Studies
in the History of Mathematics and
Physical Sciences

2

Herman H. Goldstine

A History of Numerical Analysis from the 16th through the 19th Century

Springer-Verlag
New York Heidelberg Berlin

Herman H. Goldstine
IBM Research, Yorktown Heights, New York, New York 10598/USA
and
Institute for Advanced Study, Princeton, New Jersey 08540/USA

AMS Subject Classifications: 01A40, 01A45, 01A50, 01A55, 65-03

Library of Congress Cataloging in Publication Data: Goldstine, Herman Heine, 1913– . A history of numerical analysis from the 16th through the 19th century. (Studies in the history of mathematics and physical sciences; 2) Bibliography: p. . Includes index. 1. Numerical analysis—History. I. Title. II. Series.
QA297.G64 519.4'09'03 77-5029

Printed in the United States of America.

9 8 7 6 5 4 3 2 1

ISBN 0-387-90277-5 Springer-Verlag New York
ISBN 3-540-90277-5 Springer-Verlag Berlin Heidelberg

To Ellen

"Shall I compare thee to a summer's day?"

Preface

In this book I have attempted to trace the development of numerical analysis during the period in which the foundations of the modern theory were being laid. To do this I have had to exercise a certain amount of selectivity in choosing and in rejecting both authors and papers. I have rather arbitrarily chosen, in the main, the most famous mathematicians of the period in question and have concentrated on their major works in numerical analysis at the expense, perhaps, of other lesser known but capable analysts. This selectivity results from the need to choose from a large body of literature, and from my feeling that almost by definition the great masters of mathematics were the ones responsible for the most significant accomplishments. In any event I must accept full responsibility for the choices.

I would particularly like to acknowledge my thanks to Professor Otto Neugebauer for his help and inspiration in the preparation of this book. This consisted of many friendly discussions that I will always value. I should also like to express my deep appreciation to the International Business Machines Corporation of which I have the honor of being a Fellow and in particular to Dr. Ralph E. Gomory, its Vice-President for Research, for permitting me to undertake the writing of this book and for helping make it possible by his continuing encouragement and support. I should also like to acknowledge the kindness of the Institute for Advanced Study in sustaining me intellectually through this task and for providing me with its facilities. I have been considerably helped by watching my colleagues here at their labors. They have served as exemplars of the highest standards of science and scholarship, and I hope this book reflects to some extent their inspiration. Since I bear the onus of responsibility for the contents of this work, I do not enumerate the names of these colleagues except to thank Professor Marshall Clagett for his many courtesies and kindnesses and Professor Bengt Strömgren of the University of Copenhagen for opening that university's libraries to me and providing me with other facilities as well.

In closing I wish especially to express my deep gratitude to Janet Sachs for her many kindnesses in helping to improve this book's style.

Finally my thanks are due to Springer–Verlag for its splendid work and help in making this material available in attractive form.

July 1977 HERMAN H. GOLDSTINE

Table of Contents

Introduction

In Chapter 1, I have attempted to indicate to a small extent the resurgence of interest by the Western world in science. During the sixteenth and early seventeenth century mathematical notation began to improve quite markedly. The rapid emergence of reasonable symbolisms contributed greatly to the development of mathematics and allied sciences such as mathematical astronomy.

I begin with a rather full account of both Bürgi's and Napier's discovery of logarithms in which I have tried to show in detail how Napier carried out out the laborious calculations he made in order to construct his table of logarithms by means of tables of geometrical progressions. It is fascinating to contrast this with the much more elegant and sophisticated techniques of his English successor, Henry Briggs. This man and a predecessor of his at Oxford, Thomas Harriot (1560–1621), were mainly responsible for the early developments of finite difference methods. They both understood and used interpolation formulas in general and subtabulation ones in particular. For some reason the accomplishments of these two men have not been sufficiently appreciated, perhaps because so many honors have been heaped upon Napier, who was certainly deserving of them.

Briggs noted, firstly that $\log(1 + x)$ is proportional to x for sufficiently small x; secondly that given any number y it is possible by repeated extractions of square roots to reduce it to a related number y' of the form $1 + x$ with x small; and thirdly that the logarithms of only relatively few numbers need be calculated since the others can be obtained either as sums of known logarithms or as subtabulated values. In fact Briggs not only devised subtabulation schemes, he also worked out a very ingenious difference method to eliminate some of the work of forming the square roots mentioned above.

In any case the ideas of Napier and Briggs spread rapidly across Europe, and we shall see Kepler calculating his own tables as soon as he heard of Napier's idea. From this point onwards the theory of finite differences was to be further developed with great artistry by such men as Newton, Euler, Gauss, Laplace, and Lagrange, among others. In fact we shall see that virtually all the great mathematicians of the seventeenth and eighteenth centuries had a hand in the subject.

Among Newton's predecessors one of the most extraordinary was François

Vieta. He had a considerable influence upon the young Newton and upon numerical analysis. I have given some illustrations of his work, in particular his solution of algebraic equations.

To illustrate the rapid advance of mathematics between Kepler and Newton, I have closed the chapter with the solution of the same problem in plane geometry as given by Kepler and Newton. This illustrates the progress in mathematical notation that took place over 75 years between the two men, and how mathematical analysis became easier as a result of that progress.

Chapter 2 is entitled "The Age of Newton". There I have discussed small portions of Newton's works: notably, his contributions to numerical techniques such as his method for solving equations iteratively, his interpolation and numerical integration formulas as well as his ideas on calculating tables of logarithms and of sines and cosines, In the part on interpolation there is included a brief account of how Wallis found the value of π by interpolation and how Newton generalized this by making the upper limit of an integral variable. As we shall see, this immediately led him to the binomial theorem.

Newton's friends and contemporaries quickly took up his ideas and published a great deal of work which is of interest in our field. Thus we find Halley developing series expansions for calculating logarithmic tables, and Roger Cotes systematically working out Newton's ideas on numerical integration formulas. Stirling and Maclaurin used Newton's techniques in developing important results on sums of series, and their names are still well known today and associated with fundamental series and approximation methods.

About the same time de Moivre and James Bernoulli worked at building the foundations for probability theory and the latter, while working in many other directions, was estimating the sums of powers of successive integers. This topic was also considered by Maclaurin and later by Euler. Their work resulted in a considerable body of literature important to numerical analysis under the general title of Bernoulli and Euler numbers and polynomials. This includes summation of functions and difference equations.

There is also some discussion of the extensive research of James Gregory, a Scot who worked in Edinburgh, more or less independently of Newton. Gregory had a method of using an interpolation formula involving finite differences and then of passing to the limit to find series expansions for a considerable variety of functions, essentially by Taylor's theorem. In passing we should note that Gregory's successor in the mathematics chair at Edinburgh was Maclaurin, a protégé of Newton.

The contribution of Euler, who did at least the ground work on virtually every topic in modern numerical analysis, is examined next. This work included the basic notions for the numerical integration of differential equations. Moreover, his development of lunar theory made possible the accurate calculation of the moon's position and the founding of the Nautical Almanac in Great Britain.

Lagrange worked on linear difference equations and introduced his now famous method of variation of parameters in this connection. He published extensively on the subject and must be considered as one of the founders of our field. He was very interested in interpolation theory, and he wrote several papers on the subject following up on Briggs's ideas. He introduced some quite elegant formalistic procedures which enabled him to develop many important results. He not only considered the more classical methods of interpolation, but he and Clairaut seem to have discovered trigonometric interpolation independently. He was deeply concerned with finding hidden periodicities in astronomical data and devised some interesting means for finding these periods.

Laplace used and developed the method of generating functions to study difference equations which came up in his work on probability theory. Using this apparatus, he was also able to develop various interpolation functions and to produce a calculus of finite differences. Out of his work on probabilities Laplace developed an elegant treatment of least squares. The subject, of course, was discovered by Gauss and later by Legendre. However, it is probably fair to attribute to Gauss and Laplace the real developments of the subject. But Gauss's treatment was both simpler and more elegant than Laplace's, which depended upon the Law of large numbers.

Gauss wrote much on numerical matters and obviously enjoyed calculating. He took the Newton–Cotes method of numerical integration and showed that by viewing the positions of the ordinates, taken to form the finite approximation to the integral, as parameters to be chosen he could materially improve the convergence. Jacobi reconsidered this result and gave a very elegant exposition of it. This was followed up later by Chebyshev who used another scheme to assign equal weights to the ordinates. In the Gaussian case the weights are unequal, and Gauss calculated a considerable number of them. He wrote penetratingly on interpolation and particularly on trigonometric interpolation. In fact he developed the entire subject of finite Fourier series, including what we now call the Cooley–Tukey algorithm or the fast Fourier transform.

Jacobi interested himself in a number of aspects of numerical analysis, including, as we mentioned above, the Gaussian method of numerical integration. He also gave an elegant analysis of the Euler–Maclaurin algorithm in the course of which he developed the Bernoulli polynomials. He wrote a paper on finding the characteristic values of a symmetric matrix which has given rise to the modern Jacobi method and its variants.

Cauchy was yet another great mathematician who worked on numerical methods. One of his most significant discoveries was a method for finding a rational function which passed through a sequence of given points. This idea of approximation by rational, rather than polynominal, functions is still important and in another connection — Padé approximations — is used today. In his usual way Jacobi gave a first-class exposition and analysis of this method of Cauchy. Cauchy also interested himself in trigonometric interpolation, as did Hermite, apparently in ignorance of Gauss's results.

In the course of investigating the Newton–Raphson method Cauchy came upon the well-known Cauchy–Schwarz inequality. He also made very skillful use of operational methods for solving both difference and differential equations. But probably his most important contribution to our field was made in the field of summation of functions. He based his beautiful results on his famous Residue theorem which precisely relates an integral and a sum. The exploitation of his ideas by Lindelöf and later by Nörlund has resulted in an elegant theory of considerable depth and beauty.

Another great advance Cauchy made in our subject was his method for showing the existence of the solutions of differential equations. This so-called Cauchy–Lipschitz method, as well as that of Picard, form the basis for some very important techniques for numerically integrating such equations. These theoretical methods were exploited by John Couch Adams, the astronomer, who used a successive approximation method numerically in a work with Bashforth on capillary action. This work was followed by that of F. R. Moulton who considerably improved upon it.

In a quite different direction K. Heun, W. Kutta and C. Runge developed a very pretty method for numerical integration of differential equations; and in fact one of the very first problems run on the ENIAC was done using Heun's method. Their ideas are current today.

Both J. C. Adams and Hermite wrote on Bernoulli's numbers making use of their exact form as given by Clausen and von Staudt, and Adams tabulated many of them. Hermite also studied the Bernoulli polynomials and the Euler–Maclaurin formula. Hermite was also one of the first to appreciate how Cauchy's Residue theorem could be used to obtain polynominal approximations to a function. In this he followed up on a remark of Cauchy that interpolation formulas properly should come out of the Residue theorem.

We are now on the threshold of the twentieth century, where I have quite arbitrarily decided to terminate this work. The closing section deals very superficially with the results of Cauchy and Lindelöf on the summation of functions and an asymptotic theorem of Poincaré and Perron on the relation of the zeros of an algebraic equation and certain quotients of solutions of a related linear difference equation.

1. The Sixteenth and Early Seventeenth Centuries

1.1. Introduction

One of the great discoveries of the sixteenth century was that of logarithms made independently by Bürgi and Napier. This marked a state in the development of mathematics where sufficiently sophisticated methods were finally made available for the understanding of exponentials and their inverses.

Before this time there had been considerable study of the trigonometric functions, made possible by the fact that their analysis can be undertaken purely geometrically. Greek mathematicians had already noted the importance of the relationship of a chord of a circle to the arc it subtends. Hipparchus (*circa* −140), was probably the first to introduce the chord function — in modern terms

$$\text{chord } 2\alpha = 2R \sin \alpha,$$

where R is the radius of the circle and 2α is the subtended angle. But the earliest extant chord table is that in the *Almagest* of Ptolemy (*circa* +140). Using elegant geometrical theorems, Ptolemy developed formulas for computing chd $(\alpha + \beta)$, chd $(\alpha - \beta)$ and chd $(\frac{1}{2}\alpha)$, given chd α and chd β. It is of interest to us to note that he calculated chd 1° by an approximation procedure, starting from chd 0°;45 and chd 1°;30; he found these two values by repeatedly using the half-angle formula beginning with chd 12°, which he calculated from a knowledge of the chords of 72° and 60°.

The problem of improving upon Ptolemy's method of finding chd 1° engaged many mathematicians, particularly in the Arab world, until the time of al-Kāshī (*circa* 1400), who worked at the observatory in Samarqand during the reigns of Tamerlane and his son Shahrukh. Al-Kāshī devised an elegant iterative scheme for solving the cubic

$$\sin 3\alpha = 3x - 4x^3.$$[1]

It is fair to assume that the great interest that was shown for many years in

---

[1] Aaboe [1954].

the theory of equations and in iterative methods for solving algebraic equations had its genesis in this problem of calculating sin 1° given sin 3°.

However, the study of logarithms is not a development stemming from early ideas on geometry but in a sense is a precursor of modern analysis. It was largely made possible by a series of sufficient developments in the understanding of algebraic processes and improvements in notation. As we shall see, in the hands of Briggs, it led very directly into the beginnings of numerical analysis.

1.2. Napier and Logarithms

It is interesting to trace the European origins of Napier's great discovery. In this search Michael Stifel's name is prominent. Stifel (1487–1567) was a German mathematician working at Jena, a generation before Napier. He studied the properties of exponents; in fact he seems to have coined the term "exponent" in his *Die Coss Christoffs Rudolffs* [1553]. He discussed properties of both positive and negative exponents in his *Arithmetica Integra* [1544], Book III, p. 377. There he considered the series

$$\begin{array}{ccccccc} 0 & 1 & 2 & 3 & 4 & 5 & 6\ldots \\ 1 & x & x^2 & x^3 & x^4 & x^5 & x^6\ldots \end{array} .$$

He noted the intimate connection between these two sequences, the one arithmetic and the other geometric. In fact he remarked how addition (subtraction) of terms in the former corresponds to multiplication (division) in the latter. He also knew that multiplication (division) in the former corresponds to raising to a power (extracting a root). By 1600 these properties must have been fairly well understood. Compare, e.g., Simon Jacob, who was the inventor of an early geometrical or surveying instrument.[2] These were not by any means the first or only mathematicians to consider the problem of exponents.[3]

Archimedes, in his *Sand Reckoner*, was already aware of the notion of geometrical progression. He had a theorem which came very close to being a statement of one of the laws of exponents. It is given in a somewhat anachronistic form by Heath: "If there be any number of terms of a series in continued proportion, say $A_1, A_2, A_3, \ldots, A_m, \ldots, A_n, \ldots, A_{m+n-1}$ of which $A_1 = 1$, $A_2 = 10$ [so that the series forms the geometrical progression $1, 10^1, 10^2, \ldots, 10^{m-1}, \ldots, 10^{n-1}, \ldots, 10^{m+n-2}, \ldots$], and if any two terms as A_m, A_n be taken and multiplied, the product $A_m \cdot A_n$ will be a term in the same series and will be as many terms distant from A_n as A_m is distant from A_1; also

---

[2] Jacob [1600].

[3] The interested reader may wish to consult Tropfke, *GEM*, Vol. II, pp. 132–166, for an account of the early history of exponents.

it will be distant from A_1 by a number of terms less by one than the sum of the numbers of terms by which A_m and A_n respectively are distant from A_1."[4] Thus he seems to have had some feeling for the fundamental relation $a^m \cdot a^n = a^{m+n}$ in the third century B.C.

However the first two men who are major figures in the discovery of logarithms are Joost Bürgi, a Swiss (1552–1632/33), who worked in astronomy and mechanics both in Prague and Kassel, and John Napier, Laird of Merchiston (1550–1617). It was Napier who in 1614 published his work first in Edinburgh.[5] He had labored over the concept and the actual tabulations for about twenty years. He also wrote another book on the construction of these tables, which was published posthumously in 1619.[6] This is part one of the posthumous work; part two is in "Appendix as to the making of another and better kind of Logarithms"; part three contains "Propositiones for the solutions of spherical Triangles by an easier method." Part of this work, as the complete title indicates, is by Briggs. Moreover, in 1616 the *Descriptio* appeared in an English translation. It is interesting to note that in that edition Napier's name was spelled Nepair; it has been said it was a title conferred on an ancestor for his peerless bravery. It is not known if the story is true.

The translation by Macdonald is good and makes available Napier's work in an easily accessible form.[7] This contains a number of good expositions of Napier's works. (All future references to the *Constructio* by me are to this translation.) In considering Napier's text we should know that Napier prepared his so-called artificial table to make easy the calculation of products of sines. Of course it makes no difference whether we view the independent variable as x or $\sin x$ but to Napier it was the latter. Originally Napier referred to his numbers as *artificiales*; he then coined the term *logarithm* out of λόγων plus ἀριθμός, i.e., ratio number.

Napier says in his *Constructio*:

1. *A Logarithmic table is a small table by the use of which we can obtain a knowledge of all geometrical dimensions and motions in space, by a very easy calculation.*

It is deservedly called very small, because it does not exceed in size a table of sines; very easy, because by it all multiplications, divisions, and the more difficult extractions of roots are avoided; for by only a very few most easy additions, subtractions, and divisions by two, it measures quite generally all figures and motions.

It is picked out from numbers progressing in continuous proportion.[8]

---

[4] Heath, *ARC*, p. 230. A translation of the Greek text appears in Ver Eecke [1960], Vol. I, p. 366.

[5] Napier, *Descr.* (Cf. also, Napier, *NTV.*) This is usually referred to as the *Descriptio.*

[6] Napier, *Const.* Cf. also, Napier, *EC* for an English translation. This work is usually referred to as the *Constructio.*

[7] Napier, *NTV*. This contains a number of good expositions of Napier's works.

[8] Napier, *EC*, p. 7.

We find early on in this book Napier's invention of the period to signify a decimal fraction, i.e., the so-called decimal point. He also was aware of rounding errors and had a way to deal with them. He says in Propositions 5 and 6:

5. *In numbers distinguished thus by a period in their midst, whatever is written after the period is a fraction, the denominator of which is unity with as many cyphers after it as there are figures after the period.*

Thus

$$10000000.04 \text{ is the same as } 10000000\,\frac{4}{100};$$

also

$$25.803 \text{ is the same as } 25\,\frac{803}{1000};$$

also

$$9999998.0005021 \text{ is the same as } 9999998\,\frac{5021}{10000000};$$

and so of others.

6. *When the tables are computed, the fractions following the period may then be rejected without any sensible error. For in our large numbers, an error which does not exceed unity is insensible and as if it were none.*

Thus in the completed table, instead of

$$9987643.8213051, \text{ which is } 9987643\,\frac{8213051}{10000000},$$

we may put 9987643 without sensible error.

This use of the period was clearly an improvement on Stevin's notation [Stevin, e.g., wrote 8.937 as 8 ⓪ 9 ① 3 ② 7 ③], but it was not taken up for a long time by others. In fact Briggs, Napier's friend, wrote instead 9987643|8213051.[9] In Napier's Propositions 26 and 27 he defined his logarithm function. There he says:

26. *The logarithm of a given sine is that number which has increased arithmetically with the same velocity throughout as that which radius began to decrease geometrically, and in the same time as radius has decreased to the given sine.*

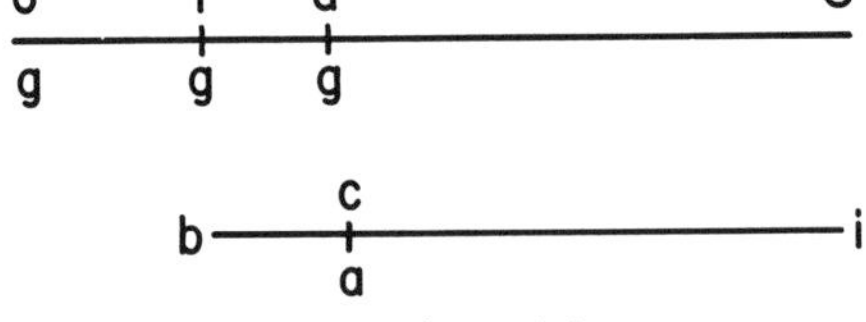

Figure 1.1.

Let the line *TS* be radius, and *dS* a given sine in the same line; let *g* move geometrically from *T* to *d* in certain determinate moments of time. Again, let *bi*

[9] Napier, *NTV*, p. 97.

be another line, infinite towards i, along which, from b, let a move arithmetically with the same velocity as g had at first when at T; and from the fixed point b in the direction of i let a advance in just the same moments of time up to the point c. The number measuring the line bc is called the logarithm of the given sine dS.

27. *Whence nothing is the logarithm of radius.*

For, referring to the figure, when g is at T making its distance from S radius, the arithmetical point d beginning at b has never proceeded thence. Whence by the definition of distance nothing will be the logarithm of radius.[10]

We can now calculate what Napier's logarithms really are in our terms. The line TS is the radius r or *sinus totus*—in this case 10^7. Then if we designate by $(r - x)$ the length Td traveled by g and by y the length bc traveled by a in the same time, dS is x and

$$\frac{d}{dt}(r - x) = x, \qquad \frac{d}{dt}y = r, \tag{1.1}$$

$$x(0) = r, \qquad y(0) = 0. \tag{1.2}$$

From these relations we see that

$$x = re^{-y/r}.$$

But Napier has defined his logarithm, Nap. Log x, as y, i.e.,

$$y = \text{Nap. Log } x,$$

and so we have

$$\log_e x = \log_e r - \frac{1}{r}\text{Nap. Log } x$$

or

$$\text{Nap. Log } x = r \log_e \frac{r}{x} = 10^r \log_e \frac{10^r}{x}. \tag{1.3a}$$

Actually the line dS represented not x for Napier but $\text{Sin } x = r \sin x = 10^7 \sin x$ and so (1.3a) is

$$\text{Nap. Log Sin } x = r \log_e \frac{r}{\text{Sin } x} = 10^7 \log_e \frac{1}{\sin x} = -10^7 \log_e \sin x. \tag{1.3b}$$

He next showed in Proposition 28 of the *Constructio* that

$$\frac{r(r - \text{Sin } x)}{\text{Sin } x} > \text{Nap. Log Sin } x > r - \text{Sin } x. \tag{1.4}$$

His proof is very nice. In Figure 1.1 he extended the line $TS = r$ backwards to a point o such that oS is to TS as TS is to dS. He then showed that bc, the logarithm of the sine dS, is greater than Td and less than oT. His proof is this: "For in the same time that g is borne from o to T, g is borne from T to d; because . . . oT is such a part of oS as Td is of TS, and in the same time

[10] Napier, *EC*, p. 19.

(by the definition of a logarithm) is a borne from b to c; so that oT, Td, and bc are distances traversed in equal times. But since g when moving between T and o is swifter than at T, and between T and d slower, but at T is equally fast with a . . .; it follows that oT the distance traversed by g moving swiftly is greater, and Td the distance traversed by g moving slowly is less, than bc the distance traversed by the point a with its medium motion in just the same moments of time; the latter is, consequently, a certain mean between the two former."[11]

(It is clear in modern terms from the relations (1.1), (1.2) that $dS = re^{-t}$, $Td = r(1 - e^{-t})$, and since $oS/TS = TS/dS$, $oS = re^{t}$ and $oT = r(e^{t} - 1)$. Therefore $oT > bc > Td$. Thus Napier's inequalities are equivalent to the statement that $r(e^{t} - 1) > rt > r(1 - e^{-t})$.)

Napier gave his Canon, i.e., his table, in the *Descriptio*. In that work he gave no explanatory text, putting that off until he brought out his *Constructio*. The detailed calculations are of some interest because of the ingenuity of his approach.

He first constructed two ancillary tables. In this preliminary work Napier was at great pains to find easy ways to carry out his task. Thus he hit on the idea of forming geometric progressions whose terms are very easy to calculate. He realized that repeated shiftings of the decimal point and subsequent subtractions are quite simple to perform. He therefore decided to tabulate the logarithms of numbers lying between his "radius and half radius" using only such operations. To do this he constructed his so-called first table, which is a tabulation of the geometrical progression whose first term is $r = 10^7$ and whose common ratio is $\rho_1 = 1 - 10^{-7}$. It is shown below. We see there that his last term is $r\rho^{100} = 9999900.0004950$.

First table.
10000000.0000000
1.0000000
99999999.0000000
.9999999
99999998.0000001
.9999998
99999997.0000003
.9999997
99999996.0000006
to be continued
up to
9999900.0004950

Thus from radius, with seven cyphers added for greater accuracy, namely, 10000000.0000000, subtract 1.0000000, you get 9999999.0000000; from this subtract .9999999, you get 9999998.0000001; and proceed in this way, as shown at the side, until you create a hundred proportionals, the last of which, if you have computed rightly, will be 9999900.0004950.

In principle he could have continued this table until he reached half radius except for the fact that the amount of work would have been completely prohibitive. He therefore shifted next to a coarser ratio and formed his second

[11] Napier, *EC*, p. 20. Notice how easily Napier handled velocities. This shows quite clearly the degree of sophistication attained in the West by 1600 in coping with non-uniform motions.

table, which is again a tabulation of the geometrical progression with first term $r = 10^7$, ratio $\rho_2 = 1 - 10^{-5}$ and last term $r\rho_2{}^{50}$.

Second table
10000000.000000
100.000000
9999900.000000
99.999000
9999700.003000
99.997000
9999600.006000
&c., up to
9995001.222927

Thus the first and last numbers of the First table are 10000000.0000000 and 9999900.0004950, in which proportion it is difficult to form fifty proportional numbers. A near and at the same time an easy proportion is 100000 to 99999, which may be continued with sufficient exactness by adding six cyphers to radius and continually subtracting from each number its own 100000th part in the manner shown at the side; and this table contains, besides radius which is the first, fifty other proportional numbers, the last of which, if you have not erred, you will find to be 9995001.222927.

Notice that the second term $r\rho_2$ in the second table is the nearest "easy" number just below $r\rho_1^{100}$, the last entry in the first table. These tables connect very nicely with only a slight roughness at the transition point. In passing we might note that Napier made a small arithmetical blunder in forming the last entry in the second table. He gave it as 9995001.222927; in fact, as Delambre and others pointed out, it should be 9995001.224804.[12] This caused an error in the last place in Napier's table of logarithms. The Canon was also affected by errors in the table of sines he used. Indeed he knew this and remarked. ". . . it would seem that the table of sines is in some places faulty. Wherefore I advise the learned, who perchance may have plenty of pupils and computers, to publish a table of sines more reliable and with larger members, in which the radius is made 100000000. . . ."[13]

Given the first and second tables Napier was now ready to form his third table which was more extensive than the others. It consisted of 69 separate geometrical progressions arranged in as many columns. Each had the same ratio $\rho = 1 - 5 \times 10^{-4}$. The rows were also geometrical progressions whose ratio was $1 - 10^{-2}$. The first term in column one was $r = 10^7$ and the last one in column 69 was 4998609.4034. Thus the third table covered the interval from radius to half radius, as Napier desired.

To assign logarithms to the quantities in the third table Napier first assigned a value to Nap. Log 9999999. This fixed all the others. By the result (1.4) above on upper and lower bounds,

$$1 + 10^{-7} + 10^{-14} + \cdots > \text{Nap. Log } 9999999 > 1.$$

He then remarks that the average of 1 and $1 + 10^{-7}$, 1.00000005, will be taken as Nap. Log 9999999. Now he has the logarithm of every term in the

[12] Napier, M. [1834]. This is an interesting biography by a descendant.

[13] Napier, *EC*, p. 46. He refers in the text to "Reinhold's common table of sines, or any other more exact." [Erasmus Reinhold (1511–1553) was a colleague of Rhaeticus at the University of Wittenberg. He was responsible for a famous table (1551) of motions of the planets based on Copernicus's *De Revolutionibus*. They were known as the *Tabulae Prutenicae* or Prussian tables after the Duke, Albert of Prussia, who patronized Reinhold.]

first table. Thus, e.g., "the logarithm of 9999998.0000001, the second sine after radius, will be contained between ...2.0000002 and 2.0000000; and the logarithm of 9999997.0000003, the third will be between the triple of the same, namely between 3.0000003 and 3.0000000." In this way he immediately found limits on the logarithms of each term in the first table.

To do the same for the second table he had first to interpolate for the logarithm of 9999900.[14] He first proved a quite interesting theorem. If Sin $a >$ Sin b, then

$$\frac{r(\text{Sin } a - \text{Sin } b)}{\text{Sin } b} > \text{Nap. Log Sin } b - \text{Nap. Log Sin } a > \frac{r(\text{Sin } a - \text{Sin } b)}{\text{Sin } a}. \quad (1.5)$$

This result is not hard to establish.[15] He did it as follows.

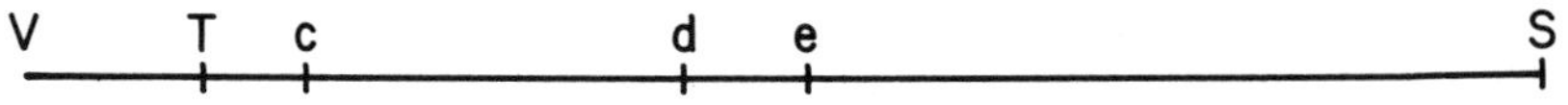

Figure 1.2.

In Figure 1.2 $TS = r$, $dS =$ Sin a, $eS =$ Sin b. Extend T backwards to V so that

$$\frac{TS}{TV} = \frac{eS}{de} = \frac{\text{Sin } b}{\text{Sin } a - \text{Sin } b}. \quad (1.6)$$

Next choose the point c so that

$$\frac{TS}{Tc} = \frac{dS}{de} = \frac{\text{Sin } a}{\text{Sin } a - \text{Sin } b}. \quad (1.7)$$

Therefore we can infer that $VS/TS = TS/cS = dS/eS$ and also

$$\begin{aligned} \text{Nap. Log } eS - \text{Nap. Log } dS &= \text{Nap. Log } cS - \text{Nap. } TS \\ &= \text{Nap. Log } cS \end{aligned} \quad (1.8)$$

since $TS = r$ and the logarithm of r is 0.

But by Napier's original result (1.4) we have

$$TV > \text{Nap. Log } cS > Tc,$$

and so with the help of (1.6), (1.7), and (1.8) we have (1.5) at once. These inequalities of Napier are quite nice and deserve a brief examination. His relations state that if $x > y > 0$

$$\frac{x}{y} - 1 > \text{Nap. Log } y - \text{Nap. Log } x = \log_e \frac{10^7}{y} - \log_e \frac{10^7}{x} = \log_e \frac{x}{y} > 1 - \frac{x}{y}.$$

It is easy to see that

$$\frac{1}{y}(x - y) = \int_y^x \frac{d\xi}{y} > \int_y^x \frac{d\xi}{\xi} > \int_y^x \frac{d\xi}{x} = \frac{1}{x}(x - y).$$

[14] Napier, *EC*, pp. 29ff.

[15] Napier, *EC*, pp. 26–27.

Thus his relations imply that the area from y to x under the curve $z = 1/x$ lies between the rectangle with height $1/y$ and width $x - y$ and the rectangle of height $1/x$ and the same width.

We can now apply the relations (1.5) to calculate the logarithm of 9999900.0000000, the second entry in the second table. We know that the logarithm of 9999900.0000000, the last one in the first table, is between 100.0000100 and 100.0000000. Then by the last theorem we see that Nap. Log 9999900 is bounded above by

$$100.0000100 + \frac{10^7 \times 0.0004950}{10^7 \times (1 - 10^{-5})} = 100.0005050$$

and below by

$$100.0000000 + \frac{10^7 \times 0.0004950}{10^7 \times (1 - 10^{-5} + 5 \times 10^{-11})} = 100.0004950.$$

Given this the logarithms of all other entries in the second table are now trivially calculable.

In the same fashion Napier proceeded to find the logarithms of the second entries in each column of the third table, and this is why all values were contaminated by his error in finding the last entry in the second table. He needed, e.g., the logarithm of 9995000.0000, the second entry in column one. To do this he used the relations (1.5), where Sin b was 9995001.222927 — instead of 9995001.224804 — and Sin a was 995000.0000. (He thereby introduced a very small error into all the logarithms of entries in the third table.) Having evaluated the logarithms of the quantities in the third table, Napier now found his so-called radical table, which was constructed by putting next to each entry in the third table its logarithm, keeping only one of the seven decimal places he previously had kept. Of this he said: "For shortness, however, two things should be borne in mind — First, that in these logarithms it is enough to leave one figure after the point, the remaining six being now rejected, which, however, if you had neglected at the beginning, the error arising thence by frequent multiplications in the previous tables would have grown intolerable in the third. Secondly, if the second figure after the point exceed the number four, the first figure after the point, which alone is retained, is to be increased by unity: thus for 10002.48 it is more correct to put 10002.5 than 10002.4; and for 1000.35001 we more fitly put 1000.4 than 1000.3. Now, therefore, continue the Radical table in the manner which has been set forth."[16]

Napier was now in a position to form his final table, his Canon of logarithms. To do this he needed not only to find an interpolatory technique so that he could form the logarithms of the sines of angles spaced 0°;1 (= 1 min.) apart, but also to find values for those outside the range of his radical table.

---

[16] Napier, *EC*, p. 35. Notice Napier's rounding-off procedure.

To accomplish his first end he gave a prescription for finding "*the logarithms of all sines embraced within the limits of the Radical table.*" It was this: "Multiply the difference of the given sine and table sine nearest it by radius. Divide the product by the easiest divisor, which may be either the given sine or the table sine nearest it, or a sine between both, however placed. By 39 there will be produced either the greater or less limit of the difference of the logarithms, or else something intermediate, no one of which will differ by a sensible error from the true difference of the logarithms on account of the nearness of the numbers in the table. Wherefore (by 35), add the result, whatever it may be, to the logarithm of the table sine, if the given sine be less than the table sine; if not, subtract the result from the logarithm of the table sine, and there will be produced the required logarithm of the given sine."[17]

To accomplish his second end he gave another prescription: namely, for finding "*the logarithms of all sines which are outside the limits of the radical table.*" He did this by writing out his so-called short table in which he recorded the logarithms of numbers (sines) in the ratios of 2, 4, 8, 10, 20, 40, 40, 80, 100, 200, . . ., 10^7. Then given any sine he multiplied it by one of these factors until the result was within the limits of the radical table.

His last result is then entitled, "*To form a logarithmic table.*"

Prepare forty-five pages, somewhat long in shape, so that besides margins at the top and bottom, they may hold sixty lines of figures. Divide each page into twenty equal spaces by horizontal lines, so that each space may hold three lines of figures. Then divide each page into seven columns by vertical lines, double lines being ruled between the fifth and sixth, but a single line only between the others.

Next write on the first page, at the top of the left, over the first three columns, "0 *degrees.*" On the second page, above, to the left, "1 degree"; and below to the right, "88 *degrees.*" On the third page above, "2 *degrees*"; and below, "87 *degrees.*" Proceed thus with the other pages, so that the number written above, added to that written below, may always make up a quadrant, less one degree or 89 *degrees.*

Then, on each page write, at the head of the first column, "*Minutes of the degree written above,*" at the head of the second column, "*Sines of the arcs to the left*"; at the head of the third column, "*Logarithms of the arcs to the left*"; at the head of the third column, "*Logarithms of the arcs to the left*"; at both the head and the foot of the third column, "*Difference between the logarithms of the complementary arcs,*" at the foot of the fifth column, "*Logarithms of the arcs to the right,*" at the foot of the sixth column, "*Sines of the arcs to the right*"; and at the foot of the seventh column, "*Minutes of the degree written beneath.*"

Then enter in the first column the numbers of minutes in ascending order from 0 to 60, and in the seventh column the number of minutes in descending order from 60 to 0; so that any of minutes placed opposite, in the first and seventh columns in the same line may make up a whole degree or 60 minutes.[18]

---

[17] Napier, *EC*, p. 36.
[18] Napier, *EC*, pp. 43–44.

To this book Napier added an Appendix, "On the construction of another and better kind of Logarithms, namely one in which the Logarithm of unity is 0." He evidently realized the awkwardness of his original system in which

$$\text{Nap. Log } xy = 10^7 \log_e \frac{10^7}{xy} = 10^7 \log_e \frac{10^7}{x} + 10^7 \log_e \frac{10^7}{y} - 10^7 \log_e 10^7$$

$$= \text{Nap. Log } x + \text{Nap. Log } y - \text{Nap. Log } 1.$$

To remedy this and also to have a system in which the logarithms of powers of 10 would be easy to calculate he proposed, but did not carry out, the construction of the logarithm where the logarithm of 1 is 0 and the logarithm of 10 (or 1/10) is 10^{10}. It is clear that the new logarithm of Napier is $\text{Log } x = 10^{10} \log_{10} x$, and he therefore discovered in a sense both the systems we know today: logarithms to the bases e and 10. In the same Appendix he gave several ways to calculate his new logarithms. To this is appended: "Some remarks by the learned Henry Briggs on the foregoing Appendix." In addition the 1616 English translation by Edward Wright of the *Descriptio* contained a graphical device for interpolating in the tables. It is shown in Figure 1.3. Wright himself died in 1615 before the work was published. It was published by Wright's son with the help of Briggs.[19] It should be remarked in passing that Edward Wright in 1599 calculated and published a text on *Certaine Errors in Navigation* .... This table corrected an error arising in the use of Gerhard Mercator's charts and is in essence a tabulation of $r \log \tan (45° - x/2)$. It gave the lengths of arcs on nautical meridians and was an important tool for navigators.[20]

The accounts of the meeting and collaboration of Napier and Briggs are so well known that they need not be repeated here. We will, however, quote Briggs's own account of their scientific interdependence in the Preface to his *Arithmetica Logarithmica* (1624).

"That these logarithms differ from those which that illustrious man, the Baron of Merchiston, published in his *Canon Mirificus* must not surprise you. For I myself, when expounding their doctrine publicly in London to my

---

[19] Henderson [1926], pp. 26–28. There we find how to use the graph: clearly $DC/ED = DH/FH = AB/BG$; thus if we know DE, EC, and DH, we can read off BG. Henderson gives as an example $60/x = 5/3$.

[20] Wright [1599] and Cajori, "Algebra in Napiers' Day and Alleged Prior Inventions of Logarithms," Napier, *NTV*, pp. 93–109. What Wright did was to approximate the value of the integral $\int \sec \theta \, d\theta$ by adding up the successive values of $\sec \theta$ starting at $\theta = 0$ and going by 0°;0,1 steps. The relation between the logarithmic tangents and the values in Wright's table was first noticed by a Henry Bond in 1645 but an actual proof was not given until 1668 when it was done by James Gregory in his *Exercitationes Geometricae*, p. 7. Cf. also, Gregory, *GTV*, p. 463. There is an account of this in Bourbaki [1960], pp. 203–204. (I am indebted to Prof. A. Weil for this reference.) It is also discussed in Cajori, Napier, *NTV*, pp. 189–190. Barrow, Wallis, and Halley all gave later proofs of the relation.

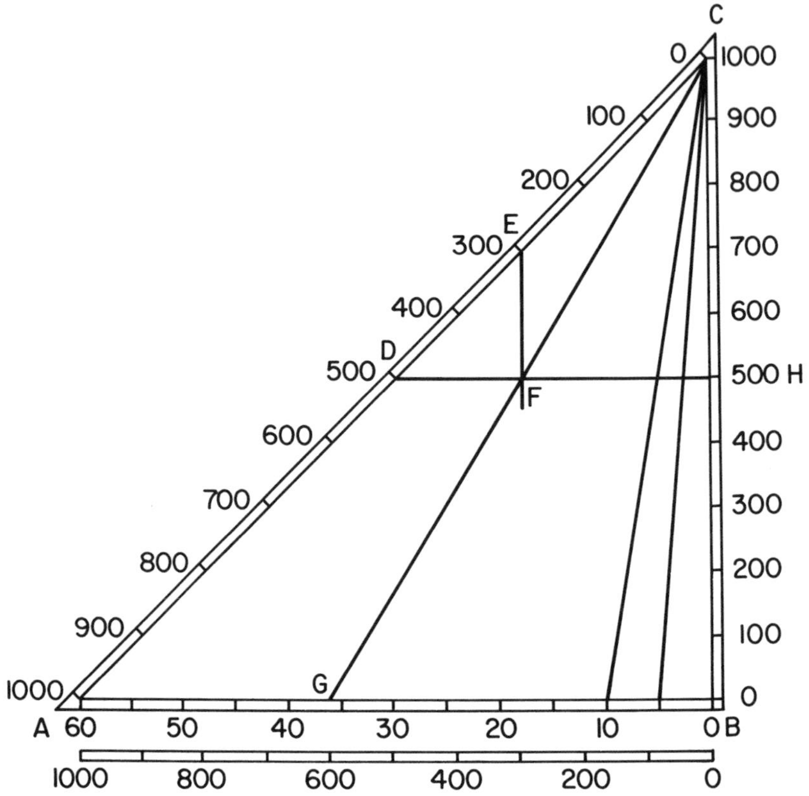

To determine x such that 60:x or 1000:x is any given ratio

Figure 1.3.

auditors in Gresham College, remarked that it would be much more convenient that 0 should be kept for the logarithm of the whole sine (as in the *Canon Mirificus*) but that the logarithm of the tenth part of the same whole sine, that is to say, 5 degrees 44 minutes and 21 seconds, should be 10000000000. And concerning that matter I wrote immediately to the author himself; and as soon as the season of the year and the vacation of my public duties of instruction permitted I journeyed to Edinburgh, where being most hospitably received by him, I lingered for a whole month. But as we talked over the change in the logarithms, he said that he had for some time been of the same opinion and had wished to accomplish it; he had however published those he had already prepared until he could construct more convenient ones if his affairs and his health would admit of it. But he was of opinion that the change should be affected in this manner, that 0 should be the logarithm of unity and 10000000000 that of the whole sine; which I could not but admit was by far the most convenient (*longe commodissimum*). So, rejecting those which I had previously prepared, I began at his exhortation to meditate seriously about the calculation of these logarithms; and in the following summer I again journeyed to Edinburgh and showed him the principal part of the logarithms I here submit. I was about to do the same in the third summer also, had it pleased God to spare him to us so long."[21]

1.3. Briggs and His Logarithms

Briggs published in 1617 a table called *Logarithmorum Chilias Prima.* (This is the first appearance of logarithms to the base 10.) This was followed in 1624 by his *Arithmetica Logarithmica.*[22] These tables of Briggs were published to 14 places but may be in error in the last place. His original idea, as we saw above, had been to make the logarithm of the *sinus totus*, the radius = 10^{10}, zero and of $10^9 \sim 10^{10} \cdot \sin(5°;44,21)$, 10^{10}. Thus the original Briggsian logarithm was $10^{10}(10 - \log_{10} x)$. He changed his ideas and put out his tables so that with $r = 10^{10}$

$$\text{Bri. Log}\, x = 10^9 \log_{10} x,$$

hence Bri. Log $1 = 0$ and Bri. Log $r = 10^{10}$.

Before discussing his work, let us say just a word about the man. Henry Briggs was born in Yorkshire in 1556 and died in 1630. He was at first a professor in Gresham's College, London, and then in 1619 he was called to one of the two chairs established by Sir Henry Savile (1549–1622) at Oxford. Much of Briggs's list was spent on the problem of making navigation safer and faster. This activity was evidently of the highest importance to England, particularly at this period when seapower was playing such a role in English history. It is therefore not surprising that Briggs took such a great interest in logarithms.

As we shall see, Briggs must be viewed as one of the great figures in numerical analysis. His ideas were far in advance of his time, and he has never been accorded the honor which is his due. This is probably because of the fallacious theory which grew up that he was merely the slavey or drudge who carried out the ideas of his master, Napier. Briggs's techniques were purely arithmetical and indicate that he must have been one of the very first, if not the first, to use the calculus of finite differences with great facility. His work is, however, difficult to read since he gave no proofs.

The main idea Briggs used (foreshadowed by a remark of Napier) was that for any number $a > 1$

$$a^{2^{-n}} \to 1.$$

Moreover for the numbers he was dealing with the convergence was not too

[21] Napier, *NTV*, pp. 126–127.

[22] Briggs, *LOG* and *ARITH*. The *Arithmetica* contains 88 pages of explanation and application.

slow. He first prepared a preliminary table for $a = 10$ and n ranging from 1 to 54, keeping up to 32 decimal places. A copy of part of this is given in Figure 1.4.[23] Let us examine the last row, which is of considerable interest.

1° Système logarithmique de Briggs
(Extrait)

	Nombres	Logarithmes
	10, 0000 00	1
1	3, 1622.77660.16387.93319.98893.54	0,5
2	1, 7782 79410 03892 28011 97304 13	0,25
3	1, 3335 21432 16322 40256 65389 308	0,125
4	1, 1547 81984 68945 81796 61918 213	0,0625
16	1, 0000 35135 27746 18566 08581 37077	0,00001 52587 89062 5
17	1, 0000 17567 48442 26738 33846 78274	0,00000 76293 94531 25
18	1, 0000 08783 70363 46121 46574 07431	0,00000 38146 97265 625
47	1, 0000 00000 00001 63608 51112 96427 283	»
48	1, 0000 00000 00000 81804 25556 48210 295	»
49	1, 0000 00000 00000 40902 12778 24104 311	»
52	1, 0000 00000 00000 05112 76597 28102 947	0,00000.00000.00002.2204.46059.25031
53	1, 0000 00000 00000 02556 38298 64006 470	0,00000.00000.00001.1102 .23024.62515
54	1, 0000 00000 00000 01278 19149 32003 235	0,00000.00000.00000.05561.11512.31257
	1, 0000 00000 00000 01	0,00000.00000.00000.04342.94481.90325.1804

Figure 1.4.

Briggs had somehow noticed that the decimal parts of the second column bear an interesting relation to each other. If $a = 1 + x$ with $x \ll 1$, then $a^{1/2} \sim 1 + x/2$. Briggs exploited this relationship most effectively, and indeed devised schemes which were used until fairly recently for making logarithmic tables.

He first calculated the function

$$10^{2^{-n}} \qquad (n = 1, 2, \ldots, 54),$$

as indicated in Figure 1.4, and noted that for n near 54

$$\log_{10} 10^{2^{-n}} = \log_{10}(1 + x_n) \sim k \cdot x_n.$$

That is, he observed that the logarithm of a number of the form $1 + x$ with x very small is essentially proportional to x. We recall that

$$\log_{10}(1 + x) = (\log_{10} e) \log_e (1 + x)$$

$$= (\log_{10} e)\left(x - \frac{1}{2}x^2 + \frac{1}{3}x^3 - \frac{1}{4}x^4 + \cdots\right).$$

[23] This is extracted from an interesting work on logarithms; Naux [1966]. (The logarithmic entry corresponding to number 54 in that figure is in error; the digit 6 should be a 5.) A very good discussion of the history of the subject is contained in the "Large and Original History of the Discoveries and Writings Relating to these Subjects." This is by way of being an introduction to Hutton [1801] and is worth reading.

Thus if x is sufficiently small — say, $x < 10^{-16}$ — its higher powers are less than 10^{-32} and so up to the first power of x

$$\log_{10}(1 + x) = x \log_{10} e = k' \cdot x; \tag{1.9}$$

where

$$k' = \log_{10} e = 0.43429448190325182765\ldots \quad .$$

Actually Briggs calculated the value of k by forming the proportion

$$\frac{2^{-53} - 2^{-54}}{10^{2^{-53}} - 10^{2^{-54}}} = \frac{0.00000\,00000\,00000\,01}{10^{-16}k};$$

and he then found

$$10^{-16}k = 0.00000\,00000\,00000\,04342\,94481\,90325\,1804.$$[24]

Notice that the two values k, k', given above, differ in the 17th decimal place. This is due to the fact that formula (1.9) is not correct to as many places as we have kept.

Clearly the more nearly correct formula is

$$\log_{10}(1 + x) = x(\log_{10} e)\left(1 - \frac{x}{2}\right). \tag{1.10}$$

Now for $x = 10^{-16}$ this gives for the expression $(\log e) \cdot (1 - x/2)$ the value

$$\begin{aligned} &0.4342\,9448190325182765 - 2.1715 \times 10^{-17} \\ &= 0.4342\,9448190325180594, \end{aligned}$$

which more nearly corresponds to Briggs's value.

To form a table of logarithms Briggs needed to consider only the prime numbers, for obvious reasons. If p is such a number, then for some n he would have

$$p^{2^{-n}} = 1 + x \tag{1.11}$$

with $x \sim 10^{-16}$. Thus by the formula (1.9) he had

$$\log_{10} p = 2^n \log_{10}(1 + x) = 2^n kx. \tag{1.12}$$

Thus he reduced his task to finding the successive square roots of a number down to the point where it was expressible in the form (1.9) above; then the relation (1.12) immediately gave the desired logarithm.

It is worth remarking on the high degree of ingenuity he displayed in finding the logarithms of the primes. Thus, e.g., he calculated $\log 2$ by starting with the fact that $2^{10}/1000 = 1.024$. Then 47 extractions of square roots gave him his result; $\log 3$ was formed by noting that $6^9 = 10077696$

---

[24] Delambre, *MOD*, Vol. I, pp. 536–537. (There is a systematic mistake in this text: one zero too many appears in some of the relevant formulas in this part of the discussion.)

(*Arithmetica Logarithmica*, p. 16). After 46 extractions of square roots he had his logarithm.

In practice this involved a tremendous amount of work, so Briggs ingeniously invented a number of labor-saving devices of a mathematical kind. Perhaps the most important is his discovery of the calculus of finite differences to expedite the extractions of square roots. He noticed in effect that when a number was of the form $1 + y$, its square root was expressible as

$$(1 + y)^{1/2} = 1 + \frac{1}{2}y - \frac{1}{8}y^2 \pm \cdots,$$

and he developed differences especially tailored to this situation. I have illustrated how it goes in the following table where superscripts indicate the order of the Briggsian difference, and where $1 + u_{i+1} = (1 + u_i)^{1/2}$.

ξ	B^1	B^2	B^3
$1 + u_1$			
	$\frac{1}{2}u_1 - u_2 = B_1^1$		
$1 + u_2$		$\frac{1}{4}B_1^1 - B_2^1 = B_1^2$	
	$\frac{1}{2}u_2 - u_3 = B_2^1$		$\frac{1}{8}B_1^2 - B_2^2 = B_1^3$
$1 + u_3$		$\frac{1}{4}B_2^1 - B_3^1 = B_2^2$	
	$\frac{1}{2}u_3 - u_4 = B_3^1$		
$1 + u_4$			

He then used these differences to extrapolate forward. Let us see how he did this by considering the example

n	ξ_n	B^1	B^2	B^3	B^4
1	1.00757 13453 69831				
		71386 59690			
2	1.00377 85340 25226		33 63766		
		17813 01156		693	
3	1.00188 74857 11457		4 19778		0
		04449 05511		43	
4	1.00094 32979 50217		52429		1
		01111 73949		4	
5	1.00047 15378 01159		6550		
		00277 86937			
6	1.00023 57411 13643				

We notice that the fourth Briggsian differences are very small. Proceed to find the entries for line 7 by means of the relations

$$B^4_{j+1} = \frac{1}{32} B^4_j - B^5_j, \qquad B^3_{j+2} = \frac{1}{16} B^3_{j+1} - B^4_{j+1}, \qquad B^2_{j+3} = \frac{1}{8} B^2_{j+2} - B^3_{j+2},$$

$$B^1_{j+4} = \frac{1}{4} B^1_{j+3} - B^2_{j+3}, \qquad B^0_{j+5} = u_{j+5} = \frac{1}{2} B^0_{j+4} - B^1_{j+4} = \frac{1}{2} u_{j+4} - B^1_{j+4}.$$

With their help, and the assumption that the $B_j{}^5$ are 0, we find

$$B^4_3 = 0.03, \qquad B^3_4 = 0.22, \qquad B^2_5 = 818.53, \qquad B^1_6 = 6945915.72,$$
$$B^0_7 = u_7 = 1178636\,10905.78.$$

This gives us 1.00011 78636 10906 as the ξ entry in line 7. Since B^4_3 is essentially zero and $B^4_4 = B^4_3/32$, it is even less work to find u_8. In fact we have $B^2_6 = 102.38$, $B^1_7 = 1736376.62$, $u_8 = 589300\ 69076$,

$$\xi_8 = 1.00005\ 89300\ 69076.$$

We may proceed in the same way doing only divisions by low powers of 2 to achieve the successive square roots beyond this point. Let us extend the table further to see what happens:

n	ξ_n	B^1	B^2	B^3
		00069 45916		
7	1.00011 78636 10906		102	
		00017 36377		0
8	1.00005 89300 69076		13	
		4 34081		0
9	1.00002 94646 00457		2	
		1 08518		0
10	1.00001 47321 91710		0	
		27130		
11	1.00000 73660 68725		0	
		6782		
12	1.00000 36830 27580			
		1696		
13	1.00000 18415 12094			
		424		
14	1.00000 09207 55623			
		106		
15	1.00000 04603 77705			
		26		
16	1.00000 02301 88826			
		7		
17	1.00000 01150 94406			
		2		
18	1.00000 00575 47201			
		0		
19	1.00000 00287 73600			

We note that the various differences decrease very rapidly so that it becomes increasingly easy to extend the table.

Suppose we have already found the logarithms of the primes through 101 (there are 26 primes involved), and that we wish to find the logarithm of the prime 173. We have 173/170 = 1.01764 70588 23529 and

$$1.01764\ 70588\ 23529/1.01 = 1.00757\ 13453\ 69831.$$

But this is the entry for ξ_1 in our table above. If we tentatively take the entry for ξ_{10} to evaluate $\log_{10} \xi_1$, we find

$$\log_{10} \xi_1 = 2^9 \cdot \log_{10} e \cdot 1.4732\ 19171 \cdot 10^{-5} = 0.00327\ 5832097$$

and therefore

$$\begin{aligned}\log_{10} 173 &= \log_{10} 170 + \log_{10} \xi_1 + \log 1.01\\ &= 2.230448921 + .00327\ 5832 + .004321374\\ &= 2.238046127.\end{aligned}$$

Actually $\log_{10} 173 = 2.23804610$. Let us next try with ξ_{14}; we then find $\log_{10} 173 = 2.2384610$. Note the value of the first Briggsian difference B^1 in row 14 as compared to earlier ones. In fact Briggs's basic relation (1.9), $\log_{10}(1 + x) = x \log_{10} e$, requires the higher powers of x to vanish to the number of places involved.

Let us stop for a moment to see another aspect of Briggs's differences for the square-root function.[25] As before let the ith Briggsian difference be written as B_j^i where $j = 1, 2, \ldots$ indicates the row in which B_j^i appears. We recall that

$$B_j^0 = u_j, \qquad B_0^0 = u \tag{1.13}$$

$$B_j^{i+1} = \frac{1}{2^{i+1}} B_j^i - B_{j+1}^i \qquad (i, j = 0, 1, \ldots), \tag{1.14}$$

where $1 + u$ is the quantity whose successive square roots we wish to determine. Now Briggs recorded his differences as powers of u.[26]

We do this as follows:

$$u_{j+1} = \frac{1}{2} u_j - B_j^1, \ldots, \qquad B_{j+1}^i = \frac{1}{2^{i+1}} B_j^i - B_j^{i+1}, \ldots;$$

this may be written as

$$u_{j+1} = \frac{1}{2} u_j - \frac{1}{2^2} B_{j-1}^1 + \frac{1}{2^3} B_{j-2}^2 - \frac{1}{2^4} B_{j-3}^3 + \cdots + \frac{(-1)^i}{2^{i+1}} B_{j-i}^i + \cdots \ . \tag{1.15}$$

---

[25] Whiteside, *Patterns*, p. 234. This is a very elegant paper and well worth reading in its entirety. Also Hutton [1801], pp. 67–68. The Briggsian differences appear in Briggs, *ARITH*, p. 16 of his Introduction.

[26] He went up to B_{j-9}^9. Whiteside, *Patterns*, p. 234.

However, Briggs was well aware that eventually his differences became very small — *perexiguus* — and may be neglected. Thus for some i, say I, we have

$$B_{j+1}^{I} = \frac{1}{2^{I+1}} B_j^I \qquad (I < j),$$

and the series (1.15) may be terminated when $i = I$. Briggs chose $I = 9$.

Finally, Briggs evaluated the successive B_{j-1}^i for $i = 1, 2, \ldots, 9$ with the help of the fact that $(1 + u_j) = (1 + u_{j-1})^{1/2}$ and found quite correctly that, if $x = u_j$,

$$B_{j-1}^1 = \frac{1}{2} B_{j-1}^0 - B_j^0 = \frac{1}{2} x^2$$

$$B_{j-2}^2 = \frac{1}{4} B_{j-2}^1 - B_{j-1}^1 = \frac{1}{2} x^3 + \frac{1}{8} x^4$$

$$B_{j-3}^3 = \frac{1}{8} B_{j-3}^2 - B_{j-2}^2 = \frac{7}{8} x^4 + \frac{7}{8} x^5 + \frac{7}{16} x^6 + \frac{1}{8} x^7 + \frac{1}{64} x^8$$

$$\vdots$$

$$B_{j-9}^9 = \frac{1}{512} B_{j-9}^8 - B_{j-8}^8 = \qquad 2805527x^{10}.$$

If these are now substituted into (1.15), there results — *mirabile dictu* — the Binomial theorem for $n = \frac{1}{2}$, $u_j = x$, namely:

$$(1 + x)^{1/2} = 1 + \frac{1}{2} x - \frac{1 \cdot 1}{2 \cdot 4} x^2 + \frac{1 \cdot 1 \cdot 3}{2 \cdot 4 \cdot 6} x^3 - \frac{1 \cdot 1 \cdot 3 \cdot 5}{2 \cdot 4 \cdot 6 \cdot 8} x^4 + \cdots .$$

This must be regarded as the first time the Binomial theorem was developed for a noninteger exponent. There are two relevant comments: first, the calculating labor involved in finding the Briggsian differences as functions of u_j is not at all trivial and one can only wonder at Briggs's prowess; second, it is curious that he perceived the essential value of the formula *a priori*. Whiteside points out the curious fact that the first use of a series approximation to find logarithms was not of the logarithm but of the square root.[27]

Not only did Briggs use these Briggsian differences, he also was facile with ordinary differences and used them to subtabulate his tables. In fact his general *modus operandi* was this: firstly he found the logarithms of the first 25 primes, 2 through 97, using his method of repeated square roots plus his Briggsian differences together with other clever devices to simplify the calculation of these roots; secondly he tabulated the logarithms of about 20 percent of his table with the help of these primes; and thirdly he subtabulated, i.e., he filled in intermediate values, to find the rest of his entries. To do this he had to discover one of the now well-known interpolation formulas, which we discuss in Section 1.5.[28]

---

[27] Whiteside, *Patterns*, p. 234.

[28] Whittaker, *WR*, p. 11.

For some obscure reason his ideas were not taken up by his colleagues and remained neglected for half a century. In 1672 Leibniz claimed for himself, at a meeting of English scientists, the discovery of the calculus of finite differences.[29]

1.4. Bürgi and His Antilogarithms

So far we have not done justice to Bürgi, the Swiss. He was born in a village in St. Gall in 1552 and eventually became connected with the courts at Prague and Kassel.[30] His chief interest was in astronomical instruments and astronomy generally. He made several astronomical clocks, did research in trigonometry and also was reputed by Kepler to be a first-class observer. He calculated a table, not of logarithms, but of antilogarithms. They are contained in Bürgi, *ANT*.[31] The body of his table was printed in two colors: red for the logarithms and black for the numbers themselves. Thus in the sample page (Figure 1.6) the listings along the top row and left column are *red numbers* $10N$. The tabular listings in black are of the geometrical progression whose Nth term is $10^8(1 + 10^{-4})^N$. Thus, e.g., we see that corresponding to the red number 2620 (10×262) he has tabulated 102654489, which is correct — his entry is in row 13 and column 6. The title page to his book (Figure 1.5) is a synopsis of his tables starting with 5000 ($= 10 \times 500$) corresponding to 105126407, and proceeding around to 230,270, corresponding to 10^8. Actually 5000 corresponds to 105126847 and 230,270 to 999999800. However, all the other entries on the circle are either correct or in a couple of cases out by 1 in the last place. Notice in Figure 1.5 that someone has inserted by hand a corrected value 84 corresponding to 5000 on the innermost circle. Random checks of Bürgi's tables by me show them to be quite accurate.

The explanatory text as given in Gieswald [1856] shows that Bürgi had a precise notation for the decimal point. It appears not only there but also in the Figure 1.5 where $23027\overset{\circ}{0}\,022$ appears. It is to be read as 230270.022. Thus $1.0001^{230270.022} = 10^8 \times 10^2$ with an error of about 8 in 10^8.

To see his usage of the table consider his example of finding the fourth proportional D to the numbers $A = 945919848$, $B = 100160120$, $C = 880122800$. Their *red numbers* are $\alpha = 22471\overset{\circ}{0}$, $\beta = 16\overset{\circ}{0}$, $\gamma = 21750\overset{\circ}{0}$, respectively. Bürgi now proceeds to form $D = (BC)/A$ as follows: he notes that $\beta + \gamma < \alpha$ and therefore forms $\delta' = \beta + \gamma + \varepsilon - \alpha$, where $\varepsilon = 23027\overset{\circ}{0}\,022$.

---

[29] Naux [1966], p. 117. Cf. p. 20 below for Hofmann's account of this.

[30] Naux [1966], pp. 92–98.

[31] For some unknown reason the "complete explanation" promised in Bürgi's title did not appear in the 1620 edition. In fact it was not published until 1856. Gieswald [1856].

PLATE XII

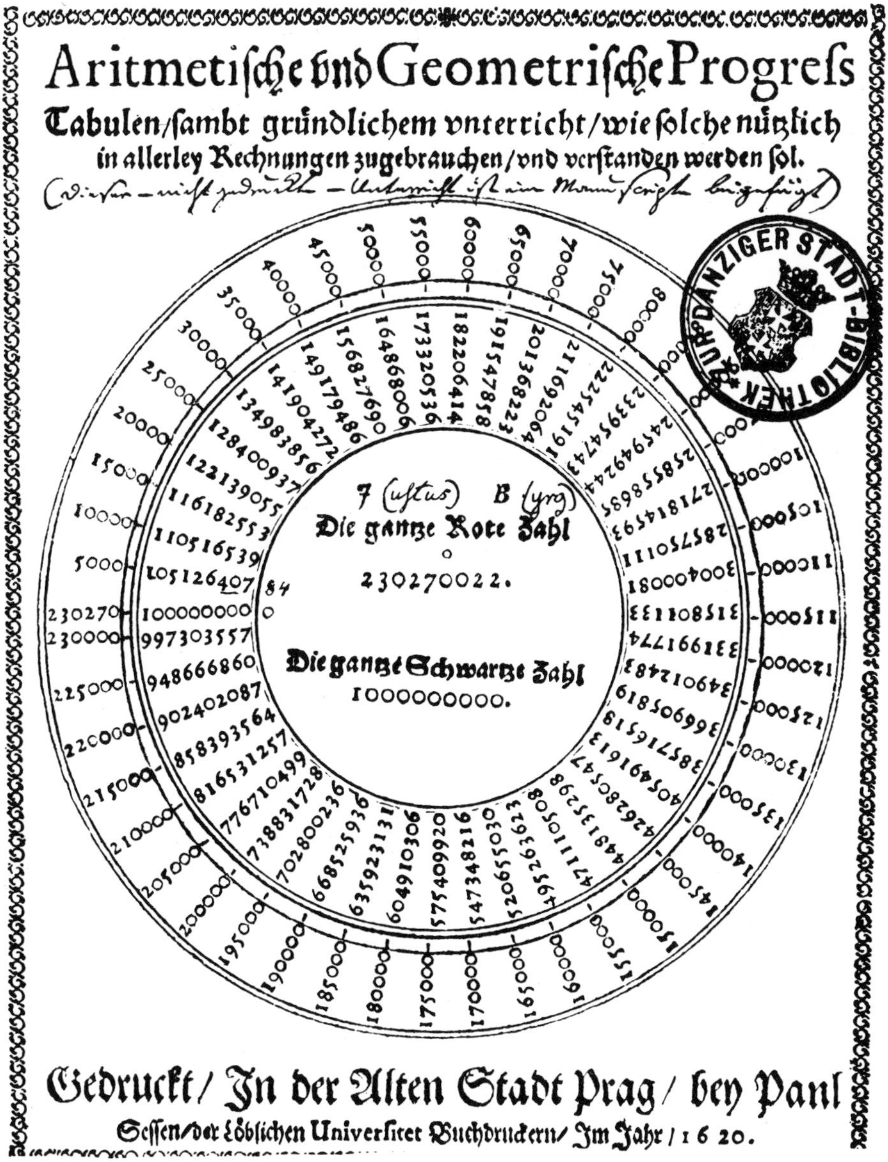

Figure 1.5.

This gives him $\delta' = 22322\mathring{0}\,022$, which corresponds to 931931024. Thus the fourth proportional is 93193102.4.[32]

Whether Bürgi's work actually was independent of Napier's and precedent to it is difficult to answer. That it was independent seems clear to me. But

[32] Gieswald [1856], p. 327.

PLATE XIII

	0	500	1000	1500	2000	2500	3000	3500
0	100000000	100501227	101004966	101511230	102020032	102531384	103045299	10356179[illegible]
10	····10000	····11277	····15067	····21381	····30234	····41637	····55603	····72140
20	····20001	····21328	[illegible]	····31524	····40437	····51891	····65909	····8250[illegible]
30	····30003	····31380	····35271	····41687	····50641	····62146	····76216	····92861
40	····40006	····41433	····45374	····51841	····60846	····72402	····86523	103603221
50	····50010	····51487	····554[illegible]	····61[illegible]06	····71052	····8266[illegible]	····9683[illegible]	····13581
60	····60015	····61543	····65584	····72158	····81259	····92918	1031071[illegible]	····23942
70	····70021	····71599	····75691	····82309	····91467	102603177	····1745[illegible]	····34305
80	····80028	····81656	····85799	····92468	102101676	····13438	····27064	····4466
90	····90036	····91714	····95907	101602627	····11887	····23699	····38077	····5503[illegible]
100	100100045	100601773	101106017	····12787	····22098	····33961	····48391	····6539[illegible]
110	····10055	····11834	····161[illegible]	····[illegible]049	····32310	····44725	····587[illegible]5	····7576[illegible]
120	····20066	····21895	····26239	····33111	····42523	····54489	····69021	····86137
130	····30078	····31957	····36352	····43274	····52738	····64755	····7933[illegible]	····96501
140	····40091	····4[illegible]	····46165	····5[illegible]38	····62953	····75021	····89656	1037[illegible]687[illegible]
150	····50105	····52084	····56580	····63604	····73169	····85289	····99975	····17241
160	····60120	····62150	····66696	····73770	····83386	····95557	103210295	····27613
170	····70136	····72216	····76812	····83938	····93606	102705827	····[illegible]0616	····[illegible]986
180	····80153	····82283	····86930	····94106	102203824	····16097	····3093[illegible]	····4836[illegible]
190	····90171	····92351	····97045	101704275	····14045	····26365	····41261	····58734
200	100200190	1007[illegible]	1012[illegible]168	····14446	····24266	····3664[illegible]	····515[illegible]	····6911[illegible]
210	····10210	····12491	····17289	····24617	····34488	····46915	····61910	····7[illegible]47
220	····20231	····22562	····27411	····34790	····44717	····57190	····72237	····8[illegible]865
230	····30253	····32634	····37522	····44963	····54936	····67466	····82564	103800244
240	····40276	····42707	····47657	····55138	····65162	····77742	····9289[illegible]	····10624
250	····50300	····52782	····57782	····65313	····75388	····88020	103303221	····21[illegible]05
260	····60325	····6285[illegible]	····679[illegible]	····75490	····85616	····98209	····1355[illegible]	····31387
270	····70351	····72933	····78095	····8566[illegible]	····95845	102808579	····23883	····41770
280	····80378	····83011	····88162	····95846	102306274	····18860	····34210	····5215[illegible]
290	····90406	····93189	····98291	101806025	····1630[illegible]	····[illegible]9142	····44[illegible]	····6[illegible]54
300	100300435	100803168	101308421	····16206	····26536	····39425	····54883	····729[illegible]
310	····10465	····13248	····18552	····26387	····36765	····49708	····65219	····8331[illegible]
320	····20496	····23330	····28684	····36570	····47003	····59993	····755[illegible]	[illegible]
330	····30528	····33412	····38817	····46754	····57237	····70279	····85893	103904091
340	····40562	····43496	····48950	····56939	····67473	····80566	····96232	····14481
350	····50596	····53580	····59085	····67124	····77710	····[illegible]0855	103406571	····2487[illegible]
360	····60631	····63665	····69221	····77311	····87947	102901144	····1691[illegible]	····35265
370	····70667	····73752	····79358	····87499	····98186	····11434	····27254	····45659
380	····80704	····83839	····89496	····97687	102408426	····21725	····37506	····5605[illegible]
390	····90742	····93927	····99635	101907877	····18667	····32017	····4794[illegible]	····6644[illegible]
400	100400781	100904017	101409775	····1806[illegible]	····28905	····42310	····58285	····76846
410	····10821	····14107	····19916	····2826[illegible]	····39152	····52604	····6[illegible]63[illegible]	····87243
420	····20862	····24199	····30058	····3845[illegible]	····49396	····62900	····7897[illegible]	····9[illegible]64[illegible]
430	····30904	····34291	····40201	····48640	····59641	····73196	····89326	10400804[illegible]
440	····40948	····44384	····50345	····58841	····69887	····83493	····99674	····1[illegible]44[illegible]
450	····50991	····54479	····60489	····69037	····80133	····9379[illegible]	103510024	····2884[illegible]
460	····61037	····64574	····70636	····79234	····90381	103004091	····20375	····3924
470	····[illegible]083	····74671	····80783	····89432	102500630	····1429[illegible]	[illegible]	····106[illegible]
480	····[illegible]113[illegible]	····8476[illegible]	····90931	····99631	····10880	····2469[illegible]	····4108[illegible]	····600[illegible]
490	····91178	····94867	101501080	102009831	····2113[illegible]	····34995	····51435	····7046[illegible]
500	100501227	101004966	····11230	····20032	····31384	····45299	····61790	····80816

Figure 1.6.

precisely who first conceived of the idea is much more difficult to ascertain, and I can not. Kepler wrote: "... [this] led the way for Justus Byrgius to the very same logarithms many years before Napier's edition, although this man, a procrastinator and guardian of his secrets, abandoned his child at birth and did not rear it for publicity."[33] Whatever the facts are, Bürgi's tables did appear after Napier's. It seems to be more or less unknowable who was the first inventor and also unimportant for our purposes. Possibly Bürgi started before Napier but his tables are not nearly as convenient as are the latter's. Both should share, along with Briggs, very great credit.

The fame of logarithms spread very quickly from Britain across the continent. By 1618 there appeared a five-place version of Napier's table put out in Cologne by Benjamin Ursinus — his German name was Behr — under the title *Cursus Mathematici Practici*; he later put out a more ambitious work.[34] This latter work had one more decimal place than did Napier's table and tabulated results every 0°;0,10 instead of every 0°;1. In 1624 Kepler brought out his tables which we discuss later. Quite soon after this, we find other authors in various countries bringing out tables, some following Napier but most following Briggs. The interested reader should consult Henderson's careful bibliographic account of the subject.

1.5. Interpolation

Al-Bīrūnī, a great mathematician writing in Arabic, attempted to use quadratic interpolation, and it is probable that there were many imitators and followers in later times. But it was not until the seventeenth century that we find an extensive study and development of the calculus of finite differences. Briggs played a very important part in this as we have already mentioned, and used sophisticated interpolation formulas to subtabulate his logarithmic tables. However, we need to go back a little in time to a colleague of his, Sir Thomas Harriot (1560–1621) of Oxford, to find the real inventor of the calculus of finite differences. Harriot was Walter Raleigh's teacher and was deeply interested with Briggs in the problems of oceanic navigation.[35] (We should remember this was the period of the great Elizabethan sailors: Drake, Frobisher, Hawkins, and Raleigh.) In fact he went with a group Raleigh sent out to colonize Virginia and then returned.

Harriot was evidently a deeply learned man but is now a neglected figure, perhaps because so little of his work has as yet been carefully studied. It consists of manuscripts in the British Museum and in the archives of Petworth House in Sussex. From our parochial point of view his development of

[33] "...qui..., Justo Byrgio multis annis ante editionem Neperianam viam praeiverunt ad hos ipsissimos logarithmos. Etsi homo cunctator et secretorum custos foetum in partu destituit, non ad usos publicos educavit." For a discussion of this topic cf. Tropfke, *GEM*, Vol. II, pp. 210*n*–211*n*. The tables of Bürgi come from Napier, *NTV*, Plates XII, XIII, following p. 210.

[34] Henderson [1916], p. 33.

[35] Lohne [1965] and Newton, *Papers*, Vol. IV (1674–1684), pp. 4–5, pp. 43–45.

interpolation and subtabulation are noteworthy. Lohne gives us his general interpolation formula, thus:

$$
\begin{array}{l|l|l|l|l}
G + \begin{array}{c} k\\ \hline 1,n \end{array} F + \begin{array}{r} k-1,n\\ k\\ \hline 1,n\\ 2,n \end{array} & D + \begin{array}{r} k-2,n\\ k-1,n\\ k\\ \hline 1,n\\ 2,n\\ 3,n \end{array} & C + \begin{array}{r} k-3,n\\ k-2,n\\ k-1,n\\ k\\ \hline 1,n\\ 2,n\\ 3,n\\ 4,n \end{array} & B + \begin{array}{r} k-4,n\\ k-3,n\\ k-2,n\\ k-1,n\\ k\\ \hline 1,n\\ 2,n\\ 3,n\\ 4,n\\ 5,n \end{array} & A
\end{array}
\tag{1.16}
$$

etc.

Here the symbols $A, B, \ldots, G$ stand for the 5th, 4th, ..., 0th order differences. Moreover, the curious symbols in front of these letters are

$$\frac{k}{n}, \frac{(k-n)k}{1\cdot n\cdot 2\cdot n}, \ldots, \frac{(k-4n)(k-3n)(k-2n)(k-n)k}{1\cdot n\cdot 2\cdot n\cdot 3\cdot n\cdot 4\cdot n\cdot 5\cdot n}.$$

Thus Harriot's formula becomes in more modern form

$$f\left(x - \frac{k}{n}\right) = f(x) + \frac{k}{n}\Delta f(x) + \frac{(k/n)(k/n-1)}{2!}\Delta^2 f(x) + \cdots$$
$$+ \frac{(k/n)(k/n-1)\cdots(k/n-4)}{5!}\Delta^5 f(x),$$

where, of course,

$$\Delta f(x) = f(x+1) - f(x), \ldots, \Delta^5 f(x) = \Delta(\Delta^4 f(x)).^{36}$$

Not only is this result very elegant, but his method of finding it and using it to subtabulate is also quite nice. Lohne comments that Harriot proceeded from special cases to general rules in a very conscious way. In Figure 1.7(a) we show his table for the function tabulated in column g. The columns f, d, c, b, a are the differences. Then on the right in column G he tabulated his function for every other value of his independent variable and showed its differences in columns F, D, C, B, and A. In the little insets he displayed similar schemes. E.g., for the case $n = 3$ he displayed every third entry in column f, together with the associated differences, etc.

He applied these ideas to subtabulation in Figure 1.7(b), where we see the function in column C — $25x^2 + 20x + 3$ — tabulated at a unit spacing and then subtabulated to $\frac{1}{5}$ spacing in the column c with the help of the relation

[36] It was he who invented the symbols >, <. (He also wrote on the colony of Virginia.)

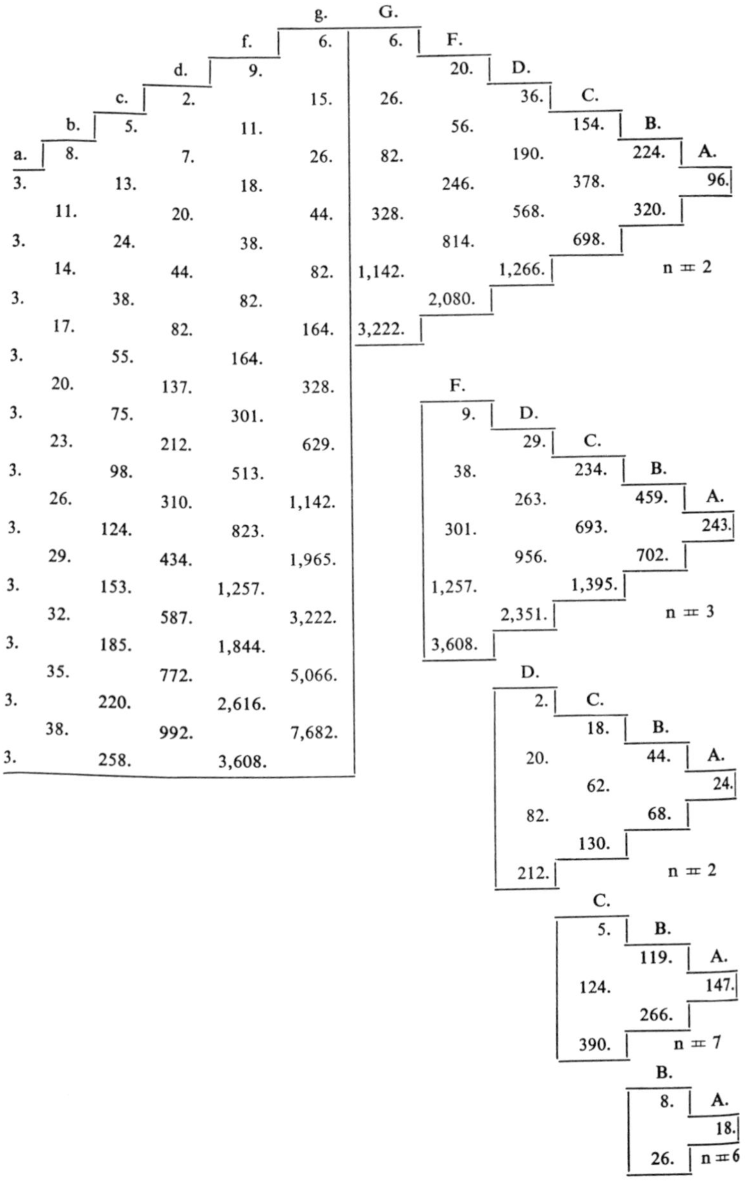

Figure 1.7(a).

(1.16); there we set $n = 5, k = 1, 2, 3, 4$, and use the values of the differences in the boxes, i.e., we really want the relation

$$c = C + \frac{k}{n} B + \frac{(k/n)(k/n - 1)}{2!} A.$$

This material of Harriot is in a text which Lohne dates as 1611 or perhaps a

k	C	B	A
0	3		
		45	
1	48		50
		95	
2	143		50
		145	
3	288		50
		195	
4	483		50
		245	
5	728		

$\frac{k}{n}$	c	b	a
0	3		
		5	
$\frac{1}{5}$	8		2
		7	
$\frac{2}{5}$	15		2
		9	
$\frac{3}{5}$	24		2
		11	
$\frac{4}{5}$	35		2
		13	
$\frac{5}{5}$	48		

Figure 1.7(b).

little earlier. This dating suggests strongly that Briggs, upon his arrival at Oxford, learned of this interpolation formula from Harriot, since the two overlapped by about two years.

In any event Briggs shows in his *Arithmetica Logarithmica*, *ARITH*, how to subtabulate by several methods. These results appear in his Chapters XII and XIII but unfortunately without proof or suggestion of how he found his results. A number of suggestions have been offered as to how he may have arrived at his rules.[37] No one can now be sure how he arrived at his formulas, but both Legendre and Lagrange have high praise for him.[38] The interested reader should also consult Lagrange, *Oeuvres*, Vol. VII, pp. 535–553. It was Lagrange who was perhaps the first to recognize Briggs's great contributions to the field. He says that the French honor Mouton, canon of Lyon, as the discoverer of interpolation . . . "I find, however, that Henry Briggs, who calculated the logarithms of the numbers from 1 to 20,000 and from 90,000 to 100,000, proposed a method of interpolation to fill this gap based upon a consideration of successive differences which he said had previously been successfully used to construct a trigonometrical table for sines and tangents of every hundredth of a degree. (See his *Arithmetica Logarithmica*, Chap. XIII and his *Trigonometria Britannica*, Chap. XII.)" Compare also, Legendre [1817], pp. 219–222, for additional work by Legendre on interpolation and preparation of large tables. Compare as well, Collins [1712], pp. 32–37, for a letter from Leibniz to Oldenburg dated February 3, 1672/3.

The first of Briggs's schemes is a method of interpolating nine values between any two entries in a table where second differences are constant or

[37] Delambre, *MOD*, Vol. II, pp. 407ff. and pp. 83–84 as well as Maurice [1847]. This latter paper is an amplification by Maurice of Legendre's analysis of Briggs's methods. This appears in the *Connaissance des Temps* 1817; see pp. 219ff. for Legendre's views.
[38] Turnbull [1932] and Newton, *Papers*, Vol. IV, pp. 43*n*–45*n*.

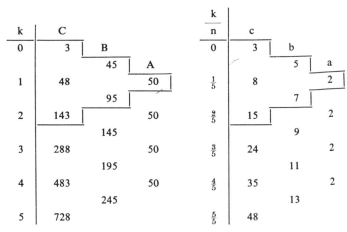

nearly so. This problem is studied in Chapter XII of *ARITH*. Suppose it is desired to put these nine values between A and $A + a$. Then Briggs handled the matter as in the following table where

$$f(x) = A + xa + \frac{x(x-1)b}{2}:$$

	First Differences
A	
	$\frac{1}{10}a - \frac{9}{200}b$
$A + \frac{1}{10}a - \frac{9}{200}b$	
	$\frac{1}{10}a - \frac{7}{200}b$
$A + \frac{2}{10}a - \frac{16}{200}b$	
	$\frac{1}{10}a - \frac{5}{200}b$
$A + \frac{3}{10}a - \frac{21}{200}b$	
	$\frac{1}{10}a - \frac{3}{200}b$
$A + \frac{4}{10}a - \frac{24}{200}b$	
	$\frac{1}{10}a - \frac{1}{200}b$
$A + \frac{5}{10}a - \frac{25}{200}b$	
	$\frac{1}{10}a + \frac{1}{200}b$
$A + \frac{6}{10}a - \frac{24}{200}b$	
	$\frac{1}{10}a + \frac{3}{200}b$
$A + \frac{7}{10}a - \frac{21}{200}b$	
	$\frac{1}{10}a + \frac{5}{200}b$
$A + \frac{8}{10}a - \frac{16}{200}b$	
	$\frac{1}{10}a + \frac{7}{200}b$
$A + \frac{9}{10}a - \frac{9}{200}b$	
	$\frac{1}{10}a + \frac{9}{200}b$
$A + a$	

The table is borrowed from Henderson [1926], p. 46. (Cf. also, Hutton [1801], p. 72.) If the second difference b varies, Briggs chose for b the average value of the second differences over the interval.

This formula certainly shows Briggs's acquaintance with the standard interpolation procedures, but his next methods show much deeper understanding. In Chapter XIII, Briggs took up a different and very unusual set of formulas for subtabulating, again without any indication of how he was led to it. Let us define what Briggs called *mean* differences by means of operator methods somewhat in imitation of Legendre. We let

$${}_n\Delta^1 = \frac{E^n - I}{n} \qquad (Ef(x) = f(x+1)), \tag{1.17}$$

i.e.,

$${}_n\Delta^1 f(x) = \frac{f(x+n) - f(x)}{n};$$

and in general let

$${}_n\Delta^p = \frac{{}_n\Delta^{p-1}E^n - {}_n\Delta^{p-1}}{n} \qquad (p = 2, 3, \ldots). \tag{1.17'}$$

Thus we have

$${}_n\Delta^p = ({}_n\Delta^1)^b = \left(\frac{E^n - I}{n}\right)^p. \tag{1.18}$$

Finally let us set

$${}_n\underline{\Delta}^p = n^p\,{}_n\Delta^p$$

so that ${}_n\Delta^p$ is our ordinary pth difference. Thus Briggs says: "Take the first, second, third, fourth, &c., differences . . . and divide the first differences by 5, the second by 5^2 . . . and these . . . will be what I call the *mean* differences of the corresponding orders." The problem Briggs solved was this: "Given a series of equidistant numbers and their logarithms, to find the logarithms of the four numbers interpolated at equal intervals between each adjacent two of the given numbers."

He showed how to relate the mean differences ${}_n\Delta^p$ ($n = 3, 5$) to the corresponding ${}_1\Delta^p$ by certain "corrections." By means of these, e.g., Briggs proposed filling in the hiatus (from 20,000 to 90,000) in his logarithmic tables. He also used them in his construction of his trigonometric tables.[39] He was particularly incensed with his Dutch editor Adrian Vlacq, who brought out the *Arithmetica Britannica* in Gouda in 1628, because Vlacq omitted Chapter XIII from this edition.[40]

Briggs included his methods in his *Trigonometria* and wrote there: "The method of corrections has been treated by me in Chapter 13 of the *Arithmetica Logarithmica*, London edition. This chapter together with the following one were omitted from the Dutch edition without my consent or knowledge. Moreover, the producer of that edition, a man in other respects industrious and not unlearned, seems not to have followed my intentions in all matters. Therefore lest anything be missing to anyone who wishes to produce the whole Canon I have judged that certain very necessary things need to be added."[41]

Let us suppose given a function f_x at the values $x = 0, n, 2n, \ldots$, where $n = 2m + 1$; we wish to find the remaining values $f_1, f_2, \ldots, f_{n-1}, f_{n+1}, \ldots$. Accordingly we turn to the relations (1.18) and express them as

$$ {}_n\Delta^p = \frac{1}{n^p}(E^{n-1} + E^{n-2} + \cdots + E + I)^p(E - 1)^p. $$

Note that

$$ E + E^{-1} = E^{-1}(E + I)^2 + 2I = \Delta^2_{-1} + 2I, $$

and hence that

$$ E^2 + E^{-2} = (E + E^{-1})^2 - 2I = (\Delta^2_{-1} + 2I)^2 - 2I = \Delta^4_{-2} + 4\Delta^2_{-1} + 2I; $$

---

39 Briggs, *TRIG.*

40 Vlacq, *AL.*

41 "Modus correctionis à me traditus est Arithmeticae Logarithmicae Capite 13, in Editione Londinensi: Istud autem Caput unâ cum sequenti in Editione Batava me inconsulto & inscio omissum fuit: nec in omnibus, Editionis illius Author Vir alioqui industrius & non indoctus meam mentem videtur assequutus: Ideoque ne quicquam desit cuiquam, qui integrum Canonem conficere cupiat, quaedam maximè necessaria illinc huc transferenda censui."

since $(\Delta^2_{-1})^2 = \Delta^2_{-1}\Delta^2_{-1} = \Delta^4_{-2}$. (We could express $E^i + E^{-i}$ for $i > 2$ as well, but we do not need to.) Then, symbolically,

$$
\begin{aligned}
{}_3\Delta^p_{pq} &= \frac{1}{3^p}(E + E^{-1} + I)^p(E - I)^p E^{p(q+1)} \\
&= \frac{1}{3^p}(3I + \Delta^2_{-1})^p \Delta^p_{p(q+1)} \\
&= \left(\Delta^1_{q+1} + \frac{1}{3}\Delta^3_q\right)^p,
\end{aligned}
\tag{1.19}
$$

and

$$
\begin{aligned}
{}_5\Delta^p_{pq} &= \frac{1}{5^p}(E^2 + E^{-2} + E + E^{-1} + I)^p(E - I)^p E^{p(q+2)} \\
&= \frac{1}{5^p}(5I + 5\Delta^2_{-2} + \Delta^4_{-2})^p \Delta^p_{p(q+2)} \\
&= \left(\Delta^1_{q+2} + \Delta^3_{q+1} + \frac{1}{5}\Delta^5_q\right)^p.
\end{aligned}
\tag{1.20}
$$

Briggs evaluated these formulas for $p = 1, 2, \ldots, 20$.[42] His results appear in tabular form. He considered the basic relation to be, e.g., ${}_1\Delta^4 = {}_5\Delta^4$ minus certain corrections. Briggs gives these "corrections" in the following table:

20								
19								
18	18 (20)							
17	17 (19)							
16	16 (18)	1232 (20)						
15	15 (17)	1080 (19)						
14	14 (16)	938 (18)	4004 (20)					
13	13 (15)	806 (17)	3174 (19)					
12	12 (14)	684 (16)	2464 (18)	62964 (20)				
11	11 (13)	572 (15)	1870 (17)	43120 (19)				
10	10 (12)	470 (14)	1380 (16)	28380 (18)	43440 (20)			
9	9 (11)	378 (13)	984 (15)	17784 (17)	23688 (19)			
8	8 (10)	296 (12)	672 (14)	10472 (16)	11872 (18)	101248 (20)		
7	7 (9)	224 (11)	434 (13)	5684 (15)	5320 (17)	36680 (19)		
6	6 (8)	162 (10)	260 (12)	2760 (14)	2040 (16)	10760 (18)	4080 (20)	
5	5 (7)	110 (9)	140 (11)	1140 (13)	620 (15)	2280 (17)	560 (19)	
4	4 (6)	68 (8)	64 (10)	364 (12)	128 (14)	272 (16)	032 (18)	0016 (20)
3	3 (5)	36 (7)	22 (9)	72 (11)	12 (13)	008 (15)		
2	2 (4)	14 (6)	4 (8)	04 (10)				
1	1 (3)	2 (5)						
A	B	C	D	E	F	G	H	I

[42] Newton, *Papers*, Vol. IV, pp. 44*n*–45*n*. (There Whiteside notes that Cotes also worked on this material in his *Canontechnia*.) Cf. also, Delambre, *MOD*, Vol. II, pp. 416–418.

Thus we read in row 4 that

$$ {}_1\Delta^5 = {}_5\Delta^4 - [4_1\Delta^6 + 6.8_1\Delta^8 + 6.4_1\Delta^{10} + 3.64_1\Delta^{12} + 1.28_1\Delta^{14} + 0.272_1\Delta^{16} + 0.032_1\Delta^{18} + 0.0016_1\Delta^{20}]. $$

These results can be verified from (1.20) with $p = 4$.[43]

The remaining nontrivial problem facing Briggs was how to solve the relations (1.19) and (1.20) for the ${}_1\Delta^p$ given the ${}_n\Delta^p$. He solved this problem by noting that if ${}_1\Delta^p$ is constant for some p, then all higher order differences vanish. Thus in the formula above if ${}_1\Delta^p$ $(p > 5)$ vanishes, then ${}_1\Delta^4 = {}_5\Delta^4$. In fact in this case

$$ {}_1\Delta^5 = {}_5\Delta^5, \qquad {}_1\Delta^4 = {}_5\Delta^4, \qquad {}_1\Delta^3 = {}_5\Delta^3 - 3_1\Delta^5, $$

$$ {}_1\Delta^2 = {}_5\Delta^2 - 2_1\Delta^4, \qquad {}_1\Delta^1 = {}_5\Delta^1 - {}_1\Delta^3 - \frac{1}{5}\,{}_1\Delta^5. $$

These follow directly out of (1.20) or, equally well, from an inspection of Briggs's table.

To illustrate the use of his method Briggs included in his *Trigonometria*, as well as in his *Arithmetica*, a number of examples. I give his illustration based on the function $y = x^6$ in Figures 1.8(a), 1.8(b). In this example Briggs has tabulated $y = x^6$ for $x = 50, 55, 60, 65$, and 70, together with its differences. This is evidently the first table in the Figures 1.8. In the second table he has formed the mean differences by dividing the ordinary ones by the powers of 5. Thus ${}_5\Delta^6$ is 11250000 divided by 5^6, etc. In the third table Briggs formed his ${}_1\Delta$ for several values. (Notice his notation as well as his remark that the sixth and fifth differences are not to be corrected.) His fourth and final table shows the subtabulated results with dashes in front of the given quantities and differences. (These appear in *Trigonometria*, Chap. XII.[44]

This is essentially how Briggs filled in his table, proceeding from quantities he knew. It shows the very remarkable progress made in the very few years from Napier to Briggs. On the one hand we see Briggs calculating little beyond the logarithms of 25 primes by a comparatively few square roots, with the help of his elegant difference schemes and calculating the rest of his table by subtabulations. On the other hand, we saw how laboriously Napier evaluated his sequences.[45]

---

[43] Briggs, *TRIG*, p. 38. (Note that ${}_1\Delta^p = \Delta^p$).

[44] Briggs, *TRIG*, pp. 40ff.

[45] It should be noted that Briggs published his great *Arithmetica Logarithmica* in 1624. It contained the logarithms of 30,000 numbers to 14 places each, together with their differences. The tables go from 1 to 20,000 and 90,000 to 100,000. The missing portions were filled in by Adrian Vlacq of Gouda who republished the work in 1627/28 to ten decimal places, but he included the logarithms of the sine, cosine, and tangent at an interval of 4 minutes of arc. Vlacq also published in 1633 for Briggs in Gouda the logarithms of the sine and tangent for 1/100th of a degree to 14 places plus certain other trigonometric tables. Upon Briggs's death Henry Gellibrand, a friend and professor at Gresham, completed these trigonometrical works at Briggs's request and published them as *Trigonometria Britannica*. For a detailed account the reader may wish to consult some account such as Henderson's in the Bibliography.

Differentiæ Datæ

Sextæ	*Quintæ*	*Quartæ*	*Tertiæ*	*Secundæ*	*Primæ*	*Sextæ Potestates datæ. Quadratocubi*	*Latera data.*
		·564375000		4734406250		15625000000	50
	118125000		2185312500		12055640625		
11250000		682500000		6919718750		· 27680640625	55
	129375000		2867812500		18975359375		
11250000		811875000		9787531250		46656000000	60
	140625000		3679687500		28762890625		
11250000		952500000		13467218750		75418890625	65
	151875000		4632187500		42230109375		
		1104375000		18099406250		117649000000	70

Differentiæ Mediæ.

6æ.	5æ. *mediæ*	*Diff.* 4æ. *mediæ.*	*Diff.* 3æ. *med.*	*Diff.* 2æ. *med.*	*Diff.* 1æ. *med.*	*Quadratocubi.*	*Lat*
720		903000		189376250		15625000000	50
	37800		17482500		2411128125		
720		1092000		276788750		27680640625	55
	41400		22942500		3995071875		
720		1299000		391501250		46656000000	60
	45000		29437500		5752578125		
720		1524000		538688750		75418890625	65
	48600		37057500		8446021875		
720						117649000000	70

Specimen Differentiarum correctarum.

Differentiæ sextæ & quintæ non corriguntur.

4a. media . . . 903000	3a. media . . . 17482500	2a media 189376250
4or. 6æ. aufer. 2880	3. quintæ auf: 113400	2æ. 4tæ. correctæ 1800240
4a. correcta 900120	3a correcta 17369100	1 4 6æ 1008
		2a. correcta. 187575002
4a. media. 1092000	3a. media 22942500	1a. media 2411128125
4. sextæ aufer: 2880	3. quintæ auf: 124200	1. 3a. cor. auf: 17369100
4a. correcta. 1089120	3a. correcta 22818300	; quintæ aufer: 7560
		1a. correcta 2393751465

Qua-

Figure 1.8(a).

6ª.	5ª.	Differ.4ª.	Differ.3ª.	Differ.2ª.	Differ.1ª.	Quadrato cubi.	Lat.
720		− 900120		−187575002		− 15625000000	50
	36360		15459060		1971287801		
		936480		203034062		17596287801	1
	37080		16395540		2174321863		
		973560		219429802		19770609664	2
	− 37800		− 17369100		−2393751465		
		1011360		236798702		22164361129	3
	38520		18380460		2630550167		
		1049880		255179162		24794911296	4
	39240		19430340		2885729329		
720		1089120		−274609502		− 27680640625	55
	39960		20819460		3160338831		
		1129080		295128962		30840979456	6
	40680		21648540		3455467793		
		1169760		316777502		34296447249	7
	−41400		− 22818300		−3772245295		
		1211160		339595802		38068692544	8
	42120		24029460		4111841097		
		1253280		363625262		42180533641	9
	42840		25282780		4475466359		
720		-1196120		−388908002		− 46656000000	60
	43560		26578860		4864374361		
		1339680		415486862		51520374361	1
	44280		27918540		5279861223		
		1383960		443405402		56800235584	2
	−45000		− 29302500		−3723266625		
		1428960		472707902		62523502209	3
	45720		30731460		6195974527		
		1474680		503439362		68719476736	4
	46440		32206140		6699413889		
720		1521120		−535645502		− 75418890525	65
	47160		33727260		7235059391		
		1568280		569372762		82653950016	6
	47880		35295540		7804432153		
		2616160		604668302		90458382169	7
	−48600		− 36911700		−8409100455		
		1664760		641580002		98867482628	8
	49320		38576460		9050680457		
		1714080		680156462		107918163081	9
	50040		40290540		9730836919		
720		- 1764180		−720447002		−117649000000	70

Figure 1.8(b).

1.6. Vieta and Briggs

The rate of progress in mathematics during the period from Vieta onwards is staggering. Very possibly it was due in important measure to the improvements in notation that allowed mathematicians to comprehend functional relations in a way that geometrical ratios hardly do. We shall see as we proceed how great were the advances in mathematics that came with notational improvements. Notation had by now been developed sufficiently to permit deep algebraic generalizations.

Perhaps it is relevant at this point to mention a beautiful solution affected by Vieta of an equation of degree 45. In 1593 Adrianus Romanus (Adriaan van Roomen) issued a challenge to all mathematicians of the world to solve the equation arising when the expression

$$\begin{aligned}45(1) &- 3795(3) + 9{,}5634(5) - 113{,}8500(7) + 781{,}1375(9) - 3451{,}2075(11)\\ &+ 1{,}0530{,}6075(13) - 2{,}3267{,}6280(15) + 3{,}8494{,}2375(17)\\ &- 4{,}8849{,}4125(19) + 4{,}8384{,}1800(21) - 3{,}7865{,}8800(23)\\ &+ 2{,}3603{,}0652(25) - 1{,}1767{,}9100(27) + 4695{,}5700(29)\\ &- 1494{,}5040(31) + 376{,}4565(33) - 74{,}0259(35)\\ &+ 11{,}1150(37) - 1{,}2300(39) + 945(41)\\ &- 45(43) + 1(45)\end{aligned}$$

is equated to a given constant A.[46] The quantities such as (7) mean x^7.

The general problem reduces to finding the roots of the equation of degree 45 one obtains by writing $2 \sin 45\alpha$ as a polynomial in $2 \sin \alpha$ and equating it to a number A. The form of the polynomial above is that of Romanus. He also considered a variety of similar equations and gave formulas for the sides of regular polygons of 8, 15, 32, 60, 96, 120, and 192 sides.

Vieta in 1595 with his profound knowledge of mathematics took up the problem of finding the roots of the above equation in his *Ad Problema*.[47] He found not just one root, as had Romanus, but all 23 positive ones. Romanus in his challenge gave three examples, the second of which is erroneous (cf. below): the A in the right-hand member of his equation was

$$\begin{aligned}&(1)\ A = \{2 + [2 + (2 + 2^{1/2})^{1/2}]^{1/2}\}^{1/2} = 2 \sin 84°;22,30,\\ &(2)\ A = [2 + (2 - \{2 - [2 - (2 - 2^{1/2})^{1/2}]^{1/2}\}^{1/2})^{1/2}]^{1/2},\\ &(3)\ A = (2 + 2^{1/2})^{1/2} = 2 \sin 67°;30,\end{aligned}$$

and he asked for a solution when

$$(4)\ A = \left\{1\frac{3}{4} - \left(\frac{5}{16}\right)^{1/2} - \left[1\frac{7}{8} - \left(\frac{45}{64}\right)^{1/2}\right]^{1/2}\right\}^{1/2} = 2 \sin 12°.$$

---

[46] This appears in van Roomen [1593]. Cf. Vetter [1930] or Vieta, *OP*, pp. 305–306.
[47] Vieta, *OP*, pp. 305–324.

Incidentally, his solutions to cases (1), (2), (3) above are

$$(1)\ x = (2 - \{2 + [2 + (2 + 3^{1/2})^{1/2}]^{1/2}\}^{1/2})^{1/2} = 2 \sin 1°,52,30,$$

$$(2)\ x = [2 - (2 + \{2 + [2 + (2 + 3^{1/2})^{1/2}]^{1/2}\}^{1/2})^{1/2}]^{1/2} = 2 \sin 0°;56,15,$$

$$(3)\ x = \left(2 - \left\{2 + \left(\frac{3}{16}\right)^{1/2} + \left(\frac{15}{16}\right)^{1/2} + \left[\frac{5}{8} - \left(\frac{5}{64}\right)^{1/2}\right]^{1/2}\right\}^{1/2}\right)^{1/2}$$
$$= 2 \sin 1°,30.$$

Vieta points out (p. 309) that Romanus had an error in his second example. He correctly states it should be "*R* binomia 2 − *R* bin. 2 − *R* bin. 2 + *R* bin. 2 + *R* bin. 2 + *R*2" = $[2 - (2 - \{2 + (2 + 2^{1/2})^{1/2}\}^{1/2})^{1/2}]^{1/2}$ = 2 sin 42°;11,15.

Vieta's solution is both beautiful and typical of the man. Firstly Vieta solved the equation $3N - 1C = A$, i.e., $3x - x^3 = A$ where A is the challenge (4) above of Romanus — he also took $A = \sqrt{2}$ as another problem and gave its solution.[48] He calls his solutions of this $x = B$. (This equation is evidently the one used to trisect the angle A.) Secondly he solves $3y - y^3 = B$ and calls his solutions D. Thirdly he solves the quintic $5z - 5z^3 + z^5 = D$ and calls these solutions $z = G$. Notice that $45 = 3 \cdot 3 \cdot 5$ so Vieta has in effect split his equation of degree 45 into the problems

$$3x - x^3 = 2 \sin 3(15\alpha), \qquad 3y - y^3 = 2 \sin 3(5\alpha),$$
$$5z - 5z^3 + z^5 = 2 \sin 5\alpha.$$

When $A = 2 \sin 45° = \sqrt{2}$ Vieta gave as his solutions $2 \sin \varphi$ with $\varphi = 1$, 9, 17, 25, 33, 41, 49, 57, 65, 73, 81, 89, and $\varphi = 3$, 11, 19, 27, 35, 43, 51, 59, 67, 75, 83. This is shown in Figure 1.9(a) taken from his *Opera*. In that table the solutions are given to 15 decimals; the "Arabic" numbers are $2 \sin \varphi$ where φ appears in the last column in Latin. (There is a misprint in the entry for 2 sin 1°; it should be 3,490,481. . . .)

The solution to Romanus's original problem appears to nine decimals in Figure 1.9(b). There the angles are given in Roman numerals. Thus 0°;16 appears as ∴ XVI and 3°;44 as III XLIV, etc. (Note the spacing of 8° between angles.) In Figure 1.10(a) Vieta drew BC as the side of the regular decagon inscribed in a circle of radius 1, and BD as that of the hexagon. He then had arc $BD = 60°$, arc $BC = 36°$, and arc $CD = 24°$. He trisected CD and found $CE = 8°$; this he trisected and found $CF = 2°;40$. A fifth part CG of arc $CF = 0°;32$. He asserted that the line CD is Romanus's value A in (4) above and that the line CG is his solution. (Recall that the chord of 2α is $2R \sin \alpha$.)

To find the other positive roots Vieta used Figure 1.10(b) where the line CG is the same as before. Then from G he laid off the arc $G\alpha = 8°$. Next he divided the remaining arc of 352° into 22 equal parts $\alpha\beta, \beta\gamma, \ldots$. Thus his tabulation in Figure 1.9(b) gives the lengths of the chords $C\alpha, C\beta, C\gamma, \ldots$.

[48] The solution for $A = \sqrt{2}$ is on p. 311 and for the other A on p. 314 of his *Opera*.

Terminus poſterior datus. Commu-nis divi-ſor.	141,421, / 100,000,	356,237,309 / 000,000,000		PARTIVM. Quadraginta quinq;.
¶ *Claſsicus terminus prior quæſitus.*	3,490	681,287,456		Vnius.

Endecas inſuper terminorum una, de quibus ſingulis idem thema poteſt explicari.

I.	31,286	893,008,046	Quibus terminis ſimiles ſunt rectæ quæ in circulo ſubtenduntur duplo circumferentiarum.	Novem.
¶ II.	58,474	340,944,547		Septendecim.
¶ III.	84,523	652,348,139		Viginti quinque.
IV.	108,927	807,003,005		Triginta trium.
¶ V.	131,211	805,798,101		Quadraginta unius.
¶ VI.	150,941	916,044,554		Quadraginta novem.
VII.	167,734	113,589,085		Quinquaginta ſeptẽ.
¶ VIII.	181,261	557,407,329		Sexaginta quinque.
IX.	191,260	951,192,607		Septuaginta trium.
X.	197,537	668,119,027		Octoginta unius.
¶ XI.	199,969	539,031,270		Octoginta novem.

Endecas altera

				PARTIVM.
I.	10,467	191,248,599		Trium.
¶ II.	38,161	799,075,309	Quibus terminis ſimiles ſunt rectæ quæ in circulo ſubtenduntur duplo circumferentiarum.	Vndecim.
¶ III.	65,113	630,891,431		Novendecim.
IV.	90,798	099,947,909		Viginti ſeptem.
¶ V.	114,715	287,270,209		Triginta quinque.
¶ VI.	136,399	672,012,499		Quadraginta trium.
VII.	155,429	192,291,394		Quinquaginta unius.
¶ VIII.	171,433	460,140,422		Quinquaginta novẽ.
¶ IX.	148,100	970,690,488		Sexaginta ſeptem.
X.	193,185	165,257,813		Septuaginta quinq;.
¶ XI.	198,509	230,328,264		Octoginta trium.

Poſita videlicet ſemidiametro 1. circumferentia vero tota circuli partium tercentũ ſexaginta.

Figure 1.9(a).

AD ADR. ROMANI PROBLEMA

				PARTIVM.	SCRVP.
In numeris qualium AC	100,000	000		XC.	∴
Talium data CA sit ἐξηγηστα	41,582	338		XII.	∴
Clasica CG quaesita	930	839	*Qualium*	∴	XVI.
Reliquarum Endecas prima			*autem*	III.	XLIV.
Cα	13,022	572			
Cβ	40,671	389	*tota cir-*	XI.	XLIV.
Cγ	67,528	585	*culi cir-*	XIX.	XLIV.
Cδ	63,071	414	*cumfe-* *rentia est*	XXVII.	XLIV.
Cε	116,802	731	*partium*	XXXV.	XLIV.
Cζ	136,260	439	*IIICLX.* *taliũ ipsa,*	XLIII.	XLIV.
Cη	157,027	354	*quae à re-*	LI.	XLIV.
Cϑ	172,737	783	*ctis designa-* *tis subten-*	LIX.	XLIV.
Cι	185,086	061	*duntur, cir-*	LXVII.	XLIV.
Cκ	193,831	852	*cumferen-* *tia semis-*	LXXV.	XLIV.
Cλ	198,849	238	*ses sunt*	LXXXIII.	XLIV.
Endecas altera.					
Cχ	28,756	098		VIII.	XVI.
Cϕ	56,021	654		XVI.	XVI.
Cυ	82,196	811		XXIV.	XVI.
Cτ	106,772	100		XXXII.	XVI.
Cσ	129,269	199		XL.	XVI.
Cρ	149,250	207		XXVIII.	XVI.
Cϖ	166,326	235		LVI.	XVI.
Co	180,164	914		LXIV.	XVI.
Cξ	190,496	888		LXXII.	XVI.
Cν	197,121	055		LXXX.	XVI.
Cμ	199,908	485		LXXXVIII.	XVI.

Figure 1.9(b).

It is interesting to notice that in his paper Vieta discriminates between geometers and analysts, and also that he uses the word "coefficient" in the modern sense. In fact he seems to have coined many terms, only a few of which are still in the literature. He was doing such pioneering work that he was forced to introduce terms for concepts new to his time.

Briggs was able to find both the binomial coefficients and various related sequences with the help of algorithms he referred to as his "wonderfully

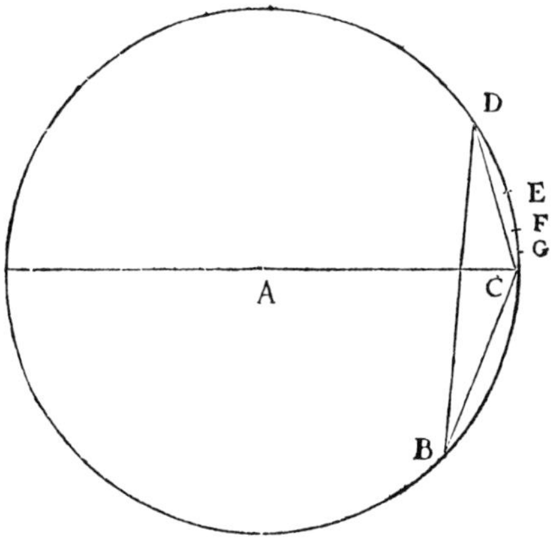

Figure 1.10(a).

useful abacus," *Abacus Panchrestus*, Πάγχρηστος (Figures 1.11(a),(b)). In his *Trigonometria* (cited below) he constructed several of these abaci. The first, and perhaps most important was the one which yielded the binomial coefficients for him. Thus, e.g., the diagonal proceeding from column 9, row 1 by unit steps to the right and down to column 1, row 9 is a tabulation of $\binom{9}{r-1}$ for $r = 1, 2, \ldots, 9$ and similarly for other parallel diagonals.

Another usage by Briggs of this abacus was to find the coefficients to express sin nI, cos nI as polynomials in sin I, cos I. Thus for n odd

$$2 \sin nI = (-1)^{(n-1)/2}\Big[(2 \sin I)^n - \frac{n}{1!}(2 \sin I)^{n-2} + \frac{n(n-3)}{2!}(2 \sin I)^{n-4}$$
$$- \frac{n(n-4)(n-5)}{3!}(2 \sin I)^{n-6}$$
$$+ \frac{n(n-5)(n-6)(n-7)}{4!}(2 \sin I)^{n-8} - \cdots\Big],$$

and the coefficients appear in the odd columns A, C, E, G, etc. Hence for $n = 9$ Briggs wrote + 1 ⑨ − 9 ⑦ + 27 ⑤ − 30 ③ + 9 ① to represent

$$2 \sin 9I = (2 \sin I)^9 - 9(2 \sin I)^7 + 27(2 \sin I)^5$$
$$- 30(2 \sin I)^3 + 9(2 \sin I).$$

In this expression the coefficient of $(2 \sin I)^9$ is the entry in column 9, row 1; of $(2 \sin I)^7$ is the sum of the entries in column 7, rows 1 and 2; of $(2 \sin I)^5$ is the sum of those in column 5, rows 2 and 3; etc. The signs of these coefficients are given in row 0. For even multiples of I an analogous procedure was

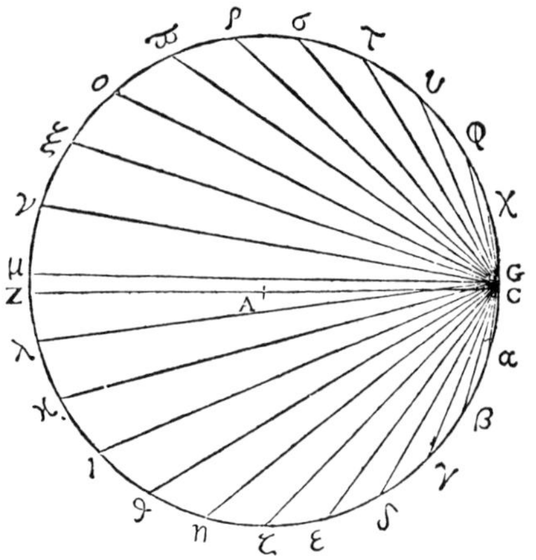

Figure 1.10(b).

followed by Briggs to express $\sin^2 2nI$ using the even columns B, D, F, H, etc. Thus

$$4 \sin^2 4I = -(2 \sin I)^8 + 8(2 \sin I)^6 - 20(2 \sin I)^4 + 16(2 \sin I)^2,$$

or in Brigg's notation -1 ⑧ $+ 8$ ⑥ $- 20$ ④ $+ 16$ ②. Notice again the coefficient of $(2 \sin I)^8$ is the entry in column 8, row 1; of $(2 \sin I)^6$ is the sum of the entries in column 6, rows 1 and 2; of $(2 \sin I)^4$ is the sum of the entries in column 4, rows 2 and 3, etc. The signs of the coefficients are again given in row 0.

He also appended another abacus for forming these same functions. I give

ABACVS ΠΑΓΧΡΗΣΤΟΣ.

M	L	K	I	H	G	F	E	D	C	B	A
−(12)	−(11)	+(10)	+(9)	−(8)	−(7)	+(6)	+( )	−(4)	−(3)	+(2)	(1)
1	1	1	1	1	1	1	1	1	1	1	1
13	12	11	10	9	8	7	6	5	4	3	2
91	78	66	55	45	36	28	21	15	10	6	3
455	364	286	220	165	120	84	56	35	20	10	4
1820	1365	1001	715	495	330	210	126	70	35	15	5
6188	4368	3003	2002	1287	792	462	252	126	56	21	6
18564	12376	8008	5005	3003	1716	924	462	210	84	28	7
50388	31824	19448	11440	6435	3432	1716	792	33[illegible]	12[illegible]	[illegible]	8
125970	75582	43758	24310	12870	6435	3003	1287	495	165	45	9
293930	167960	92378	48620	24310	11440	5005	2002	715	220	55	10
646646	352716	184756	92378	43758	19448	8008	3003	1001	286	66	11

Figure 1.11(a). The complete table contains 30 rows and is then repeated with scattered entries in rows 1 through 23, thus:

17383860											
30421755	13037895										
	21474180	8436285	3124550								18
			4686825	1562275	480700						19
					657800	177100	42504				20
	13037895						53130	10626	1771		21
	21474180	8436285	3124550						2024	253	22
			4686825	1562275	480700						23

Figure 1.11(b).

an excerpt in Figure 1.11(c).[49] For example, to express 2 sin 7I as a polynomial in (2 sin I) he used the abacus and found

$$2 \sin 7I = 7(2 \sin I) - 14(2 \sin I)^3 + 7(2 \sin I)^5 - (2 \sin I)^7.^{50}$$

M	L	K	I	H	G	F	E	D	C	B	A
−(12)	−(11)	+(10)	+(9)	−(8)	−(7)	+(6)	+(5)	−(4)	−(3)	+(2)	(1)
1	1	1	1	1	1	1	1	1	1	1	1
14	13	12	11	10	9	8	7	6	5	4	3
104	90	77	65	54	44	35	27	20	14	9	5
546	442	352	275	210	156	112	77	50	30	16	7
2275	1729	1287	935	660	450	294	182	105	55	25	9
8008	5733	4004	2717	1782	1122	672	378	196	91	36	11
24752	16744	11011	7007	4290	2508	1386	714	336	140	49	13
68952	44200	27456	16445	9438	5148	2640	1254	540	204	64	15
176358	107406	63206	35750	19305	9867	4719	2079	825	285	81	17
419900	243542	136136	72930	37180	17875	8008	3289	1210	385	100	19
940576	520676	277134	140998	68068	30888	13013	5005	1716	506	121	21
					51272	20384	7371	2366	650	144	23
						30940	10556	3185	819	169	25
							14756	4200	1015	196	27
								5440	1240	225	29
									1496	256	31
										286	33
	34512075									324	35
			7811375							361	37
					1138500					400	39
							95634			441	41
									3795	484	43
										529	45

Figure 1.11(c).

[49] Briggs, *TRIG*, pp. 21–23.

[50] The entries in the table are arranged so as to be reached by knight's moves. Thus, e.g., start with 1 in column 7, go down one and to the right two to the 7 in column 5; iterate this, proceeding to the 14 in column 3 and finally to the 7 in column 1. (There are 45 rows in this abacus. It is discussed, e.g., in Hutton [1801], pp. 75ff.)

This is, in Briggsian terminology,

$$-1\,⑦ + 7\,⑤ - 14\,③ + 7\,①$$

Vieta made use of an abacus for finding these coefficients, but his was quite differently arranged. The coefficients appear along each row of the table in Figure 1.12, in which the odd numbered columns are negative and the even ones positive. Thus to express the relation

$$2\cos 3I = 1\left(2\cos\frac{1}{2}I\right)^6 - 6\left(2\cos\frac{1}{2}I\right)^4 + 9\left(2\cos\frac{1}{2}I\right)^2 - 2$$

he used the coefficients -6, $+9$, -2 in row 5. The general inductive rule as given by Whiteside is clearly

$$2\cos\frac{1}{2}(n+1)I = \left(2\cos\frac{1}{2}I\right)\left(2\cos\frac{1}{2}nI\right) - 2\cos\frac{1}{2}(n-1)I.$$

Prima Negata.	Secunda affirmata.	Tertia negata.	Quarta affirmata.	Quinta negata.	Sexta affirmata.	Septima negata.	Octava affirmata.	Nona negata.
2								
3								
4	2							
5	5							
6	9	2						
7	14	7						
8	20	16	2					
9	27	30	9					
10	35	50	25	2				
11	44	77	55	11				
12	54	112	105	36	2			
13	65	156	182	91	13			
14	77	210	294	196	49	2		
15	90	275	450	318	140	15		
16	104	552	660	672	336	64	2	
17	119	442	935	1122	714	204	17	
18	135	546	1287	1782	1386	540	81	2
19	152	665	1729	2717	2508	1254	287	19
20	170	800	2275	4604	4290	2640	825	100
21	189	952	2940	5733	7007	5148	1079	385

Figure 1.12.

The rule of formation of this table of Vieta is not difficult to establish.[51]

The table above appears in Vieta, *OP*, p. 295. It is in his *Ad Angulares Sectiones*. . . .

There is a discussion of the methods of Briggs, Mouton, and Newton by Maurice.[52] Gabriel Mouton (1618–1694) was a French astronomer who also

[51] Newton in his youth was much influenced by Vieta, and he annotated various texts of Vieta's. Cf. Newton, *Papers*, Vol. I, pp. 63–88.

[52] Maurice [1847]. This paper gives an account of the interpolation methods of Gabriel Mouton (1618–1694). Cf. Mouton [1670].

made important contributions to the theory of finite differences independently of Briggs. The presentation of Mouton is, as we can easily understand, clearer than is that of Briggs. After all, his work was done about fifty years later. Since Mouton was not a mathematician we do not find many references to him in the literature. However Delambre, *MOD*, Vol. 2, does have a very complete discussion of his work. He felt that the most important part of this material of Mouton was his method of interpolation by differences; and he mentions a number of cases considered by Mouton without proof. In particular Delambre was deeply impressed by Problem IV where, again without proof, Mouton shows how to intercalate into a table enough entries to achieve one with constant differences. Mouton acknowledges that this subtabulation scheme was due to a friend named François Regnaud who undertook the problem at his request.[53]

1.7. Kepler

Amongst those who produced tables of logarithms was Johann Kepler. His first table was very like that of Napier and appeared in 1624.[54] As indicated in the title the tables were published in Marburg under the aegis of Philip, Landgrave of Hesse. Although Kepler refers to a *chilias* he actually gives 1036 logarithms. The first 36 are the logarithms of the numbers 1, 2, ..., 10; 20, 30, ..., 100; 200, 300, ..., 1000; 2000, 3000, ..., 9000. The *chilias* proper then starts with the logarithms of 10,000, 20,000, ..., 10,000,000. A sample from a page is given in Figure 1.13(a). The values recorded are Keplerian logarithms

$$\text{Kep. log } N = 10^5 \log 10^5/N,$$

and, as we see, the results are given to 8 decimal places. The first column contains the angle whose sine is the number in the second column. Thus the fourth quantity in column 1 in Figure 1.13(a) is 0;13,45 = arcsin 0.0039947 $\sim$ 0.004 = 0;0,14, the value in the fifth column. The third column gives the sine in hours, minutes and seconds; thus, $24 \cdot \sin 0;13,45 = 0.09599 \sim 0.096$ in sexagesimal form; i.e., 0,5,46.[55]

In 1625 Kepler brought out a Supplement to his table containing precepts for their use — *praecepta de eorum usu*. Typical of its contents is an example of a spherical triangle with vertices P, V, S and sides $PV = 38°;30$, $VS = 40°;0$, $V\hat{P}S = 31°;34$. The problem is to find $V\hat{S}P$ and side VP. Another topic he discusses is how to find the logarithm of a number outside the range of his table.

---

[53] Delambre, *MOD*, Vol. 2, p. 360.

[54] Kepler IX *CHIL*, pp. 275–426.

[55] Kepler IX *CHIL*, p. 319.

	ARCUS *Circuli cum differentiis*	SINUS *seu Numeri absoluti*	*Partes vicesimae quartae*	LOGARITHMI *cum differentiis*	*Partes sexagenariae*
	— 3. 26			10536. 05	
	0. 3. 26	100.00	0. 1. 26	690775. 54	0. 4
	3. 27			69314. 72	
	0. 6. 53	200.00	0. 2. 53	621460. 82-+	0. 7
	— 3. 26			40546. 51	
	0. 10. 19	300.00	0. 4. 19	580914. 31	0. 11
	3. 26			28768. 21	
	0. 13. 45	400.00	0. 5. 46	552146. 10-+	0. 14
	— 3. 27			22314. 35	
H 2	0. 17. 12	500.00	0. 7. 12	529831. 75—	0. 18
	3. 26			18232. 16	
	0. 20. 38	600.00	0. 8. 38	511599. 59	0. 22
	— 3. 26			15415. 07	
	0. 24. 4	700.00	0. 10. 5	496184. 52-+	0. 25
	3. 26			13353. 14	
	0. 27. 30	800.00	0. 11. 31	482831. 38-+	0. 29
	— 3. 26			11778. 30	
	0. 30. 56	900.00	0. 12. 58	471053. 08	0. 32
	3. 27			10536. 05	
	0. 34. 23	1000.00	0. 14. 24	460517. 03	0. 36
	— 3. 26			9531. 02	
	0. 37. 49	1100.00	0. 15. 50	450986. 01	0. 40
	3. 26			8701. 14	
	0. 41. 15	1200.00	0. 17. 17	442284. 87	0. 43
	— 3. 27			8004. 27	
	0. 44. 42	1300.00	0. 18. 43	434280. 60	0. 47
	3. 26			7410. 80	
	0. 48. 8	1400.00	0. 20. 10	426869. 80-+	0. 50
	— 3. 26			6899. 28	
	0. 51. 34	1500.00	0. 21. 36	419970. 52—	0. 54
	3. 26			6453. 85	
	0. 55. 0	1600.00	0. 23. 2	413516. 67—	0. 58
	— 3. 27			6062. 47	
	0. 58. 27	1700.00	0. 24. 29	407454. 20	1. 1
	3. 26			5715. 84	
	1. 1. 53	1800.00	0. 25. 55	401738. 36	1. 5
	— 3. 27			5406. 72	
	1. 5. 20	1900.00	0. 27. 22	396331. 64	1. 8
	3. 26			5129. 33	
	1. 8. 46	2000.00	0. 28. 48	391202. 31	1. 12
	— 3. 26			4879. 02	
	0. 12. 12	2100.00	0. 30. 14	386323. 29	1. 16
	3. 26			4652. 00	
	1. 15. 38	2200.00	0. 31. 41	381671. 29	1. 19
	— 3. 27			4445. 17	
	1. 19. 5	2300.00	0. 33. 7	377226. 12—	1. 23
	3. 26			4255. 97	
	1. 22. 31	2400.00	0. 34. 34	372970. 15-+	1. 26
	— 3. 26			4082. 20	
H2v	1. 25. 57	2500.00	0. 36. 0	368887. 95-+	1. 30
	3. 27			3922. 07	
	1. 29. 24	2600.00	0. 37. 26	364965. 88	1. 34
	— 3. 26			3774. 03	
	1. 32. 50	2700.00	0. 38. 53	361191. 85—	1. 37
	3. 26			3636. 77	
	1. 36. 16	2800.00	0. 40. 19	357555. 08-+	1. 41
	— 3. 27			3509. 13	
	1. 39. 43	2900.00	0. 41. 46	354045. 95	1. 44
	3. 26			3390. 15	
	1. 43. 9	3000.00	0. 43. 12	350655. 80—	1. 48
	— 3. 26			3278. 99	

Figure 1.13(a).

HEPTACOSIAS LOGARITH-

Arcus Quadrantis. *Cum differentiis.*	Sexagesima scrupula.	Logarithmi *Cum differentiis.*	Quadrivicenæ part et scru.	Partes et Sexagesi: privativorum.
P. ′ ″	′ ″		P. ′	P. ′ ″
0. 0. 0	0. 0	Infinitum.	0. 0	Infinitum.
4.47		*Infinitum.*		
0. 4.47	0. 5	657925.14	0. 2	720. 0. 0
4.46		69314.72		
0. 9.33	0.10	588610.42	0. 4	360. 0. 0
4.46		40546.51		
0.14.19	0.15	548063.91	0. 6	240. 0. 0
4.47		28768.21		
0.19. 6	0.20	519295.70	0. 8	180. 0. 0
4.46		22314.35		
0.23.52	0.25	496981.35	0.10	144. 0. 0
4.47		18232.16		
0.28.39	0.30	478749.19	0.12	120. 0. 0
4.46		15415.07		
0.33.25	0.35	463334.12	0.14	102.51.26
4.47		13353.14		
0.38.12	0.40	449980.98	0.16	90. 0 0
4.46		11778.30		
0.42.58	0.45	438202.68	0.18	80. 0. 0
4.47		10536.05		
0.47.45	0.50	427666.63	0.20	72. 0. 0
4.46		9531.03		
0.52.31	0.55	418135.60	0.22	65.27.16
4.47		8701.14		
0.57.18	1. 0	409434.46	0.24	60. 0. 0
4.47		8004.26		
1. 2. 5	1. 5	401430.20	0.26	55.23. 5
4.46		7410.80		
1. 6.51	1.10	394019.40	0.28	51.25.43
4.47		6899.28		
1.11.38	1.15	387120.12	0.30	48. 0. 0
4.46		6453.86		
1.16.24	1.20	380666.26	0.32	45. 0. 0
4.47		6062.46		
1.21.11	1.25	374603.80	0.34	42.21.11
4.46		5715.86		
1.25.57	1.30	368887.94	0.36	40. 0. 0
4.47		5406.71		
1.30.44	1.35	363481.23	0.38	37.53.41
4.47		5129.33		
1.35.31	1.40	358351.90	0.40	36. 0. 0
4.46		4879.02		
1.40.17	1.45	353472.88	0.42	34.17. 9
4.47		4652.00		
1.45. 4	1.50	348820.88	0.44	32.43.38
4.47		4445.17		
1.49.51	1.55	344375.71	0.46	31.18.16
4.46		4255.97		
1.54.37	2. 0	340119.74	0.48	30. 0. 0
4.47		4082.19		
1.59.24	2. 5	336037.55	0.50	28.48. 0
4.47		3922.07		
2. 4.11	2.10	332115.48	0.52	27.41.32
4.47		3774.03		
2. 8.58	2.15	328341.45	0.54	26.40. 0
4.46		3636.77		
2.13.44	2.20	324704.68	0.56	25.42.52
4.47		3509.14		
2.18.31	2.25	321195.54	0.58	24.49.39
4.47		3390.14		
2.23.18	2.30	317805.40	1. 0	24. 0. 0
4.47		3278.99		
2.28. 5	2.35	314526.41	1. 2	23.13.33
4.46		3174.87		
2.32.51	2.40	311351.54	1. 4	22.30. 0
4.47		3077.17		
2.37.38	2.45	308274.37	1. 6	21.49. 5
4.47		2985.29		
2.42.25	2.50	305289.08	1. 8	21.10.35
4.47		2898.75		
2.47.12	2.55	302390.33	1.10	20.34.17
4.45		2817.10		
[illegible]	[illegible]	[illegible]	[illegible]	[illegible]

Arcus Quadrantis. *Cum differentiis.*	Sexagesima scrupula.	Logarithmi *Cum differentiis.*	Quadrivicenæ part. et scru.	Partes et Sexagesi: privativorum.
P. ′ ″	′ ″		P. ′	P. ′ ″
2.51.58	3. 0	299573.23	1.12	20. 0. 0
4.47		2739.88		
2.56.45	3. 5	296833.35	1.14	19.27.35
4.47		2666.83		
3. 1.32	3.10	294166.52	1.16	18.56.52
4.47		2597.55		
3. 6.19	3.15	291568.97	1.18	18.27.43
4.46		2531.77		
3.11. 5	3.20	289037.20	1.20	18. 0. 0
4.47		2469.27		
3.15.52	3.25	286567.93	1.22	17.33.39
4.47		2409.76		
3.20.39	3.30	284158.17	1.24	17. 8.34
4.47		2353.05		
3.25.26	3.35	281805.12	1.26	16.44.39
4.47		2298.95		
3.30.13	3.40	279506.17	1.28	16.21.49
4.46		2247.28		
3.34.59	3.45	277258.89	1.30	16. 0. 0
4.47		2197.90		
3.39.46	3.50	275060.99	1.32	15.39. 8
4.47		2150.62		
3.44.33	3.55	272910.37	1.34	15.19. 9
4.47		2105.35		
3.49.20	4. 0	270805.02	1.36	15. 0. 0
4.47		2061.93		
3.54. 7	4. 5	268743.09	1.38	14.41.38
4.47		2020.26		
3.58.54	4.10	266722.83	1.40	14.24. 0
4.48		1980.27		
4. 3.42	4.15	264742.56	1.42	14. 7. 4
4.47		1941.81		
4. 8.29	4.20	262800.75	1.44	13.50.47
4.47		1904.81		
4.13.16	4.25	260895.94	1.46	13.35. 6
4.48		1869.21		
4.18. 4	4.30	259026.73	1.48	13.20. 0
4.47		1834.91		
4.22.51	4.35	257191.82	1.50	13. 5.26
4.47		1801.86		
4.27.38	4.40	255389.96	1.52	12.51.25
4.48		1769.96		
4.32.26	4.45	253620.00	1.54	12.37.53
4.47		1739.17		
4.37.13	4.50	251880.83	1.56	12.24.50
4.48		1709.45		
4.42. 1	4.55	250171.38	1.58	12.12.12
4.47		1680.71		
4.46.48	5. 0	248490.67	2. 0	12. 0. 0
4.48		1652.93		
4.51.36	5. 5	246837.74	2. 2	11.48.12
4.47		1626.06		
4.56.23	5.10	245211.68	2. 4	11.36.47
4.48		1600.03		
5. 1.11	5.15	243611.65	2. 6	11.25.43
4.48		1574.83		
5. 5.59	5.20	242036.82	2. 8	11.15. 0
4.47		1550.42		
5.10.46	5.25	240486.40	2.10	11. 4.37
4.48		1526.75		
5.15.34	5.30	238959.65	2.12	10.54.33
4.47		1503.78		
5.20.21	5.35	237455.87	2.14	10.44.47
4.48		1481.51		
5.25. 9	5.40	235974.36	2.16	10.35.18
4.48		1459.90		
5.29.57	5.45	234514.46	2.18	10.26. 5
4.47		1438.86		
5.34.44	5.50	233075.60	2.20	01.17. 9
4.48		1418.46		
5.39.32	5.55	231657.14	2.22	10. 8.27
4.48		1398.64		
5.44.20	6. 0	230258.51	2.24	10. 0. 0

Figure 1.13(b).

Kepler also includes two logarithmic tables in his *Rudolphine Tables*. The first, which is entitled *Heptacosias logarithmorum logisticorum* (700 logistical logarithms), is a tabulation of logarithms to base e (Figure 1.13(b)). The argument of the table is in sexagesimal form, ranging from 0°;0,5 to 0°;1 at an interval of 0°;0,5. This appears in column two. In column three we find

$$-10^5 \log_e x$$

where x is the argument. Thus opposite 0′10″ in Kepler's table we find the entry 588610.42; now 0′10″ = 0.002777... and $\log_e 0.002777 = -5.8861040$. In column one he recorded the angle in degrees, minutes, seconds such that the Keplerian logarithm of its sine is the corresponding entry in the table. Thus in column one opposite to 0°;57,18 we read 409434.46 — this corresponds in column two to 0°;1,0. Now sin 0°;57,18 is 0.016667 and log 0.016667 = −4.0943172. In column four Kepler has recorded the time in the sense that 0°;60 has been divided into 24 hours = 1440 minutes. Thus 0°;0,5, the second entry in column two is $0^h;2^m$ and 0°;4 is $1^h;36^m$. The last column, number five, is intimately related to column two. To see the relation, let x be the entry in column two in minutes; then the entry y in column four is the reciprocal of x expressed in sexagesimal form. Thus if x is 0°;3,50, then the entry y in column four is 15;39,08, as we see.[56]

His other table is called *CANON Logarithmorum et Antilogarithmorum ad singula Semicirculi Scrupula* (cf. Figure 1.13(c)). Along the top of each sheet are two rows: the upper of these increases by 1° steps going from 90° to 179°; the lower decreases by 1° steps, going from 0° to 89°. Down the first column are entries in minutes which correspond to the lower row; and up the last column are comparable entries for the upper row. The actual tabular entries are

$$-10^5 \log_e \sin x$$

Thus if we look at x = 4°;07 we find 263404. Now sin 4°;07 = 0.071788 and log 0.071788 = −2.63404.

Kepler remarks in the Introduction to the *Rudolphine Tables* (p. 62) that antilogarithms are the logarithms of the sine of the complementary angle, and he goes on to say these sines have been named "cosines" by the Englishman Gunther. Edmund Gunther (1581–1626) also invented a forerunner of the slide rule.

The frontispiece to the *Rudolphine Tables* itself, shown in Figure 1.14, is worthy of inspection. The temple itself represents Astronomy, and the columns holding it up are named after various great figures: we see Aratus and Meton on the extreme edges, Hipparchus and Ptolemy, as well as Copernicus and Brahe. At the base we see, *inter alia*, a map of Brahe's island of Hveen in the sound between Denmark and Sweden. Notice also the panel to the left of the map, where we see poor, unhappy Kepler in a nightcap with

[56] Kepler X *RT*.

CANON Logarithmorum et Antilogarithmo-

Scr. Partes	90		91		92		93		94		95		96	Anti
	0		1		2		3		4		5		6	Log
	Pro 10″	*Decre.*		*Dec.*		*Dec.*		*Dec.*		*Dec.*		*Dec.*	*Decr.46*	
0	Infinitum.		404828	275	335528	139	295007	92	266274	69	244006	56	225830	60
1	814257	11553	3175	271	4699	137	4454	92	265859	69	243674	55	554	59
2	744942	6758	401549	267	3876	136	3903	91	446	69	343	55	278	58
3	704396	4795	399949	263	3060	135	3356	91	265034	69	243013	55	225003	57
4	675627	3719	8374	259	2251	134	2811	90	4624	68	242684	55	224729	56
5	653313	3039	6824	255	1448	133	2270	90	4216	68	357	55	456	55
6	635081	2569	5298	251	330651	132	1731	89	263809	68	242031	54	224183	54
7	619666	2229	3794	247	329861	131	1195	89	404	68	241705	54	223911	53
8	606313	1963	2313	243	9077	130	290663	88	263001	67	380	54	640	52
9	594535	1756	390853	240	8299	129	290133	88	2599	67	241057	54	369	51
10	583999	1555	389414	236	7527	128	289606	87	2199	67	240735	54	223100	50
11	574468	1450	7996	235	6761	127	9081	87	261801	67	414	53	222831	49
12	565766	1334	6598	230	6001	126	8559	87	404	66	240094	53	563	48
13	557762	1236	5219	227	5247	125	8040	86	261008	66	239775	53	295	47
14	550351	1150	3858	224	4498	124	7524	86	260614	66	457	53	222029	46
15	543452	1075	2516	221	3755	123	7011	85	260222	65	239140	53	221763	45
16	536998	1010	381192	218	3018	122	6500	85	259832	65	238824	53	498	44
17	530936	953	379885	215	2285	121	5991	84	443	65	509	52	221233	43
18	525220	901	8595	212	1559	120	5485	84	259055	65	238195	52	220969	42
19	519814	855	7321	210	320837	120	4982	83	8669	64	237881	52	706	41
20	514684	813	6063	207	320120	119	4482	83	8284	64	569	52	444	40
21	509805	776	4821	204	319409	118	3984	83	257901	64	237258	52	220182	39
22	5153	741	o 3595	202	8703	117	3488	82	519	64	236948	52	219921	38
23	500708	708	2383	200	8001	116	282995	82	257139	63	639	51	660	37
24	496452	680	1185	197	7305	116	504	82	256760	63	331	51	401	36
25	492370	654	370002	195	6613	115	282015	81	383	63	236024	51	219142	35
26	488448	629	368833	193	5926	114	1529	81	256007	63	235718	51	218884	34
27	4674	606	7677	191	5244	113	1045	80	5633	63	413	51	626	33
28	481038	585	6534	188	4567	112	280564	80	5260	62	235108	51	369	32
29	477529	565	5404	186	3894	112	280085	80	254888	62	234804	50	218113	31
30	4139	547	4287	184	3225	111	279608	79	518	62	501	50	217857	30
31	470860	529	3183	182	2561	110	9134	79	254149	62	234200	50	602	29
32	467685	513	2090	180	1902	109	8662	78	253781	61	233899	50	348	28
33	4608	498	361009	178	1246	109	8192	78	415	61	599	50	217094	27
34	461623	483	359940	176	310595	108	7724	78	253050	61	300	50	216841	26
35	458724	469	8882	175	309948	107	7258	77	2686	61	233002	50	589	25
36	5907	457	7835	173	9306	106	6795	77	2324	60	232705	49	337	24
37	3167	445	6799	171	8667	106	6334	76	251963	60	409	49	216086	23
38	450500	433	5774	169	8033	105	5875	76	604	60	232114	49	215835	22
39	447903	422	4759	168	7402	104	5418	76	251246	60	231820	49	585	21
40	5371	412	3754	166	6776	104	274963	75	250889	60	526	49	336	20
41	2902	402	2759	164	6153	103	510	75	533	59	231233	49	215088	19
42	440493	392	1774	162	5534	102	274059	75	250178	59	230941	48	214840	18
43	438140	383	350800	161	4919	102	3610	74	249825	59	650	48	593	17
44	5841	375	349833	159	4308	101	3163	74	473	59	360	48	346	16
45	3594	367	8877	158	3701	100	2718	74	249122	59	230071	48	214100	15
46	431396	359	7929	156	3097	100	2276	73	248773	58	229783	48	213854	14
47	429245	351	6990	155	2497	99	1835	73	425	58	495	48	609	13
48	7140	344	6060	153	1900	99	1396	73	248078	58	229208	48	365	12
49	5078	337	5139	152	1307	98	270959	73	247732	58	228922	47	213122	11
50	3058	330	4226	151	300718	97	524	72	387	58	637	47	212879	10
51	421078	324	3321	149	300132	97	270091	72	247044	57	353	47	636	9
52	419136	317	2425	148	299549	96	269660	72	246702	57	228069	47	394	8
53	7232	312	1536	147	8970	95	9230	71	361	57	227786	47	212153	7
54	5363	306	340655	146	8394	95	8802	71	246021	57	504	47	211912	6
55	3528	300	339782	144	7822	94	8376	71	245682	57	227223	47	672	5
56	411726	295	8917	143	7252	94	267952	71	344	56	226943	47	433	4
57	409956	290	8059	142	6686	93	530	70	245008	56	664	46	211194	3
58	8217	285	7208	141	6123	93	267110	70	244673	56	385	46	210955	2
59	6508	280	6365	140	5564	93	6691	70	339	56	226107	46	717	1
60	404828		335528		295007		266274		244006		225830		210480	0
	Pro 10	*Increm.*		*Incre.*		*Incre.*		*Incr.*		*Incr.*		*Incr.*	*Incre.39*	Scr. Partes
Log	179		178		177		176		175		174		173	
Anti	89		88		87		86		85		84		83	

Figure 1.13(c).

Figure 1.14.

his glasses, figures and a replica of the temple's dome on his nightstand. On the roof we see a set of allegorical figures; one of whom has 6931472 on her halo. This is the logarithm of sine 30°. In her hands are two rods: one presumably representing the numbers and the other their logarithms. We also see Kepler's ellipse, a scale or balance with the Sun at the fulcrum, Galileo's telescope, a lodestone, and a compass as well as the Sun shining on a sphere and casting its shadow.[57]

To solve his famous equation

$$\alpha = \beta + e \sin \beta$$

for β as a function of α Kepler produced an iterative scheme in a work he styled an epitome of Copernican astronomy.[58] He calls his method *regula positionum* or the rule of position. To describe his algorithm, suppose that $e = 0.09265$ and note that for $0 \leq \beta \leq \pi$, $\alpha \geq \beta$. For this example Kepler chose his angle α to be 50°;9,10 and hence knew that

$$\sin \beta \leq \sin 50°;9,10 \sim 0.76776.$$

Knowing this, Kepler picked β_0 to be arcsin 0.7 ~ 44°;25 and remarked that e in degrees is 5°;18,30 or in seconds 19,110, i.e., 0.09265 radians = 5°;18,30. (Actually Kepler used 11,910″ = 3°;18,30 in this example.) Then

$$\alpha_0 = \beta_0 + e \sin \beta_0 = 46°;44,$$

which is about 3°;25 too small. He then chose β_1 to be β_0 + 3°;25 = 47°;50. Actually, for ease of calculation, he took β_1 = arcsin 0.74 ~ 47°;44 — and recalculated finding α_1 = 50°;10,59 which is too large by 0°;1,49. He modified his guess for β to $\beta_2 = \beta_1$ − 0°;1,49 = 47°;42,17. This yielded α_2 = 50°;9,7 which he said was essentially the same as the given 50°;9,10.[59]

In concluding our discussion of Kepler's work, let us illustrate how he handled algebraic notations and equations. See Figure 1.16, in which the circle $\gamma\upsilon\nu$ with center at β is given, as are the points ε and ρ. The geometrical

---

[57] Kepler X *RT*, pp. 31 ff. for a discussion of the entire structure.

[58] Kepler VII *EPIT*, pp. 387–388; this is in Chap. 4 of Book V. (A part of the text, Books IV and V, has been translated by C. G. Wallis and appears in Vol. 16 of the *Great Books of the World* Chicago, 1952, pp. 998–999. This work represents truly a uniting of all his ideas into beautiful and complete text on astronomy. It is in truth not an epitome of Copernican but of Keplerian astronomy. However he says "I admit that this formulation of the hypotheses is not Copernican. But because the part concerning the eccentric circle is subordinate to the general hypothesis which employs the annual movement of the Earth and the stillness of the sun; therefore the name comes from the more important part of the hypothesis...and so this part has a good title for being referred to Copernicus." (VII *EPIT*, pp. 364–365. This is p. 967 of the English translation.)

[59] In this example Kepler gives as the value for e not 19,110″ but 11,910″. This latter value he uses throughout the example. The original value of 19,110″ is determined on pp. 387–388 of the Latin text and p. 992 of the English. In an earlier calculation Kepler started with β = 46°;18,51 and determined α to be 50°;9,10 using e = 19,110″. In the latter example he started with α = 50°;9,10 and found β = 47°;42,17 using e = 11,910″.

problem is to pass a circle $\nu\varepsilon\rho$ through those points, which is tangent to the given circle. In the figure o is the midpoint of $\varepsilon\rho$, $o\alpha$ is perpendicular to $\rho\varepsilon$ and is parallel to $\iota\beta$. The point ν is the point of tangency and ψ is the center of the circle to be found. Thus Kepler reduced the problem to solving the equation

$$\{[(o\psi^2 + o\varepsilon^2)^{1/2} + \beta\nu]^2 - \alpha\beta^2\}^{1/2} + \alpha o = o\psi. \tag{1.21}$$

Both Newton and Kepler solved the same problem and it is of interest to contrast their methods. Newton's drawing is reproduced in Figure 1.15.[60]

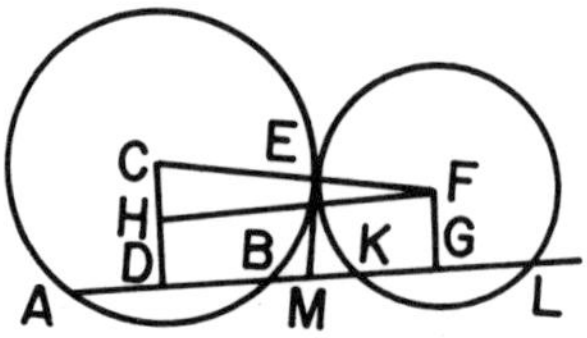

Figure 1.15.

Newton formulated and solved the problem as follows:

Problem XLV. *To describe a Circle through two given Points, which shall touch another Circle given in Position.*

Let A, B be the two Points given, EK the Circle given in Magnitude and Position, F its Center, ABE the Circle sought, passing through the Points A and B, and touching the other Circle in E, and let C be its Center. Let fall the Perpendiculars CD and FG to AB being produced, and draw CF cutting the Circles in the Point of Contact E, and draw also FH parallel to DG, and meeting CD in H. These being thus constructed, make AD or $DB = a$, DG or $HF = b$, $GF = c$, and EF (the Radius of the Circle given) $= d$, and $DC = x$; and CH will be $(= CD - FG) = x - c$, and $CFq\,(= CHq + HFq) = xx - 2cx + cc + bb$, and $CBq\,(= CDq + DBq) = xx + aa$, and consequently CB or $CE = \sqrt{xx + aa}$. To this add EF, and you will have $CF = d + \sqrt{xx + aa}$, whose Square $dd + aa + xx + 2d\sqrt{xx + aa}$, is equal to the Value of the same CFq found before, *viz.* $xx - 2cx + cc + bb$. Take away from both Sides xx, and there will remain $dd + aa + 2d\sqrt{xx + aa} = cc + bb - 2cx$. Take away moreover $dd + aa$, and there will come out $2d\sqrt{xx + aa} = cc + bb - dd - aa - 2cx$. Now, for Abbreviation sake, for $cc + bb - dd - aa$, write $2gg$, and you will have $2d\sqrt{xx + aa} = 2gg - 2cx$, or $d\sqrt{xx + aa} = gg - cx$. And the Parts of the Equation being squared, there will come out $ddxx + ddaa = g^4 - 2ggcx + ccxx$. Take from both Sides $ddaa$ and $ccxx$, and there will remain $ddxx - ccxx = g^4 - ddaa - 2ggcx$. And the Parts of the Equation being divided by $dd - cc$, you will have

$$xx = \frac{g^4 - ddaa - 2ggcx}{dd - cc}.$$

[60] Newton, *Works*, Vol. 2, p. 152. (I am indebted to an observation by Smart [1953], pp. 322–324, for the reference to Newton's Problem XLV in his *Universal Arithmetic*, which was first published in 1707 at Cambridge by Newton's successor, William Whiston, in the face of Newton's most emphatic objections.) The *Arithmetica Universalis* consists of lectures by Newton in the period 1673–1683. Cf. Newton, *Works*, Vol. 2, pp. xxii ff.

And by Extraction of the affected Root

$$x = \frac{-ggc + \sqrt{g^4dd - d^4aa + ddaacc}}{dd - cc}.$$

Having found therefore x, or the Length of DC, bisect AB in D, and at D erect the Perpendicular

$$DC = \frac{-ggc + d\sqrt{g^4 - aadd + aacc}}{dd - cc}.$$

Then from the Center C, through the Point A or B, describe the Circle ABE; for that will touch the other Circle EK, and pass through both the Points A, B. Q.E.F. (Note that Newton wrote, e.g., CFq for CF^2.)

A comparison of this with Kepler's solution helps to demonstrate how rapidly mathematics advanced in three-quarters of a century. In that solution all work is numerical. As we noted just above the equation (1.21) the given circle in Kepler's figure, Figure 1.16, is $\gamma\upsilon\nu$, the two given points are ε and ρ. The desired circle is then $\nu\varepsilon\rho$ with center at ψ. In Newton's notation $\varepsilon o = a = o\rho$, $o\iota = b = \alpha\beta$, $\beta\iota = c = \alpha o$, $\beta\nu = d$, $\psi o = x$. Then Kepler's relation (1.21) above becomes

$$\{[(x^2 + a^2)^{1/2} + d]^2 - b^2\}^{1/2} + c = x.$$

Kepler had

$$a = 476\tfrac{1}{2}, \qquad b = 90668, \qquad c = 13971, \qquad d = 65656\tfrac{1}{2}.$$

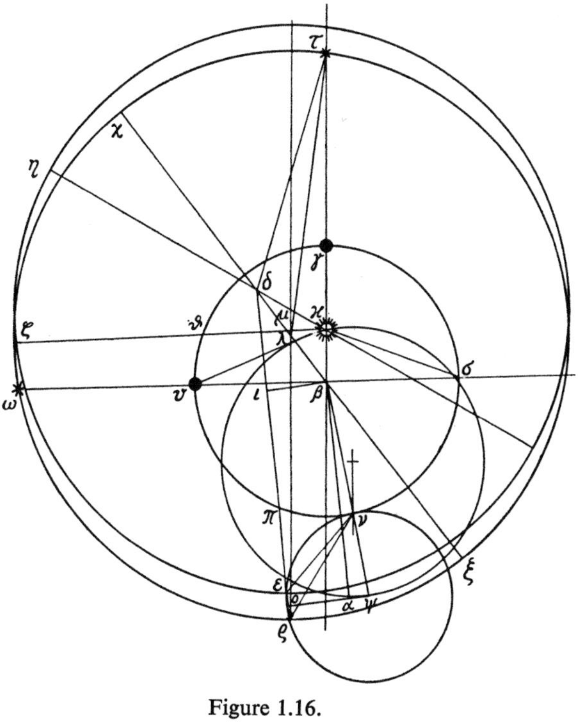

Figure 1.16.

Kepler's solution was this (there are a number of arithmetical errors in Kepler's solution):

I designate ψo as the unknown and note that its square is also an unknown. To that square add 227,052, the square of εo. This gives the square of $\psi \varepsilon$ or $\psi \nu$. The square of $\beta \nu$ is 4,310,747,477. Add this to the square of $\psi \nu$ and complete the square [by adding twice the area of the rectangle of sides $\beta \nu$ and $\psi \nu$.] (He has formed the expression $\beta \nu^2 + \psi \nu^2 + 2\beta \nu \cdot \psi \omega$). This gives the square of $\beta \psi$. The area of the rectangle is the square root of 4,310,747,475Z + 978,763,835,636,363. This gives a first expression for the square of $\beta \psi$.

Since however αo is equal to 13,971, $\psi \alpha$ is equal to the unknown less 13,971. Its square is 1Z − 27,942R + 195,188,841. To this square add 8,220,686,224, the square of $\beta \alpha$, so that a second expression for the square of $\beta \psi$ is 1Z − 27,942R + 8,415,875,065. This is now to be equated to 1Z + 4,310,974,527 plus twice the square root of 4,310,974,475Z + 97,876,383,536,363. Deduct 1Z and 4,310,974, 529 from both members of the equation. This yields −27,942R + 4,104,900,538 equal to twice the root of 4,310,747,475Z + 978,763,835,536,363 or the radical is equal to −18,971R + 2,052,450,269. The square of this expression is 195,188,841Z − 57,349,565,416,398R + 421,252,106,718,172,361. This is equal to 4,310,974,527Z + 978,763,835,536,363.

Deduct 195,188,841Z and 978,763,835,536,363 from both sides of the equation and add to them 57,349,565,416,398R. The equation remains valid, and it becomes 4,115,558,634Z + 57,349,565,416,398R equal to 4,211,573,342,882,635, 998. Reduced to lowest terms this becomes 1Z + 13,934R equal to 1,023,329,690. The solution of this equation gives for the unknown $o\psi$ the value 25,772.[61]

[61] Kepler, *AN*, pp. 34–36. The R stands for *res* and meant the variable x; Z for zenzus or x^2. (The word *zenzus* is derived from *census* meaning wealth, which was a translation of the Arabic *māl* meaning the same thing; this word was used by al-Khwārizmī and was borrowed from the Indians.) Here is the Latin text of the last paragraph so we can see the original notation. "*Abjice utringue* 195, 188, 841 $\mathfrak{Z}$ *et* 978, 763, 835, 536, 363, *et adde utrinque* 573, 349, 565, 416, 398 $\mathfrak{R}_x$. *Stabunt utrinque aequalia; illinc* 4, 115, 558, 634 $\mathfrak{Z}$ + 57, 349, 565, 416, 398 $\mathfrak{R}_x$; *hinc vero* 4, 211, 573, 342, 882, 635, 998. *Et in minimis numeris* 1 $\mathfrak{Z}$ + 13934 $\mathfrak{R}_x$ *aequant* 1,023,329,690. *Peracta aequatione prodit* $o\psi$ *unitatis figuratae valor* 25772."

2. The Age of Newton

2.1. Introduction

In the latter half of the seventeenth century we find a sudden burst of mathematical output in western Europe. One can speculate that this phenomenon followed the developments in notational expression which occurred shortly after Kepler's time and which permitted a much more profound understanding of many mathematical concepts. Kepler's notational capacity was quite limited, and was probably typical of his contemporaries. However, somewhat younger men such as Descartes were already facile in notational matters, and with this facility came an understanding of functional relationships. Descartes's *La Géométrie*, for example, played an important part in the developments of the second half of the century.[1]

Descartes's elegant treatment of geometry by algebraic means was very largely responsible for the almost explosive development that soon took place. While it is not germane to recount the work of Descartes in detail, some facets are discussed here since they play a key role in the art of computing. The Geometry consists of three small books: the first concerns itself with geometrical constructions involving only straight lines and circles, the second with the *nature des lignes courbes* and the third with what we now call the elementary theory of equations.

In Book II we find not only the introduction of analytical geometry with its concomitant functional relationships for curves but also the germ of the differential calculus. Descartes showed very clearly how to find the angle between two curves. In his 1637 edition he did this by means of the normal to a curve. The problem he posed was this: "Let CE be the given curve, and let it be required to draw through C a straight line making right angles with CE."[2] To do this he proceeded as follows: in Figure 2.1 he imagined CP to be the normal to the curve CE at C, and he put $MA = CB = y$, $CM = BA = x$. He also set $PC = s$, $PA = v$ and noted that $PM = v - y$ and that in his notation

$$x = \sqrt{ss - vv + 2vy - yy}, \tag{2.1}$$

[1] Descartes [1637].
[2] Descartes [1637], pp. 94–95.

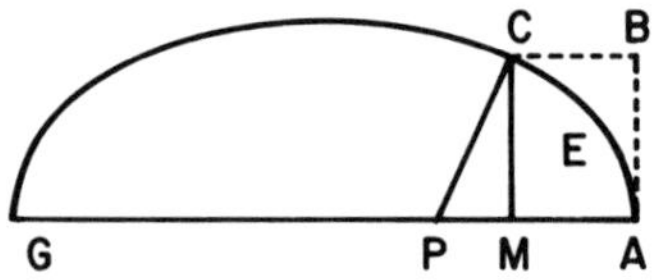

Figure 2.1.

or inversely (in our notation)

$$y = v + \sqrt{s^2 - x^2}. \tag{2.2}$$

He next remarked that either of these relations together with the equation $F(x, y) = 0$ of the curve enabled him to eliminate x or y and to find an equation in one unknown. Thus for the ellipse

$$xx = ry - \frac{r}{q}yy,$$

he found

$$yy + \frac{qry - 2qvy + qvv - qss}{q - r} = 0. \tag{2.3}$$

Descartes now noted that this equation was not primarily to be used to find y, e.g., but rather to impose a constraint on v or s. He imposed the condition that the circle about P as center, which passes through C, "touchera la ligne courbe CE, sans la coupper" — i.e., touch but not cut the curve. If it were to cut the curve, then the equation given by (2.3) above would have two unequal roots. He went on to argue that for equal roots the quadratic (2.3) had to be of the form

$$y^2 - 2ey + e^2 = 0,$$

and from this he found

$$PA = v = e - \frac{r}{q}e + \frac{1}{2}r.$$

But he pointed out that $e = y$, and hence

$$v = y - \frac{r}{q}y + \frac{1}{2}r.$$

He gave other examples and showed how to handle matters for equations of the same type as (2.3) but of degree higher than two.[3] He went on to say: "I desire rather to tell you in passing that this method, of which you have here an example, of supposing two equations to be the same form in order to compare them term by term . . . will apply to an infinity of other problems and is not the least important feature of my general method."[4]

---

[3] Descartes [1637], pp. 96–97, 104–105.
[4] Descartes [1637], pp. 112–113.

His Book III is also very interesting because he had been led from his analytical approach to a consideration of algebraic equations. Here he followed in the footsteps of Fibonacci, Cardan, and Stifel. He says: "Every equation can have as many distinct roots as the number of dimensions of the unknown quantity in the equation."[5] But later he goes on to say that often "some of the roots are false or less than nothing." He proceeds to say: ". . . that is, while we can conceive of as many roots for each equation as I have already assigned, yet there is not always a definite quantity corresponding to each root so conceived of. Thus while we may conceive of the equation $x^3 - 6xx + 13x - 10 = 0$, as having three roots, yet there is only one real root 2, while the other two, however we may increase, diminish, or multiply therein . . . remain always imaginary."[6] He then gives the theorem that for an equation with integral coefficients any integral root is an exact divisor of the constant term. He used this result together with his earlier ones to solve equations with fractional coefficients. With the help of this he discussed ruler and compass constructions.

In leaving our very short discussion of Descartes we must not overlook his great intellectual rival Fermat. The latter made important contributions to the subtangent problem as he did to so many fields. Both men worked on the same problem and each solved it by novel means. Let us examine very briefly Fermat's approach.[7] Consider the parabolic curve $y^2 = kx$ in Figure 2.2 and look for the subtangent PT to the curve at the point $P = (x, y)$.

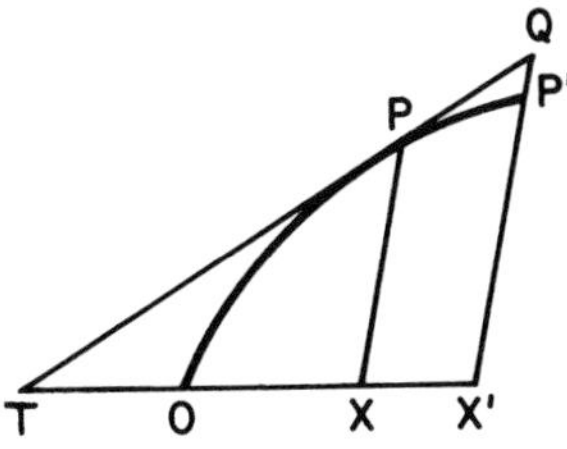

Figure 2.2.

Fermat took a neighboring point $P' = (x', y')$ and considered the point Q where the tangent PT cuts $X'P'$. Then if $TX = t$, $y^2/y'^2 = x/x'$; and $X'P' \leq X'Q$, $XP/X'Q = TX/TX'$. Hence

$$\frac{x' - x}{x} \leq \frac{(t + x' - x)^2 - t^2}{t^2}$$

so that $t \leq 2x + (x' - x)/t$; now as $x' \to x$,

$$XT = t = \lim_{x' \to x} \left(2x + \frac{x' - x}{t'}\right) = 2x.$$

Descartes' approach was closer to modern methods. He drew the chord PP'

---

[5] Descartes [1637], p. 159.

[6] Descartes [1637], p. 175.

[7] Whiteside, *Patterns*, p. 356.

and let it meet the X-axis in a point T'. Then he formed XT' and found XT by a limiting process. Fermat's approach had a considerable impact on Newton who wrote, "I had the hint of this method from Fermat's way of drawing tangents and by applying it to abstract equations, directly and invertedly, I made it general. Mr. Gregory and Dr. Barrow used and improved the same method of drawing tangents."[8]

2.2. Logarithms and Finite Differences

It was during the period we are now in that great advances occurred in the development of the logarithm not only as an analytical concept but also as a calculating apparatus. Today, we take for granted the relation

$$\log x = \int_1^x \frac{dx}{x}, \tag{2.4}$$

and perhaps imagine it was always so. But remember what we saw of Briggs, Bürgi, and Napier. They knew nothing of integrals and certainly did not discuss any relationships between logarithms and hyperbolic areas. Whiteside tells us that it was a Belgian Jesuit, A. A. deSarasa, who made this discovery "reading through the *Opus Geometricum* of his friend Gregory St. Vincent . . . ," whose work appeared in Antwerp in 1647.[9] He was one of the earliest mathematicians to study integration.

Apparently this connection did not commend itself to the mathematical community until somewhat later. Thus we find William, Viscount Brouncker (1620–1684), first president of the Royal Society, writing on the subject in 1668. Whiteside tells us he did the work ten years earlier.[10] The paper is entitled, "The squaring of the Hyperbola, by an infinite series of Rational Numbers, together with its Demonstration, by that Eminent Mathematician, The Right Honourable the Lord Viscount Brouncker." His approach is quite ingenious and perhaps deserves repetition here. In Figures 2.3 and 2.4, which are similar to his but borrowed from Whiteside, consider first that the hyperbolic area $ABCE$ is bounded between the sum of parallelograms $ABCF + FKdN + MNPb + HKIf + \cdots$ and the parallelogram $ABDE$ less than the sum of the inscribed triangles $CED + CdE + dbE + Cfd + \cdots$. To evaluate the areas of these figures Brouncker divided the interval AB into 2^n intervals each of length $AB/2^n$; note that d' is the midpoint of AB, that b' is the midpoint of AB, that b' is the midpoint of Ad', that a' is the midpoint of Ab', etc. Thus the parallelograms have lengths AB, $AB/2$, $AB/2^2$, etc. Brouncker chose the asymptote OB as his x-axis, his origin at O, his equation for

[8] Turnbull [1921], p. 5.
[9] Hofmann, *St. Vincent*.
[10] Brouncker [1668].

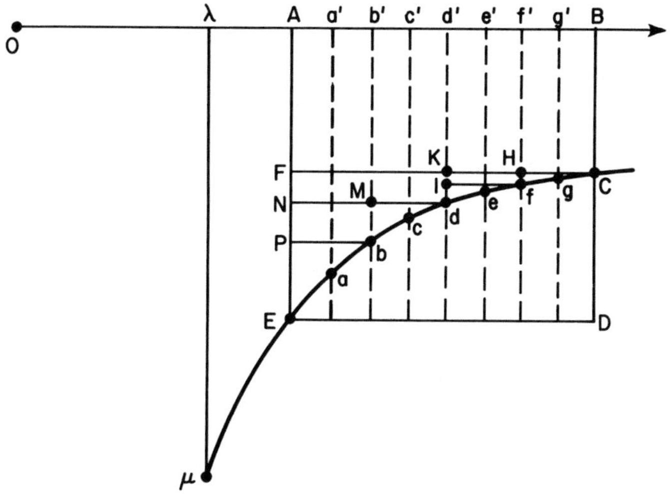

Figure 2.3.

the hyperbola as $xy = 1$ with $OA = AE = AB = 1$. Then the first parallelogram had area $1/(1\cdot 2)$, the second $(\frac{2}{3} - \frac{1}{2})\cdot\frac{1}{2} = 1/(3\cdot 4)$, the third $1/(5\cdot 6)$, etc. Thus Brouncker had a lower bound for the hyperbolic area of

$$\frac{1}{1\cdot 2} + \frac{1}{3\cdot 4} + \frac{1}{5\cdot 6} + \cdots \quad . \tag{2.5}$$

Similarly he evaluated his upper bound and found it to be

$$1 - \left(\frac{1}{2\cdot 2} + \frac{1}{2\cdot 3\cdot 4} + \frac{1}{4\cdot 5\cdot 6} + \cdots\right). \tag{2.6}$$

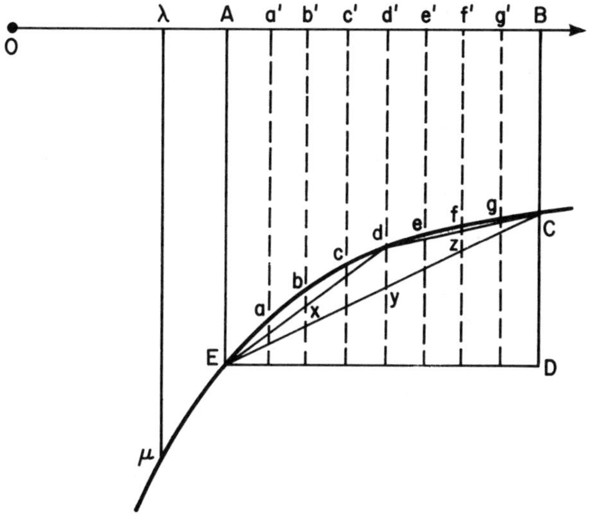

Figure 2.4.

It is not hard to see that in the limit these sums converge to log 2. His relation (2.5) above is sometimes known as Mercator's expansion of log 2. He found log 10/log 2 to be 2.302585/0.693147 = 3.321928; which is correct. The Brouncker approach, which gave at least an indication of the general case, is quite tedious. Indeed at this stage it must almost have seemed that the relation between logarithms and hyperbolic areas was of little utility. (In his paper he also gave a third sum

$$\frac{1}{2\cdot 3\cdot 4} + \frac{1}{4\cdot 5\cdot 6} + \frac{1}{6\cdot 7\cdot 8} + \cdots$$

and showed how to use it in the case $AE/BC = 5/4$ and $2/1$, keeping ten decimal places.)

In roughly the same period Pietro Mengoli (1625–1686) gave an interesting analytical formulation of the logarithm,[11] but it suffered from the general computational awkwardness of the series he obtained. The really elegant solution apparently came to several people almost simultaneously. It is usually attributed to a well-known Danish mathematician Nicolaus Mercator (1620–1687), who worked in England.[12] (Mercator was one of the founders of the Royal Society.) The result in question is the series expansion

$$\log_e (1 + x) = x - \frac{1}{2}x^2 + \frac{1}{3}x^3 - \frac{1}{4}x^4 + \cdots \qquad (-1 < x < 1).$$

This work of Mercator was picked up almost at once by John Wallis in a letter to Brouncker which appeared in August, 1668.[13] Here Wallis gives a proof of Mercator's result based on the integration techniques in his work *Arithmetica Infinitorum* (Oxford, 1656). He calculated $\log_e 1.21$ by taking terms through x^{11} in the series expansion, finding 0.190620361. Moreover, in a letter dated September 26, 1668 to Collins, Wallis says, in response to an assertion of Huygens that he had squared the hyperbola, "... for it hath been known likewise to a great many more, ever since the book of Gregorii de Sancto Vincentio, And it was in print in my Commercium Epistolicum, Epist. 39, 40... ."[14]

However, as early as 1664 Newton had already calculated $\log_e 1.1$ to 68 decimal places. At first, he made two mistakes, one in the 28th place and one in the 43rd.[15] In 1676 he revised his calculation without really introducing any new ideas.[16] By that date he fully understood subtabulation as we shall

---

[11] Whiteside, *Patterns*, pp. 224–225.

[12] Mercator [1667]. Cf. also, Hofmann, *Mercator*.

[13] Wallis [1668], pp. 753–764. The paper consists of letters dated July 8 and August 5, 1668, and of illustrative calculations by Mercator.

[14] Rigaud [1841], Vol. II, pp. 500ff.

[15] Newton, *Papers*, Vol. I, pp. 112–115 and 134–142. Cf. also, Plate I for a copy of Newton's calculation of log. 1.1 in his own hand. (This series of Whiteside's is a truly monumental and magnificent accomplishment fully in keeping with the superb man whose works he has so elegantly edited.)

[16] Newton, *Papers*, Vol. IV, pp. 22ff.

see soon. He also gave careful thought to producing a logarithmic canon and wrote at some length on the topic over a number of years. It is very likely that his great interest in the theory of finite differences came about through his work on logarithms.[17] In Problem 9 of his *Methods of Series and Fluxions*, 1670–1671 (Newton, *Papers*, Vol. III, pp. 211–236, in particular pp. 227ff. entitled, "To determine the area of any proposed curve"), Newton obtains the series expansions for ab log $((a + x)/a)$, ab log $(a/(a - x))$ as well as for ab log $((a + x)/(a - x))$. He uses these to find to 16 places, in general, the logarithms of 10/8, 10/9, 11/10, 12/10, 100/98, 100/99, 101/100, 102/100, 1000/998, 1000/999, 1001/1000, and 1002/1000. (Then the logarithms of 12/8 and 12/9 are directly calculable.) With the aid of these Newton finds $\log 2 = \log(1.2/0.8) \times (1.2/0.9) = 0.69314\,71805\,59945\,3$, which is the correct result. He then finds log 3 from $3 = 2 \times (1.2/0.8)$; log 5 from $2 \times 2/0.8$; log 10 from 2×5, log 100 from 10×10, log 1000 from 10×100, log 7 from $49 = (5 \times 10 \times 0.98)$; log 11 from 10×1.1; log 13 from $1000 \times 1.001/7 \times 11$, log 17 from $100 \times 1.02/2 \times 3$, log 37 from $1000 \times 0.999/3 \times 3 \times 3$; log 101 from 100×1.01, log 167 from $1000 \times 1.002/2 \times 3$, log 499 from $1000 \times 0.998/2$. He then goes on to say these can be converted to Briggsian logarithms upon division by $2.30258\,50929\,94045\,7 = \log 10$. From these primes he can obtain many entries in his table. He said: "Then the logarithms of all numbers in the canon which come from multiplying these together are to be hunted down (as usual) by addition of their logarithms, and the places still vacant afterwards filled in with the aid of this theorem." Whiteside points out that in later work Newton distinguished between "interpolating" and "intercalating." Newton now gives the rule previously mentioned for finding the value of the logarithm midway between two entries. He says let "n be the number whose logarithm is to be assigned, x the difference between it and the nearest numbers equally distant from it on either side whose logarithms are known, and d the half difference of their logarithms: the required logarithm of . . . n will then be obtained by adding

$$d + \frac{\frac{1}{2}dx}{n} + \frac{\frac{1}{12}dx^3}{n^3}\cdots$$

to the logarithm of the lesser number."[18]

---

[17] The interested reader may wish to consult Newton, *Papers*, Vol. II, p. 234*n*, for a brief account of Newton's knowledge of the literature. It may be relevant at this point to mention his notation for an integral. He wrote

$$\frac{aa}{x + y}$$

inside a rectangle to mean the integral $\int a^2\,dy/(x + y)$, or wrote out four letters for the corners of the area in question. As we know, it was Leibniz who introduced the $\int$ notation. The Newtonian rectangular notation is certainly much less useful in that it is difficult to denote limits for the variables.

[18] Newton, *Papers*, Vol. III, pp. 232–233.

Note first in this rule that

$$d = \frac{1}{2}\log\frac{n+x}{n-x} = \sum_{i=1}^{\infty}\frac{1}{2i-1}\left(\frac{x}{n}\right)^{2i-1};$$

then Newton says by way of proof: "The area

$$\left[\log\frac{n+x}{n-x}\right] \qquad \left(\text{that is, } \frac{2x}{x} + \frac{\frac{2}{3}x^3}{n^3} + \frac{\frac{2}{5}x^5}{n^5}\cdots\right)$$

is to the area

$$\left[\log\frac{n}{n-x}\right] \qquad \left(\text{or } \frac{x}{n} + \frac{\frac{1}{2}x^2}{n^2} + \frac{\frac{1}{3}x^3}{n^3}\cdots\right)$$

as the difference between the logarithms of the extreme numbers (or $2d$) to the difference between those of the lesser and middle ones, and this difference therefore will be

$$\frac{dx/n + \frac{1}{2}\,dx^2/n^2 + \frac{1}{3}\,dx^3/n^3}{x/n + \frac{1}{3}\,x^3/n^3 + \frac{1}{5}x^5/n^5\cdots},$$

that is, on performing the division,

$$d + \frac{\frac{1}{2}\,dx}{n} + \frac{\frac{1}{12}\,dx^3}{n^3}\cdots \quad ."$$

He goes on to say: "But I consider that the first two terms, $d + \frac{1}{2}\,dx/n$, are accurate enough for constructing a canon even though the logarithms are extended to fourteen or perhaps fifteen places of figures, provided that the number whose logarithm is to be assigned is not less than 1000."

This method is workable but it requires considerable skill and cleverness. In 1676 Newton discusses in a most elegant way how to calculate a logarithmic canon. This is in an unsent part of a letter to Leibniz.[19] He also gives a discussion of how to form a canon of sines, as we shall soon see. But first we shall discuss the construction of his logarithmic tables, "for that here affords a pleasant digression." He first calculates the hyperbolic logarithms of 10, 0.98, 0.99, 1.01, and 1.02: "this will take but an hour or two's time." From these he has the *true* logarithms of 98, 99, 100, 101, 102. He then subtabulates these by ten intervals, finding thereby the logarithms of all numbers between 980 and 1020. He next subtabulates between these to find the logarithms of all numbers between 9,800 and 10,000. Given these he can now trivially find

[19] Newton, *Papers*, Vol. IV, pp. 22ff. This part was discarded by Newton and was not included in his "Epistola Posterior" (October 1676). Exactly what was sent is discussed by Whiteside. Cf. also, Collins [1712], p. 81.

the logarithms of the 25 primes less than 100 as follows:

$$\sqrt[10]{\frac{9984 \times 1020}{9945}} = 2, \quad \sqrt[4]{\frac{8 \times 9963}{984}} = 3, \quad \frac{10}{2} = 5, \quad \sqrt{\frac{98}{2}} = 7, \quad \frac{99}{9} = 11,$$

$$\frac{1001}{7 \times 11} = 13, \quad \frac{102}{6} = 17, \quad \frac{988}{4 \times 13} = 19, \quad \frac{9936}{16 \times 27} = 23, \quad \frac{986}{2 \times 17} = 29,$$

$$\frac{992}{32} = 31, \quad \frac{999}{27} = 37, \quad \frac{984}{24} = 41, \quad \frac{989}{23} = 43, \quad \frac{987}{2[1]} = 47,$$

$$\frac{9911}{11 \times 17} = 53, \quad \frac{9971}{13 \times 13} = 59, \quad \frac{9882}{2 \times 81} = 61, \quad \frac{9849}{3 \times 49} = 67, \quad \frac{994}{14} = 71,$$

$$\frac{9928}{8 \times 17} = 73, \quad \frac{9954}{7 \times 18} = 79, \quad \frac{996}{12} = 83, \quad \frac{9968}{7 \times 16} = 89, \quad \frac{9894}{6 \times 17} = 97.$$

In this way he finds the logarithms of all numbers less than 100. He then says "it remains merely to interpolate these also once and then again by ten intervals at a time." He goes on to say: "To subtabulate the terms of any table by hundredths when, as usually happens in logarithms, first differences suffice is an obvious matter and one well known. But when second, third, and other differences come into it, a general rule for eliciting these differences and compounding them to best effect would seem desirable. The formula which I find most agreeable, though barely relevant to my purpose and wholly unworthy to be passed on in a letter to Leibniz, I shall nonetheless — since it is applicable to the computation of tables of any kind whatsoever — not hesitate to describe here somewhat lavishly for the use of calculators: to these, if you [Oldenburg] know any and the matter seems important enough, you might communicate it." Evidently he now had a table of logarithms to the base e from 1 to 10,000. That there was still much interest in how best to calculate logarithms may be seen from an excellent paper published in the spring of 1695 by Edmund Halley (1656–1742), the great astronomer and close friend of Newton.[20] By way of introduction Halley wrote:

The invention of the Logarithms is justly esteemed one of the most Useful Discoveries in the Art of Numbers, and accordingly has a Universal Reception and Applause; and the great Geometricians of this age have not been wanting to cultivate this Subject with all the Accuracy and subtility a matter of that consequence doth require; and they have demonstrated several very admirable Properties of these Artificial Numbers, which have rendered thier Construction much more facile than by those operose Methods at first used by their truly Nobel Inventer the Lord *Napeir*, and our worthy Country-man Mr. *Briggs*.

But notwithstanding all their Endeavours, I find very few of those who made constant use of Logarithms, to have attained an adequate Notion of them; or to understand the extent of the use of them: Contenting themselves with the Tables of them . . . , without daring to question them, or caring to know how to rectifie them, should they be found amiss, being I suppose under the apprehension

[20] Halley [1695].

of some great difficulty therein. For the sake of such the following Tract is principally intended, but not without hopes however to produce something that may be acceptable to the most knowing in these matters.

Halley had a very neat scheme which depended upon a definition of the logarithm close to that of Bürgi. If it is desired to find the logarithm of a number $\xi = 1 + q$, then

$$\log(1+q) = \lim_{m\to\infty} \frac{(1+q)^{1/m} - 1)}{1/m}.^{21}$$

To evaluate this expression Halley makes use of Newton's "Doctrine of Series" and writes

$$\overline{1+q}^{\frac{1}{m}} = 1 + \frac{1}{m}q + \frac{1-m}{2mm}qq + \frac{1-3m+2mm}{6m^3}q^3 + \cdots \quad .$$

Since he knows that $\log(1+q) = \lim[(1+q)^{1/m} - 1]/(1/m)$, this gives him the relation

$$\log \xi = \log(1+q) = q - \frac{q^2}{2} + \frac{q^3}{3} - \cdots \quad .$$

In the same way he finds

$$-\log(1-q) = q + \frac{q^2}{2} + \frac{q^3}{3} + \cdots \quad .$$

He then forms the expression

$$\log\frac{1+x/z}{1-x/z} = \frac{2x}{z} * + \frac{2x^3}{3z^3} * + \frac{2x^5}{5z^5} * + \frac{2x^7}{7z^7} \text{ \&c.},$$

where $q = x/z$ and he plans to set $x = b - a$, $z = b + a$. (Halley, as well as a number of others of the period, used an asterisk to indicate a missing term.) He now has $\log b/a$, and he remarks that this last series "converges twice as swift as the former, and therefore, is more proper for the Practice of making of Logarithms: Which it performs with that expedition, that where x the difference is but the hundredth part of the Sum, the first step $2x/z$ suffices to seven places of the Logarithm, and the second step to twelve; but if *Briggs's* first twenty Chiliads of Logarithms be supposed made, as he has very carefully computed them to fourteen places, the first step alone is capable to give the Logarithm of any intermediate Number true to all the places of those Tables."[22]

Halley now produces a very neat series expansion. He wants to find the logarithm of a prime p lying between two composite numbers, say a and b, whose logarithms are already known; e.g., he knows the logarithms of 22

---

[21] Whiteside, *Patterns*, pp. 230–231.

[22] Halley [1695], p. 61.

and 24 and wishes to calculate that of 23. He does this in two steps. First he finds a series expansion for the logarithm of the ratio of the geometrical to the arithmetical mean of a and b. As before, if $x = b - a$, $z = b + a$,

$$\log \frac{\sqrt{ab}}{\frac{1}{2}z} = \frac{1}{2} \log \frac{4ab}{(a+b)^2} = \frac{1}{2} \log \left(1 - \frac{x^2}{z^2}\right) = -\left(\frac{x^2}{2z^2} + \frac{x^4}{4z^4} + \frac{x^6}{6z^6} + \cdots\right).$$

(Halley describes this as a "Theorem of good dispatch to find the Logarithm of $z/2$)." But he proceeds to find his desired result for the case $b = a + 2$. This he does by setting $y^2 = ab + z^2/4 = 2a^2 + 4a + 1$ and then noting that

$$-\log \frac{\sqrt{ab}}{\frac{1}{2}z} = \frac{1}{2} \log \frac{1 + 1/y^2}{1 - 1/y^2} = \frac{1}{y^2} + \frac{1}{3y^6} + \frac{1}{5y^{10}} + \frac{1}{7y^{14}} + \frac{1}{9y^{18}} \text{ \&c.},$$

"which converges very much faster than any Theorem hitherto published for this purpose."

He points out that the series expansion of $\log(1 + q)$ was due to "*Mercator*, as improved by the Learned Dr. *Wallis*." The second result, $\log(1 + q)/(1 - q)$ he says, "was invented and demonstrated in the Hyperbolick Spaces Analogous to the Logarithms, by the Excellent Mr. *James Gregory*, in his *Exercitationes Geometricae*, and since further prosecuted by the aforesaid Mr. *Speidall*, ... But the demonstration as I conceive was never till now perfected without the consideration of the Hyperbola, which in a matter purely Arithmetical, as this is, cannot so properly be applyed. But what follows I think I may more justly claim as my own . . ." He then illustrates with a calculation of log 23. To do this he writes ($y^2 = 1057$)

$$\begin{aligned} -\log \frac{\sqrt{22.24}}{23} &= \log 23 - \frac{1}{2} \log 22 \times 24 \\ &= \frac{1}{2} \log \frac{1 + 1/1057}{1 - 1/1057} \\ &= \frac{1}{1057} + \frac{1}{3542796579} + \frac{1}{659676558485285} \text{ \&c.} \end{aligned}$$

Given the logarithms of 2, 3, and 11 he had the logarithm of 23. He then shows how to multiply the result by 0.4342944819. . . . He does this quite adroitly: first he divides 0.434. . . by 1057, the quotient by three times the square of 1057, etc. He thus find

	1.3613169612669061294500917266980 5
1057) 43429 &c. (	4108746281014681434731588636 8
3 in 1117249) 41087 &c. (	122585215441818294600 74
5/3 in 1117249) 12258 &c. (	65832351843761 75
7/5 in 1117249) 65832 &c. (	42088297 65
9/7 in 1117249) 42088 &c. (	29 30
Summa	1.3617278360175928788677771122511 7

"which is the Logarithm of thirty two places, and obtained by five Divisions with very small *Divisors*, all which is much less work than simply multiplying the *Series* into and said Multiplicator 43429 &c."[23]

He now inverts his series for the logarithm to find series expansions for exponentials, writing

$$1 \pm q = e^{\pm L} = 1 \pm L + \frac{1}{2}L^2 \pm \frac{1}{6}L^3 + \cdots \quad .$$

His "proof" is of interest. He says in effect that

$$L = \lim_{m \to \infty} \frac{(1 \pm q)1^{/m} - 1}{1/m}$$

and hence that

$$1 \pm q = \lim (1 \pm L/m)^m = 1 \pm L + L^2/2 \pm L^3/6 + \cdots \quad .$$

In concluding his paper Halley gives the approximate formulas

$$1 + \frac{L}{1 - L/2} \quad \text{and} \quad 1 + \frac{L}{1 - L/2} - \frac{L^3/12}{1 - L}$$

for e^L. He recommends using the former for numbers not exceeding fourteen places, "as are Mr. *Briggs's* large Table of Logarithms." Note that the former one is in error by about $L^3/12$ and the latter by $L^4/8$.

2.3. Trigonometric Tables

Newton had a very elegant method for finding the multiples of a given angle which he used on a variety of occasions. In his previously cited letter to Leibniz he showed how to carry it out and how to find cosines of these angles at a very small cost in computation — one multiplication per step. In Figure 2.5 let $\alpha = I\hat{A}R$ be the given angle and let $AB = BC = CD = DE = EF = \cdots$ be taken as unity. Then from the points $B, C, D, \ldots$ drop perpendiculars on AI and AR. We see at once that $B\hat{A}C = \alpha$, $C\hat{B}D = 2\alpha$, $E\hat{D}F = 3\alpha, \ldots$ and that $AK, BL, CM, \ldots$ will be their cosines. Newton

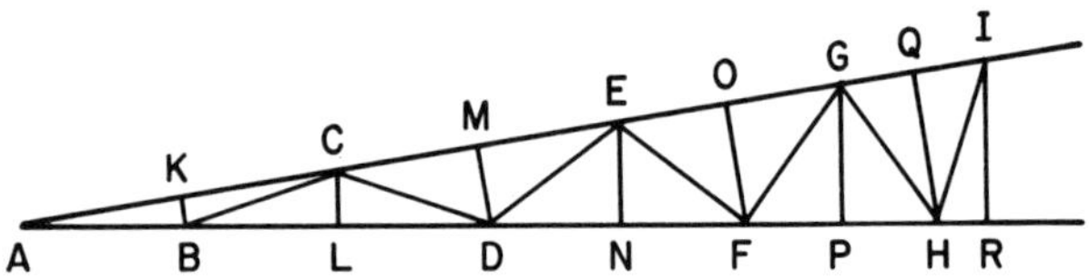

Figure 2.5.

[23] Halley [1695], p. 65. The entry 1.3617... is $\log_{10} 23$. (Note in the table above that 1.3613... is the $\log_{10} \sqrt{22.24}$.)

points out if AK is given, then $2AK = AC$, $AL = AK \cdot AC$, $AM = AK \cdot AD$, ... as can be seen from the similar triangles involved. Furthermore he has $AL - AB = BL$, $AL + BL = AD, \ldots$.[24]

Following Whiteside we may express Newton's results as recurrence relations. He sets $A = B_0$, $B = B_1$, $C = B_2$, $D = B_3, \ldots, K = C_1$, $L = C_2$, $M = C_3, \ldots$. Then $AC_1 = \cos \alpha$, $AB_n \cdot \sin \alpha = \sin n\alpha$, $B_{n-1}C_n = \cos n\alpha$. In terms of these Newton expresses the relations

$$\begin{aligned} \sin (n + 1)\alpha - \sin (n - 1)\alpha &= 2 \sin \alpha \cos n\alpha \\ \sin (n + 1)\alpha \cos \alpha - \cos (n + 1)\alpha \sin \alpha &= \sin n\alpha. \end{aligned} \tag{2.7}$$

Given $\sin m\alpha$, $\cos m\alpha$ for $m = 1, 2, \ldots, n$, the first relation (2.7) easily yields $\sin (n + 1)\alpha$ and then the second relation $\cos (n + 1)\alpha$. Newton handled the matter very neatly. Instead of using the relations (2.7) directly he wrote, in our notation:

$$\begin{aligned} B_0B_{n+1} &= B_0B_{n-1} + 2B_{n-1}C_n, \\ B_nC_{n+1} &= B_0C_1 \cdot B_0B_{n+1} - B_0B_n. \end{aligned} \tag{2.7$'$}$$

We see that he thereby finds both $B_0B_{n+1} = \sin (n + 1)\alpha/\sin \alpha$, and $B_nC_{n+1} = \cos (n + 1)\alpha$ at the cost of one multiplication. Moreover he observes that, for very small α, the work will be made still less by using the versed sine = 1 − cosine instead of the cosine of α. If we set $v = 1 - \cos \alpha$ then (2.7′) become

$$\begin{aligned} B_0B_{n+1} &= B_0B_{n-1} + 2B_{n-1}C_n, \\ B_nC_{n+1} &= 1 - vB_0B_{n+1}, \end{aligned} \tag{2.7$''$}$$

where v is a very small number.

He now writes out the series expansion for $v = 1 - \cos \alpha$ as

$$v = \frac{1}{2} z - \frac{1}{24} z^2 + \frac{1}{720} z^3 - \frac{1}{40320} z^4 \cdots,$$

where $z = \alpha^2$. Then the relations (2.7′) or (2.7″) will give *seriatim* the cosines of the multiples of α. He goes on to suggest that "... in the first instance, the cosines of every fifth or sixth degree between 0° and 60° or between 30° and 90° can be computed, and subsequently the cosines of every degree or half degree. For it is not advantageous to proceed by leaps which are too broad. Afterwards the cosines of the remaining thirty degrees are producible from these by addition or subtraction alone, seeing that the cosine of any angle N is the sum of the cosines of the angles $60° + N$ and $60° - N$, as is well known ...."[25]

[24] Newton, *Papers*, Vol. IV, pp. 22–25, and Vol. II, pp. 444–445.

[25] Newton, *Papers*, Vol. IV, p. 25.

Newton also gave Vieta's recurrence relations for finding $B_nC_n = \sin n\alpha$ as

$$B_{n+1}C_{n+1} = 2 \sin \alpha \cdot B_{n-1}C_n + B_{n-1}C_{n-1},$$

and

$$B_nC_{n+1} = 2 \cos \alpha \cdot B_{n-1}C_n - B_{n-2}C_{n-1}.^{26}$$

2.4. The Newton–Raphson and Other Iterative Methods

The Newton–Raphson procedure we know so well today "is essentially an improved version of the procedure, expounded by Viète and simplified by Oughtred . . . ."[27] According to Whiteside, Newton first published his method in the *Principia Mathematica* (Book I, Prop. 31, Scholium) as a means of solving Kepler's equation.[28] However, he certainly was concerned with the method much earlier. In 1669 Newton already had discussed the cubic $y^3 - 2y - 5 = 0$ in his work on infinite equations. [It is amusing to note that de Morgan wrote, "Invent a numerical method, neglect to show how it works on this equation, and you are a pilgrim who does not come in at the little wicket (*vide* Bunyan)."] There he shows his process as applied to this equation. He first notes that the integer part of the root is 2, and then translates his axes by means of $y = 2 + p$, obtaining $p^3 + 6p^2 + 10p - 1 = 0$. He then neglects terms higher than the first and has $p \sim 0.1$. He again translates axes by $p = 0.1 + q$ and finds $q^3 + 6.3q^2 + 11.23q + 0.061 = 0$. Again he neglects terms higher than linear and finds $q \sim -0.0054$. He continues the process one more step, finding $r \sim 0.00004853$ and $y = 2.09455147$.[29] We notice that the relation $10p - 1 = 0$ given above corresponds precisely to $p = x^2 - x_1 = -f(x_1)/f'(x_1)$, where $f(x) = x^3 - 2x - 5$ and $x_1 = 2$. Similarly, $q = x_3 - x_2 = -f(x_2)/f'(x_2)$, etc.

The procedure was systematically discussed in print by Joseph Raphson in *Analysis Aequationum Universalis* . . . , which appeared first in 1690 in London.[30] Here Raphson acknowledges Newton as the source of the pro-

---

[26] Newton, *Papers*, Vol. IV, p. 24. Here we see once again the great debt mathematics owes to Vieta. The interested reader can consult François Vieta, *OP*, pp. 298–300. This is his *Ad Angulares Sectiones*. . . .

[27] Newton, *Papers*, Vol. II, p. 218*n*. Cf. also, *Papers*, Vol. I, pp. 63ff. The procedure would seem to have originated with al-Kāshī, as we mentioned earlier. We discuss Vieta's scheme shortly.

[28] Newton, *Principia*, pp. 112–116. This is Book I, Prop. XXXI, Problem XXII and Scholium.

[29] Whiteside points out that he found only a single usage in the literature of al-Kāshī's method; this was by Briggs in his *Trigonometria Britannica* where he needed to solve both a cubic and a quintic to tri- and quinquisect a given angle. He further remarks that Newton was unaware of this book of Briggs. So Newton's method was at least a rediscovery by him.

[30] Raphson [1690].

cedure. (He did not use the function f and its derivative f', but instead used polynomials explicitly.) Newton, however, had shown his knowledge of the underlying process much earlier, as had James Gregory. There is evidence of considerable interest in iterative schemes for solving equations during this period.

In a letter of August 15, 1674, Michael Dary, writing to Newton, set up the scheme

$$z_{i+1} = (az_i^q + n)^{1/p} \qquad (i = 0, 1, \ldots)$$

for finding a root of

$$z^p = az^q + n \qquad (p > q),$$

starting with a guess z_0. This represented a small extension of a scheme Newton communicated to Dary on October 15, 1674, as Newton pointed out to Collins in a letter dated November 17, 1674. However, in two letters to Collins dated July 24 and August 27, 1675, Newton gives the formulas

$$A^{1/n} = [(n-1)B + A/B^{n-1}]/n \qquad (n = 2, 3, 4)$$

for finding square, cube, and fourth roots of a number A given a first guess B to the root. He tells Collins, in regard to making tables of roots: "... to find the cube root of A to eleven decimal places, seek the root by logarithms to five decimal places, and suppose it B. Then square B, not by logarithms, but by common arithmetic, that you may have its exact square to ten decimal places, and by this square divide A to eleven decimal places, and to the quotient add $2B$; the third part ... shall be the root cubical of A to eleven decimal places. Your surest way will be to find first the whole series of the roots B by logarithms, and try whether it be regular by differencing it; then square those roots by Napier's bones ..., and try the resulting series again by differencing it, whether it be regular ... ."[31]

The basis for the procedure is an immediate consequence of the Newton–Raphson procedure. For if we let

$$f(x) = x^n - A,$$

then that method tells us to form

$$x_{i+1} = x_i - (x_i^n - A)/nx_i^{n-1} = [(n-1)x_i + A/x^{n-1}]/n,$$

which is Newton's result. Gregory's interest in the subject dates from the same period. In a letter to Collins, a mutual friend of his and Newton, dated November 8, 1672 (see Gregory, *GTV*, p. 393), he seeks to solve for a the equation

$$b^n c + a^{n+1} = b^n a \qquad (b > 0, c > 0).$$

---

[31] Rigaud [1841], Vol. II, pp. 370–373. Cf. also, Newton, *Papers*, Vol. IV, pp. 14ff. Here Newton gave a scheme entitled "Completion of a numerical Table of Roots by a primitive Subtabulation" (May, 1675).

He does this iteratively by setting

$$a_0 = c, \qquad a_{i+1} = c + a_i^{n+1}/b^n.$$

He notes that the sequence of a_i is bounded and converges to the smaller positive root whereas if he chooses a_0 to be b, it converges to the larger one. Moreover, on April 2, 1674, Gregory sent to Collins a letter regarding the use of such a scheme on an annuity problem which is of the form $b^n c + a^{n+1} = b^{n-1}(b + c)a$, where "$b$ is an annuitie, c the present Worth, n the time in years of continuance and $b^2/a - b$ a year's interest of the annuitie."

Perhaps this is the place to mention Vieta's method for solving an equation iteratively. It first appeared in a work of his in 1600 on solving equations, where he worked out a number of numerical cases.[32] His procedure is not very different from the so-called Newton–Raphson one. It yields a digit at a time of a root of a polynomial equation, once the root has been isolated. Let the equation be $f(x) = N$, the root be $x = a_0 \cdot 10^k + a_1 \cdot 10^{k-1} + a_2 \cdot 10^{k-2} + \cdots$, and let an approximation to that root be $x_1 = a_0 \cdot 10^k + a_1 \cdot 10^{k-1} + \cdots + a_l \cdot 10^{k-l}$. To find the next digit a_{l+1}, and hence x_{l+1}, Vieta formed the auxiliary value

$$g_k(x_l) = f(x_l + 10^{k-l-1}) - f(x_l) - 10^{(k-l-1)n},$$

where n is the degree of the equation. Vieta then divided this quantity into $f(x_l) - N$ or perhaps $[f(x_l + 10^{k-l-1}) + f(x_l)]/2 - N$, and the integer part of the result gave the next digit a_{l+1} and the next approximant x_{l+1}. Thus he considered, e.g., the equation $f(x) = x^2 - 240x = 484$ — "Itaque si 1Q − 240N aequatur 484." He started with $x_0 = 200$ and formed $g_2(x_0) = f(210) - f(200) - 10^2 = 1600$. Then $(\frac{1}{2}[f(x_0) + f(x_0 + 10)] - 484)/g_2(x_0)$ has as its integer part 4; thus the tens digit is 4 and $x_1 = 240$, $g_2(x_1) = f(241) - f(240) - 1 = 240$. Hence $[(f(x_1) - 484)/g_2(x_1)]$ gives the units digit to be 2, $x_2 = 242$; Vieta found that $f(242) = 484$ exactly.[33]

In passing it is worth noting that this method was very useful and, until Newton replaced it by his own, was much employed. There are instances of its use by Harriot, Oughtred, and Wallis. In fact Oughtred made simplifications of Vieta's method in the editions of his *Clavis Mathematicae* from 1647 onwards. (Cf. Newton, *Papers*, Vol. I, pp. 63ff.: "Annotations from Viète and Oughtred. Section 1. Notes on Viète's *De numerosa potestatum ad exegesin resolutione*.") Whiteside dates this work of Newton as late 1664(?). Below I have reproduced Newton's free translation of Vieta's solution of $x^3 + 30x = 14356197$. In what follows let $f(x) = x^3 + 30x$.

---

[32] Vieta, *OP*, pp. 163–228.
[33] Vieta, *OP*, pp. 196–197.

The analysis of Cubick Equations.

The equation supposed $Lc * + 30L = 14356197$. $Lc + CqL = Pc$.

The square coëfficient		3	0	
The cube affected to be	14	356	197	(243
Sollids to be substracted	8			$= Ac$
		6	0	$= ACq$
Theire sume	8	006	0	
Rests	6	350	197	for finding y^e^ 2^d^ side.
The extraction of y^e^ seacon	d sid	e		
Coëfficient			30	or superior divisor.
The rest of y^e^ cube to be	6	350	197	resolved
The inferior divisors	1	2		$3Aq$
		6		
Their sume	1	260	30	
Sollids to be subtracted	4	8		$= 3AqE$
		96		$= 3AEq$
		64		$= Ec$
		1	20	$= ECq$
Their sume	5	825	20	

[The extraction of y^e^ 3^d^	side]		
The superior part of y^e^ divisor		30	or y^e^ square coefficient
The remainder for finding	524	997	y^e^ third side
The inferior part of y^e^ divisor	172	8	$3Aq$ that is $3 \times 24 \times 24$
		72	$3A$ or 3×24
The sume of y^e^ divisors	173	550	
Sollids to be taken away	518	4	$3AqE$
	6	48	$3AEq$
		27	Ec
		90	Ecq
Theire sume	524	997	
Remaines	000	000	

Vieta first guesses the size of the root as $x_0 = 200$. Then the schema above shows that $g_2(200) = f(200 + 10) - f(200) - 10^3 = 1{,}260{,}300$ (Whiteside erroneously states that $g_2(200)$ is ten times $f(201) - f(200) - 1$, and is also wrong in suggesting that Vieta always chose $f(A + 1) - f(A) - 1$ as his denominator.) This is the result recorded against the second "their sume." Next note that $[f(210) + f(200)]/2 - N = -5{,}719{,}547$, and that this divided by $g_2(200)$ has as its integer part 4. Thus $x_1 = 240$ and $g_2(240) = f(241) - f(240) - 1 = 173{,}550$, as we see in the schema. Then $f(240) - N = -524{,}997$ divided by this quantity gives the units digit as 3 and Vieta notes that $f(243) - N = 0$.

Yet another example of Vieta's method is the way he found the root near 300 of $x^3 - 116{,}620x = 352{,}947$. He started with $x_0 = 300$, found $g_2(300) = f(310) - f(300) - 10^3 = 1{,}623{,}800$ and then formed $[f(310) + f(300)]/2 - N = -7526547$; when this is divided by $g_2(300)$, the tens digit is found to be 4, etc.[34]

In the case when $f(x) = x^2$, Vieta's procedure is the same as Newton's:

$$g_k(x_l) = (x_l + 10^{k-l-1})^2 - x_l^2 - 10^{2(k-l-1)} = 2\cdot 10^{k-l-1}x_l = 10^{k-l-1}f'(x_l).$$

Clearly for other choices of f the two are not the same. But it is obvious why he deducted

$$f(x_l) + 10^{(k-l-1)n}$$

from $f(x_l + 10^{k-l-1})$. He hoped that this would leave him with an approximation to $10^{k-l-1}f'(x_l)$.

It is amusing to note, however, that Vieta's method was described in the 1670s as "work unfit for a Christian and more proper to one than can undertake to remove the Italian Alps into England . . . ."[35]

2.5. Finite Differences and Interpolation

Probably there is no single person who did so much for this field, as for so many others, as Newton. He apparently developed the subject in ignorance of the beautiful results of both Harriot and Briggs. The development of his ideas starts with a letter dated May 8, 1675, although a definitive publication had to wait until much later.[36]

Newton's interest in finite differences in general, and interpolation in particular, would seem to have developed in part from a desire to help an accountant acquaintance, one John Smith, and in part from a deep interest in a problem of Wallis's which we shall discuss later (p. 78). (Smith wanted to construct a table of roots of the integers and sought Newton's help. Newton's solution was to tabulate every hundredth square, cube, and fourth root and then subtabulate the balance.) This interest continued until he produced many of our ideas on finite differences.[37] Indeed, the very names of our Interpolation formulas — Gregory–Newton, Newton–Gauss, Newton–Bessel, Newton–Cotes — suggest the broad range of his important contributions to this field.[38]

---

[34] Vieta, *OP*, pp. 199–200.

[35] Whiteside, *Patterns*, p. 206.

[36] Newton, *Papers*, Vol. IV, pp. 14–73. The discussion here as everywhere by Whiteside is very good and thorough.

[37] Newton, *Papers*, Vol. IV, pp. 5ff.

[38] Whittaker, *WR*, Jordan [1929], or Steffensen [1927].

Let us first hastily review the basic notations, ideas, and results of Newton's work so that we may have a common basis available in order to examine his results. Customarily we shall deal with functions F defined on some interval $a \leq x \leq b$ and of polynomial or parabolic type, i.e., expressible in the form

$$y = F(x) = a_0 + a_1x + a_2x^2 + \cdots + a_nx^n$$

for some n. There are two types of differences that appear very frequently in the standard formulas. They are

$$\Delta_x^1 = \Delta F(x) = F(x + h) - F(x), \qquad \Delta_x^{\alpha+1} = \Delta\Delta_x^\alpha \qquad (\alpha = 1, 2, \ldots);$$
$$\delta_x^1 = \delta F(x) = F(x + h/2) - F(x - h/2), \qquad \delta_x^{\alpha+1} = \delta\delta_x^\alpha \qquad (\alpha = 1, 2, \ldots).$$

The former are called advancing or descending differences and the latter central differences. A brief examination shows these differences to be closely related. Indeed $\Delta_x^\alpha = \delta_{x+\alpha h/2}^\alpha$. A simple way to view the problem is to construct a table of differences as shown below and to compare corresponding entries. We notice that the advancing differences and the corresponding central ones are equal. The only difference between them is notational.

x	$F(x)$	Δ^1	Δ^2	Δ^3
0	$F(0)$			
		$\Delta_1^0 = \delta_{1/2}^1$		
1	$F(1)$		$\Delta_0^2 = \delta_1^2$	
		$\Delta_1^1 = \delta_{3/2}^1$		$\Delta_0^3 = \delta_{3/2}^3$
2	$F(2)$		$\Delta_1^2 = \delta_2^2$	
		$\Delta_2^1 = \delta_{5/2}^1$		
3	$F(3)$			

Looking at this table we see that the values of Δ_x^α for constant x (we take $\Delta_x = F(x)$) and advancing α all lie along a descending line. The values of $\delta_x^{2\alpha}$ for constant x all lie on a horizontal line, as do the $\delta_x^{2\alpha+1}$. Frequently the average

$$\mu\delta_x^\alpha = \frac{1}{2}(\delta_{x+h/2}^\alpha + \delta_{x-h/2}^\alpha)$$

is used. This device used for α odd puts both even and odd central differences on the same horizontal line for fixed x.

In the differential calculus the powers x^α play a key role because of the importance of Maclaurin or Taylor expansions.[39] In the difference calculus the analogous functions, often called factorials, are

$$(x)_{0,h} = 1, \qquad (x)_{\alpha,h} = x(x - h)\cdots(x - (\alpha - 1)h) \qquad (\alpha = 1, 2, \ldots),$$
$$[x]_{0,h} = 1, \qquad [x]_{\alpha,h} = x(x + (\alpha - 2)h/2)_{\alpha-1,h} \qquad (\alpha = 1, 2, \ldots).$$

---

[39] Brook Taylor (1685–1731) was among other things Halley's successor as secretary of the Royal Society. He published in 1715 a book entitled *Methodus Incrementorum Directa & Inversa* in which his expansion appears as does an account of Newton's work on finite differences. Cf. Taylor, *Methodus* and *Review*. His theorem had been announced in 1712. Colin Maclaurin (1698–1746) was something of an infant prodigy, being appointed professor of mathematics in Aberdeen at the age of 19. His expansion appeared in his *Treatise on Fluxions*; Maclaurin, *Fluxions*.

We often write for $h = 1$

$$(x)_{\alpha,1} = (x)_\alpha = \alpha! \binom{x}{\alpha} = \frac{x!}{(x-\alpha)!}.$$

It is clear that

$$\Delta(x)_\alpha = (x+1)_\alpha - x_\alpha = (x)_{\alpha-1}[(x+1) - (x - (\alpha - 1))] = \alpha(x)_{\alpha-1},$$

and similarly that $\delta[x]_\alpha = \alpha[x]_{\alpha-1}$ when $h = 1$ and $[x]_\alpha = [x]_{\alpha,1}$. With the help of these we then have for F, a polynomial of degree n, the relations

$$F(x) = \sum_{\alpha=0}^{n} \frac{1}{\alpha!}(x)_\alpha \Delta_0^\alpha, \qquad F(x) = \sum_{\alpha=0}^{n} \frac{1}{\alpha!}[x]_\alpha \delta_0^\alpha, \tag{2.8}$$

which are the exact difference analogs of Maclaurin's relation. To see how the central difference relation may be obtained, write F in the form

$$F(x) = a_0 + a_1[x]_1 + a_2[x]_2 + \cdots + a_n[x]_n,$$

and operate on this relation with δ^α. We find

$$\delta^\alpha F(x) = \sum_{\beta=0}^{n} a_\beta \delta^\alpha [x]_\beta = \sum_{\beta=\alpha}^{n} \beta!\, a_\beta \binom{\beta}{\alpha} [x]_{\beta-\alpha} \qquad (\alpha = 0, 1, \ldots).$$

Now if we set $x = 0$, we see that $\delta^\alpha F(0) = \delta_0^\alpha = \alpha!\, a_\alpha$, which is what we wished to show. The advancing difference relation is obtained similarly. The first part of (2.8) is the Harriot–Briggs relation, although it now goes by the name of Gregory–Newton since it was independently rediscovered by them.[40] (We discuss their work below.) It was also known to Mercator and Leibniz.

The other basic tool needed is the concept of divided differences. In what we have discussed above we implicitly assumed our data were tabulated on a regularly spaced lattice of points, say $x_0, x_1, \ldots, x_n$. But in point of fact experimentalists are often not able to do their work in quite such happy circumstances. For example, the observational astronomer often finds his data at quite irregular intervals and therefore needs a more complex apparatus than we have yet discussed. Let us define divided differences in this way:

$$F_1 = F(x_0, x_1) = \frac{F(x_0) - F(x_1)}{x_0 - x_1},$$

$$\begin{aligned} F_{\alpha+1} &= F(x_0, x_1, \ldots, x_{\alpha+1}) \\ &= \frac{F(x_0, x_1, \ldots, x_\alpha) - F(x_1, x_2, \ldots, x_{\alpha+1})}{x_0 - x_{\alpha+1}} \qquad (\alpha = 1, 2, \ldots); \end{aligned} \tag{2.9}$$

implicitly we have assumed that the $x_0, x_1, \ldots$ are distinct but by limiting

[40] James Gregory (1638–1675), a Scot like Maclaurin, mentioned the result in a letter of November 23, 1670 to John Collins, a mutual friend of his and Newton. Cf. Gregory, *GTV*, pp. 118ff. Gregory had many great discoveries to his credit and was clearly a superb mathematician in a heroic age. (Among these was the notion of convergence of a series.)

processes the divided differences are still well-defined when they are not, provided F has enough derivatives. We notice without difficulty that

$$F_\alpha = \sum_{\beta=0}^{\alpha} \frac{F(x_\beta)}{\prod_{\substack{\gamma\neq\beta \\ \gamma\leqq\alpha}} (x_\beta - x_\gamma)}. \tag{2.10}$$

From this we note that each F_α is symmetrical in its $\alpha + 1$ variables $x_0, x_1, \ldots, x_\alpha$. Moreover, if F is a polynomial of degree n, then F_α is a polynomial of degree $n - \alpha$ in each of its variables.

According to this, if F is of degree n, then F_{n+1} and all higher divided differences must vanish. Let $\alpha = n + 1$ in (2.10) above and let $x = x_{n+1}$. Then this relation becomes

$$F_{n+1} = 0 = \frac{F(x)}{\prod_{\gamma=0}^{n} (x - x_\gamma)} + \sum_{\beta=0}^{n} \frac{F(x_\beta)}{(x_\beta - x) \prod_{\gamma\neq\beta} (x_\beta - x_\gamma)},$$

$$F(x) = \sum_{\beta=0}^{n} \frac{\prod_{\gamma\neq\beta} (x - x_\gamma)}{\prod_{\gamma\neq\beta} (x_\beta - x_\gamma)} F(x_\beta).$$

This is known as Lagrange's Formula of interpolation.[41]

If we go back to the relations (2.9) and, so to speak, successively "unwind" them, we may write

$$\begin{aligned} F(x_0) &= F(x_1) + (x_0 - x_1)F(x_0, x_1) \\ &= F(x_1) + (x_0 - x_1)[F(x_1, x_2) + (x_0 - x_2)F(x_0, x_1, x_2)] \\ &= F(x_1) + (x_0 - x_1)F_1 + (x_0 - x_1)(x_0 - x_2)F_2 + \cdots \\ &\quad + (x_0 - x_1)(x_0 - x_2)\cdots(x_0 - x_{n-1})F_n + R_n, \end{aligned} \tag{2.11}$$

where $R_n = (x_0 - x_1)(x_0 - x_2)\cdots(x_0 - x_{n+1})F_{n+1}$.

If we replace x_0 by x, we have the result due to Newton and so named. Clearly, if F is a polynomial of degree n, $R_n = 0$.

In earlier papers Newton derived some of the basic results of interpolation theory, but in his *Regula Differentiarum*, tentatively dated by Whiteside October, 1676, he said: "Other rules of this kind might be presented, but I would prefer to embrace everything in one general rule and show how any series you wish may be intercalated in any place commanded."[42] We should understand that both Newton and Gregory realized the geometrical character of the interpolation problem and formulated it in this way: "Given any number of points, to describe a curve which shall pass through one of them."[43] First Newton took up the case where the points are equally spaced. Figure 2.6 (p. 72) is a reproduction from Whiteside of Newton's illustration showing how he set out his differences. Thus the $b, b^2, b^3, \ldots$ are his first differences,

---

[41] Lagrange VII [1795], p. 286. Fraser tells us, however, that the relation (2.9) was already known to Euler.

[42] Newton, *Papers*, Vol. IV, p. 47.

[43] Newton, *Papers*, Vol. IV, p. 61.

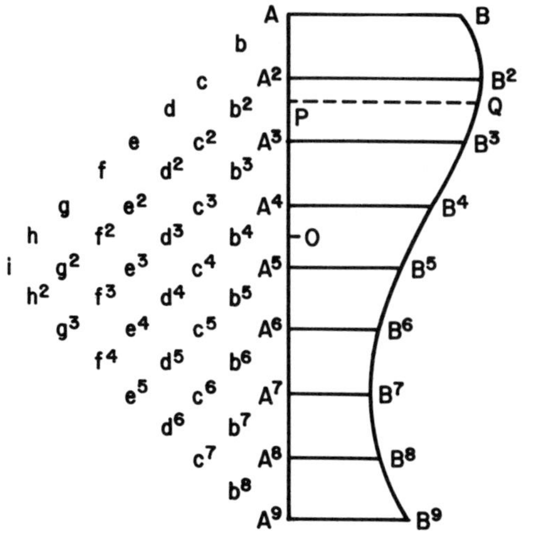

A line AA^9 is divided into equal parts AA^2, A^2A^3, A^3A^4 &c. and the parallels AB, A^2B^2, A^3B^3 &c. are drawn: to find the curve passing through the points B, B^2, B^3 &c.

Figure 2.6.

c, c^2, c^3, ... his second differences, etc. — the superscripts are not powers but are indices.[44]

Jones was a friend of Newton who was authorized to bring out his compendium containing: Newton's *De Analysi per Aequationes Infinitas*; fragments of letters from Newton to Oldenburg, Wallis and Collins; his *De Quadratura Curvarum*; his *Enumeratione*; and his *Methodus Differentialis* which contains Newton's exposition of interpolation using central differences. The preface is by Jones. (See p. 73 below.)

Perhaps Newton's words in Whiteside's translation at this point will serve to show his style.

Of the erected lines A_1B_1, A_2B_2, A_3B_3, ... seek the first differences b_1, b_2, b_3, ...; their second ones c_1, c_2, c_3, ...; their third ones d_1, d_2, d_3, ...; and so on successively until you reach the last difference i_1. Then, beginning with the last difference, take out the middle differences in alternate columns/series/ranks of differences and the arithmetic means of the two middle-most ones in the remaining ranks, proceeding as far as the series A_1B_1, A_2B_2, A_3B_3, ... of first terms. Let these be k, l, m, n, o, p, q, r, s, ..., the last of which is to denote the last difference, the last but one the arithmetic mean between the two last but one differences, the last but two the middle one of the three last but two differences, and so on in turn up to the first, which will be either the middle one of the terms $A_1[B_1]$, $A_2[B_2]$, $A_3[B_3]$, $A_3[B_3]$, ... or the arithmetic mean between the two middle-most. The former happens when the number of terms $A_1[B_1]$, $A_2[B_2]$, $A_3[B_3]$, ... is odd, the latter when it is even.

44 Newton, *Methodus* = *Works*, Vol. 2, pp. 165–173. However, it first appears in his *Waste Book* and is tentatively dated by Whiteside to 1676; it comes directly after a memorandum of October, 1676. There is also an analysis by D. C. Fraser in Newton [1927], pp. 45–69, entitled "Newton and Interpolation." Cf. especially, Newton, *Papers*, Vol. IV, pp. 36–39.

Case 1. In the former case let A_5B_5 be that middle term, that is, $A_5B_5 = k$, $\frac{1}{2}(b_4 + b_5) = l$, $c_4 = m$, $\frac{1}{2}(d_3 + d_4) = n$, $e_3 = o$, $\frac{1}{2}(f_2 + f_3) = p$, $g_2 = q$, $\frac{1}{2}(h_1 + h_2) = r$, $i_1 = s$. Then, having erected the ordinate PQ call $A_5P = x$ and multiply the terms of this progression

$$1 \times \frac{x}{1} \times \frac{x}{2} \times \frac{x^2 - 1}{3x} \times \frac{x}{4} \times \frac{x^2 - 4}{5x} \times \frac{x}{6} \times \frac{x^2 - 9}{7x} \times \frac{x}{8} \times \frac{x^2 - 16}{9x}$$

$$\times \frac{x}{10} \times \frac{x^2 - 25}{11x} \times \frac{x}{12} \times \frac{x^2 - 36}{13x} \cdots$$

one into another continually. There will arise the terms

$$1, \quad x, \quad \frac{1}{2}x^2, \quad \frac{1}{6}(x^3 - x), \quad \frac{1}{24}(x^4 - x^2), \quad \frac{1}{120}(x^5 - 5x^3 + 4x),$$

$$\frac{1}{720}(x^6 - 5x^4 + 4x^2), \quad \frac{1}{5040}(x^7 - 14x^5 + 49x^3 - 36x), \ldots,$$

and if the terms of the series $k, l, m, n, o, p, \ldots$ be respectively multiplied by these, the aggregate

$$k + lx + \frac{1}{2}mx^2 + \frac{1}{6}n(x^3 - x) + \frac{1}{24}o(x^3 - x^2) + \frac{1}{120}p(x^5 - 5x^3 + 4x) \cdots$$

of the products will be the length of the ordinate PQ.

Case 2. In the latter case let A_4B_4, A_5B_5 be the two middle terms, that is, $\frac{1}{2}(A_4B_4 + A_5B_5) = k$, $b_4 = l$, $\frac{1}{2}(c_3 + c_4) = m$, $d_3 = n$, $[\frac{1}{2}](e_2 + e_3) = o$, $f_2 = p$, $\frac{1}{2}(g_1 + g_2) = q$ and $h_1 = r$; then, having raised the ordinate PQ, bisect A_4B_5 at O and, calling $OP = x$, multiply the terms of this progression

$$1 \times \frac{x}{1} \times \frac{x^2 - \frac{1}{4}}{2x} \times \frac{x}{3} \times \frac{x^2 - \frac{9}{4}}{4x} \times \frac{x}{5} \times \frac{x^2 - \frac{25}{4}}{6x} \times \frac{x}{7} \times \frac{x^2 - \frac{49}{4}}{8x} \cdots$$

one into another continually. There will arise the terms

$$1, \quad x, \quad \frac{1}{8}(4x^2 - 1), \quad \frac{1}{24}(4x^3 - x), \quad \frac{1}{384}(16x^4 - 40x^2 + 9), \ldots,$$

and if the terms of the series $k, l, m, n, o, p, \ldots$ be respectively multiplied by these, the aggregate

$$k + lx + \frac{1}{8}m(4x^2 - 1) + \frac{1}{24}n(4x^3 - x) + \frac{1}{384}o(16x^4 - 40x^2 + 9) \cdots$$

of the products will be the length of the ordinate PQ.

But it should here be noted, first, that the intervals A_1A_2, A_2A_3, $A_3A_4, \ldots$ are in this instance taken to be units; and that the differences ought to be gathered by taking away lower quantities from upper ones (A_2B_2 from A_1B_1, A_3B_3 from A_2B_2, b_2 from b_1, and so on: in other words by making

$$A_1B_1 - A_2B_2 = b_1, \qquad A_2B_2 - A_3B_3 = b_2, \qquad b_1 - b_2 = c_1, \ldots)$$

and consequently, when those differences prove in this way to be negative, their signs are to be changed throughout.[45]

This should give us a feeling for Newton's style. Let us examine what he said in more detail. His first case in our terms involves the quantities

$$k = F_0, \quad l = \mu\delta_0^1, \quad m = \delta_0^2, \quad n = \mu\delta_0^3, \quad o = \delta_0^4, \quad p = \mu\delta_0^5, \ldots;$$

---

[45] Newton, *Papers*, Vol. IV, pp. 57–61.

and hence his relation is

$$F(x) = F_0 + x\mu\delta_0^1 + \frac{1}{2!}x^2\delta_0^2 + \frac{1}{3!}x(x^2-1)\mu\delta_0^3$$

$$+ \frac{1}{4!}x^2(x^2-1)\delta_0^4 + \frac{1}{5!}x(x^2-1)(x^2-4)\mu\delta_0^5 + \cdots$$

$$= \frac{1}{x}\sum_{\alpha=1}^{n}\frac{[x]_{2\alpha}}{(2\alpha-1)!}\mu\delta_0^{2\alpha-1} + \sum_{\alpha=0}^{n}\frac{[x]_{2\alpha}}{(2\alpha)!}\delta_0^{2\alpha},$$

which is the so-called Newton–Stirling Interpolation formula.[46] Newton's second case involves rather different quantities. He says, e.g., $k = (A_4B_4 + A_5B_5)/2$, $l = b_4, \ldots$. In our terms these may be expressed as

$$k = [F(1) + F(0)]/2 = \mu F(1/2), \qquad l = \delta_{1/2}^1, \qquad m = \mu\delta_{1/2}^2,$$
$$n = \delta_{1/2}^3, \qquad o = \mu\delta_{1/2}^4, \qquad p = \delta_{1/2}^5, \ldots;$$

and hence his relation becomes

$$F\left(x+\frac{1}{2}\right) = \frac{1}{x}\sum_{\alpha=0}^{n-1}\frac{[x]_{2\alpha+1}}{(2\alpha)!}\mu\delta_{1/2}^{2\alpha} + \sum_{\alpha=0}^{n-1}\frac{[x]_{2\alpha+1}}{(2\alpha+1)!}\delta_{1/2}^{2\alpha+1},$$

where $\mu\delta_{1/2}^0 = [F(1) + F(0)]/2$. This is the so-called Newton–Bessel Interpolation formula.[47]

Newton then proceeded to remove the restriction that the points be uniformly spaced and found generalizations of his preceding results in terms of divided differences.

In his *Principia* he introduced into Book III, Lemma 5 the Harriot–Briggs relation for equal intervals and then for divided differences.[48] This was brought in so that the reader would have the apparatus available to interpolate cometary positions.

No better way to evaluate Newton's work on interpolation can be found than to quote Stirling in his own *Methodus Differentialis* of 1730. He said in the Scholium to Proposition XX:

> Concerning the description of a curve of a parabolic kind, thro' any number of given points, several acute geometers since *Newton* have treated; but all their solutions are the same with these exhibited, which differ very little from *Newton's*, as will be manifest from the fifth lemma of the third Book of his *Principia*, and his *Differential Method* published by the universally learned *W. Jones*, Esq; *Newton* indeed describes a parabola through given points, others have considered the assignation of terms from given differences; but how it may be conceived, or under what form it may be exhibited, the problem is the same. And indeed the invention of the terms, which the values of the ordinate *T* have, is extremely ingenious, and worthy [of] the excellent author; and after the forms are had, the

---

[46] Whittaker, *WR*, p. 43. The result also appears in Stirling [1719], where he considers three cases, of which our formula is case two; and also in Stirling [1730], Prop. XX = Stirling [1739]; the Newton–Bessel result is case ii of that proposition.

[47] Whittaker, *WR*, p. 47, or Steffensen [1950], pp. 30–31.

[48] Newton, *Principia*, Book III, Lemma V, pp. 499–500.

investigation of the problem is easy, in which nothing else is required but the resolution of simple equations.

But it is to be observed, that the form of the ordinate composed of the powers $A + Bz + Cz^2 + Dz^3 +$ &c. which *Newton* assumes in demonstrating the foundation of his method, is badly designed for this purpose; for the value of every coefficient comes out into an infinite series: but if any one will assume the terms here used, he will with very little trouble get the above conclusion.[49]

Gregory's development of his interpolation formula is contained in a lengthy letter from him to Collins dated November 23, 1670, which is presumably a response to the letter of Collins of December 30, 1668, in which he says, ". . . and therefore intreate you to vouchsafe to communicate what Compendiums you have about it, as likewise how you demonstrate Briggs his interpositions by correcting the 1st 2nd 3rd 4th 5th differences of Log^{mes}." Turnbull tells us this is an illusion by Collins to the thirteenth chapter of the 1624 edition of Briggs's *Arithmetica Logarithmica*.[50]

Gregory wrote in this letter, "I remember you did once desire of me my method of finding the proportional parts in tables." He then describes his method of interpolation. It amounts to the use of the formula

$$f(x) = f(0) + \frac{x}{1!}\Delta_0 + \frac{x(x-1)}{2!}\Delta_0^2 + \cdots$$

where his function f has $f(0) = 0$ and $x = a/c$. He calls his general ordinate $f(x)$ $\alpha\gamma$, and his first, second, etc. differences $d, f, \ldots$. Finally he writes b/c for $(a/c)(a - c)/2c$, h/c for $(a/c)\cdot((a - c)/2c)\cdot((a - 2c)/3c)$, etc. In these terms he writes his formula as

$$\alpha\gamma = \frac{ad}{c} + \frac{bf}{c} + \frac{kh}{c} + \frac{li}{c} + \text{etc. in infinitum,}$$

and goes on to say, "This method, as I apprehend, is both more easie and universal than either Briggs or Mercator's, and also performed without tables."[51]

It has been suggested by several people, including Turnbull, that Gregory developed out of this interpolation formula the general Taylor expansion and hence the Binomial theorem — all this independently of Newton. (As we shall note on p. 81 he carried out 16 series expansions, apparently by a passage to the limit in his interpolation result.)

---

[49] Stirling [1749], pp. 98–99, or Newton [1927], p. 69 of Fraser's article.

[50] As we saw above, Briggs gave there his scheme for subtabulation to three or five parts in his logarithmic tables, but without proof. Cf. Gregory, *GTV*, pp. 58–59 and 118–122.

[51] Gregory, *GTV*, pp. 119–133. He also showed that the integral of sec x is log (sec x − tan x). This was a long-unsolved problem relating to Edward Wright's tables. Cf. p. 11. Collins wrote that Wright "made a table of Logarithms before Logarithms were invented and printed, but did not know he had done it." Gregory had also shown in 1668 the reciprocal natures of differentiation and integration in his *Geometriae pars universalis*, *GU*.

He illustrates his formula by the following: "Given b, $\log b = e$, $b + d$, $\log (b + d) = e + c$, it is required to find the number whose logarithm is $e + a$." To do this he forms the expression

$$b + xd + x(x - 1)d^2/2b + x(x - 1)(x - 2)d^3/6b^2 + \cdots;$$

if in this we set $p = d/b$, then it becomes

$$b\left[1 + xp + \binom{x}{2}p^2 + \binom{x}{3}p^3 + \cdots\right] = b(1 + p)^x. \tag{2.12}$$

This follows from the fact that the differences of the function $F(x) = (1 + p)^x$ are

$$\Delta^1 F(x) = (1 + p)^{x+1} - (1 + p)^x = p(1 + p)^x,$$
$$\Delta^2 F(x) = (1 + p)^{x+2} - 2(1 + p)^{x+1} + (1 + p)^x = p^2(1 + p)^x, \ldots,$$

and so $\Delta_0^\alpha = p^\alpha$. The logarithm of the right-hand member of (2.12) is $\log b + x \log (1 + d/b) = \log b + x \log (b + d) - x \log b = e + (a/c)(e + c) - (a/c)e = e + a$, provided $x = a/c$, as was to be shown.[52] Gregory uses this to calculate the daily interest rate at six percent *per annum*, finding the value of $100(1 + .06)^{1/365}$ (see Footnote 53).

Gregory gave another interesting application. He wished to find 23^3 given the table

Num.	Cubi	Diff. 1ae	2ae	3ae
10	1000			
		2375		
15	3375			
		4625	2250	
20	8000			750
		7625	3000	
25	15625			750
		11375	3750	
30	27000			

$$c = 5$$
$$a = 23 - 10 = 13$$
$$d = 2375$$
$$f = 2250$$
$$h = 750.$$

To do this he noted that $x = 13/5$, which means he fixed his origin at 10. He wrote

$$23^3 = 1000 + \frac{13}{5} \times 2375 + \frac{13 \times 8}{25 \times 2} \times 2250 + \frac{13 \times 8 \times 3}{2 \times 3 \times 5^3} \times 750 = 12167.$$

Both Newton and Gregory understood how to integrate numerically, as we see in Proposition VI and Scholium of the *Methodus*.[54] The problem is this: "To find the approximate area of any curve a number of whose ordinates can be ascertained." Newton remarks in his Scholium that "rules applicable to any given number of ordinates can be derived and recorded for reference. For example: If there are four ordinates at equal intervals, let A be the sum of the first and fourth, B the sum of the second and third, and R the interval

---

[52] Gregory, *GTV*, p. 133. We note that in effect Gregory had the Binomial theorem in his result (2.12).

[53] Rigaud [1841], Vol. II, pp. 210–211.

[54] Newton, *Works*, Vol. II, p. 172. *Principia*, pp. 499–500 (this is a corollary to Lemma V referred to above); cf. also, Gregory, *GTV*, pp. 119–128.

between the first and fourth; then the central ordinate will be $(9B - A)/16$, and the area between the first and fourth ordinates will be $(A + 3B)R/8$."

This last result of Newton's is one of the so-called Newton–Cotes Integration formulas (the *Three-eighth rule*). It may be expressed as

$$\int_a^{a+3h} F(x)\,dx = \frac{3h}{8}[F(0) + 3F(1) + 3F(2) + F(3)].$$

This result plus some additional remarks by Newton in the *Methodus* came into Roger Cotes's hands in 1711 when he saw Jones's edition (cf. p. 72 above) of the *Analysis*. According to Whiteside he incorporated these results systematically as a postscript to his "De Methodo Differentiali *Newtoniana*."[55] The results also appear in Stirling's article in the *Philosophical Transactions*, Stirling [1719]. He gives the Cotes formulas for 1, 3, 5, 7, 9, and 11 ordinates and, as an illustration, computes the area under the curve $(1 + z^2)^{-1}$ from $z = 0$ to $z = 1$ using the 9-ordinate result, which is

$$\frac{h}{28350}[989(f_0 + f_8) + 5888(f_1 + f_7) - 928(f_2 + f_6) + 10496(f_3 + f_5) - 4540f_4].$$

He found the area to be .785398187. It should be $\pi/4 = .78539816$.

However, numerical integration techniques did not originate with Newton. As early as 1639 Cavalieri, one of the early pioneers in the field of integration, had found in geometrical form the so-called Simpson's rule, which is Cotes's formula for $n = 2$.[56] It was also known to Gregory, who published it in his *Exercitationes Geometricae*.[57] According to Turnbull, Gregory worked out Simpson's rule in order to form tables of

$$\log \sec x = \int_0^x \tan t\,dt, \qquad \log \tan (x/2 + \pi/4) = \int_0^x \sec t\,dt$$

by approximating to the integrals indicated above. Simpson himself rediscovered the result in 1743.[58] This by no means was the limit of Gregory's interest in numerical integration. In the November 23, 1670, letter to Collins he also discussed the so-called Gregory formula for numerical integration.[59] This result is now often stated as

$$\frac{1}{c}\int_a^{a+\alpha c} F(x)\,dx = \sum_{\beta=0}^{\alpha-1} F(a + \beta c) + \sum_{\beta=0} b_\beta(\Delta^\beta_{a+\alpha c} - \Delta^\beta_\alpha),$$

where the b_β are the coefficients of the Bernoulli polynomials of the second

[55] Newton, *Papers*, Vol. IV, 73*n*–74*n*.

[56] Whittaker, *WR*, p. 156*n*.

[57] The book consists of only 26 pages of material. Cf. Gregory, *GTV*, p. 459.

[58] Simpson [1743], p. 109.

[59] Rigaud [1841], Vol. II, pp. 208–209, or Gregory, *GTV*, pp. 118ff.

kind.[60] They are defined by the relations

$$b_\beta = \int_0^1 \binom{x}{\beta}\, dx \qquad (\beta = 0, 1, 2, \ldots);$$

and the Bernoulli polynomial of the second kind of degree β is

$$\psi_\beta(x) = b_0\binom{x}{\beta} + b_1\binom{x}{\beta - 1} + \cdots + b_\beta.$$

Actually Gregory's statement is given in terms of the area ABP under a curve between the points $x = A$ (where the curve vanishes) and $x = P$ (where the ordinate is BP). He then says:

$$\text{"}\ldots ABP = \frac{dc}{2} - \frac{fc}{12} + \frac{hc}{24} - \frac{19ic}{720} + \frac{3kc}{160} - \frac{863lc}{60480} + \text{\&c. infinitum,"}$$

where $c = AP$, $d = PB$, f, h, i, k, l, &c. are the successive differences of the function. Thus Gregory's result is

$$\frac{1}{c}\int_a^{a+c} F(x)\, dx = \sum_{\beta=0} b_\beta \Delta_\alpha^\beta = b_0 F(a) + \sum_{\beta=z} b_\beta[\Delta_{a+c}^{\beta-1} - \Delta_a^{\beta-1}]$$

since, by definition,

$$\Delta_{a+c}^{\beta-1} - \Delta_a^{\beta-1} = \Delta_\alpha^\beta.$$

His formula follows directly by integrating his interpolation formula quoted earlier for αy. If the integrals from a to $a + c$, $a + c$ to $a + 2c, \ldots$ are summed up, the general formula above is obtained at once. An equivalent form is

$$\frac{1}{c}\int_a^{a+\alpha c} F(x)\, dx = \left[\frac{1}{2}F(a) + F(a + c) + F(a + 2c) + \cdots + \frac{1}{2}F(a + \alpha c)\right]$$
$$- \frac{1}{12}(\Delta_{a+(\alpha-1)c}^1 - \Delta_a^1) - \frac{1}{24}(\Delta_{a+(\alpha-1)c}^2 + \Delta_a^2) - \cdots \quad .$$

We shall see how his successor Maclaurin found, along with Euler, the comparable result where derivatives instead of differences appear.

Newton had another, quite different, interest in interpolation techniques, which arose from his reading of Wallis's work. The basic problem was finding the area of a quadrant of a circle

$$\Box = a_{1/2} = \int_0^1 (1 - x^2)^{1/2}\, dx$$

by integration. Gregory St. Vincent had attempted the demonstration but Huygens in his *Cyclometria* (1651) showed the difficulty with that proof.

[60] Jordan [1929], pp. 265ff. (Gregory gives the first seven coefficients explicitly.)

Wallis then took up the problem and sought to solve it by interpolating $a_{1/2}$ from the known values

$$a_n = \int_0^1 (1 - x^2)^n \, dx \qquad (n = 0, 1, 2, \ldots).^{61}$$

What Wallis did was to interpolate $a_{1/2}$ out of the array of values

$$b_{\alpha,\beta} = \frac{1}{\int_0^1 (1 - x^\alpha)^\beta \, dx} \qquad (\alpha, \beta = 1, 2, \ldots)$$

essentially by an appeal to "continuity principles." By ingenious arguments Wallis found his well-known result that if $a = \int_0^1 (1 - x^2)^{1/2} \, dx$,

$$a \text{ is} \begin{cases} \text{less y}^{\text{n}} & \dfrac{9 \times 25 \times 49 \times 81 \times 121 \times 169}{2 \times 16 \times 36 \times 64 \times 100 \times 144 \times 14} \sqrt{\dfrac{14}{13}} \qquad \text{\&c.} \\[2ex] \text{greater y}^{\text{n}} & \dfrac{3 \times 3 \times 5 \times 5 \times 7 \times 7 \times 9 \times 9 \times 11 \times 11 \times 13 \times 13}{2 \times 4 \times 4 \times 6 \times 6 \times 8 \times 8 \times 10 \times 10 \times 12 \times 12 \times 14} \sqrt{\dfrac{15}{14}}. \end{cases}$$

Wallis asserted, correctly, that the result could be carried to as many factors as one wished. (Incidentally the first use of the symbol ∞ for "infinity" occurs in this argument of Wallis.)

Newton, in analyzing this result of Wallis's in 1664/5, was led to the Binomial theorem. He did it, in effect, by replacing Wallis's fixed upper limit of 1 in the integral above by a variable upper limit, x. This elegant trick gave him the "elbow-room" he needed. Thus he considered

$$\int_0^x (1 - t^2)^{\alpha/2} \, dt \qquad (\alpha = 0, \pm 2, \pm 3, \ldots),$$

and wrote for $\alpha = 2, 4, 6, \ldots, 14$:

$$x\colon \quad x - \frac{1}{3}x^3\colon \quad x - \frac{2}{3}x^3 + \frac{1}{5}x^5\colon \quad x - \frac{3}{3}x^3 + \frac{3}{5}x^5 - \frac{1}{7}x^7\colon$$

$$x - \frac{4}{3}x^3 + \frac{6}{5}x^5 - \frac{4}{7}x^7 + \frac{1}{9}x^9\colon \quad x - \frac{5}{3}x^3 + \frac{10}{5}x^5 - \frac{10}{7}x^7 + \frac{5}{9}x^9 + \frac{1}{11}x^{11}\colon$$

$$x - \frac{6}{3}x^3 + \frac{15}{5}x^5 - \frac{20}{7}x^7 + \frac{15}{9}x^9 - \frac{6}{11}x^{11} + \frac{1}{13}x^{13}\colon$$

$$x - \frac{7}{3}x^3 + \frac{21}{5}x^5 - \frac{35}{7}x^7 - \frac{35}{9}x^9 - \frac{21}{11}x^{11} + \frac{7}{13}x^{13} - \frac{1}{15}x^{15}. \text{ \&c.}$$

---

[61] Whiteside, *Patterns*, pp. 236ff. Wallis's work is in his *Arithmetica infinitorum*; cf. Wallis, *OP*, Vol. I, pp. 462ff. Cf. also, Newton, *Papers*, Vol. I, pp. 96ff, (especially Whiteside's excellent commentary), where there are annotations on Wallis by Newton.

And if y^e^ meane termes be inserted it will bee

$$x\colon\ x - \colon\ x - \frac{1}{3}x^3\colon\ x - \frac{3}{6}x^3 + \colon\ x - \frac{2}{3}x^3 + \frac{1}{5}x^5\colon\ x - \frac{5}{6}x^3 + \frac{2}{5}x^5 -$$

The first letters x run in this progression 1. 1. 1. 1. 1. &c. y^e^ 2^d^ x^3 in this

$$\frac{-1}{3}.\frac{0}{3}.\frac{1}{3}.\frac{2}{3}.\frac{3}{3}.\frac{4}{3}.\frac{5}{3}\ \&c.$$

y^e^ 3^d^ x^5 in this 6. 3. 1. 0. $0 + 1 = 1$. $1 + 2 = 3$. $3 + 3 = 6$. $6 + 4 = 10$. $10 + 5 = 15$. y^e^ 4th x^7 [in] this.

[Note that Newton's last coefficient 2/5 is wrong; it should be 3/8.]

Here Newton broke off and recorded his results as shown on pages 82–83. We see here his discovery of the Binomial theorem. It is interesting to compare this to what he said a decade later. Turnbull and Whiteside each quote from a letter Newton wrote Oldenburg on October 24, 1676 on this matter:[62]

I considered . . . that the denominators [$2i + 1$] were in arithmetical progression, and so only the numeral [the binomial] coefficients remained to be investigated. But these [for even powers of λ] were the figures which represent powers of the number 1, 1 namely $(11)^0$, $(11)^1$, $(11)^2$ . . ., that is, . . . , 1; 1, 1; 1, 2, 1; 1, 3, 3, 1; 1, 4, 6, 4, 1.

And so I sought how in these sequences, given the first two figures, the rest might be derived, and I found that, assuming the second figure to be m, the rest could be produced by continued multiplication of the terms of this sequence:

$$\frac{m-0}{1} \times \frac{m-1}{2} \times \frac{m-2}{3} \times \cdots \text{ etc.}$$

So I applied this rule to interpolate the sequence . . . And since, for the Circle, the second Term was

$$\frac{\frac{1}{2}x^3}{3}$$

I put $m = \frac{1}{2}$ . . . .

And the like Method may be made use of to interpolate other series . . . . And thus I perceived, e.g., that $(1 - xx)^{1/2}$ was equivalent to

$$1 - \frac{1}{2}x^2 - \frac{1}{8}x^4 - \frac{1}{16}x^6 \cdots .$$

Having discovered this, I entirely neglected the Interpolation of Series, and made use of these operations only, as a more genuine foundation . . . .[63]

It is interesting to contrast with Whiteside the derivations of the Binomial theorem by Gregory and Newton. That of the former is possibly more

---

[62] Turnbull [1945], p. 13, and Whiteside, *Patterns*, pp. 256–257.

[63] Turnbull [1945], pp. 15–16, Whiteside has remarked that this letter ". . .written over ten years after the actual event, tends to touch up the crudities of the original discovery." This is in his *Patterns*, p. 258.

elegant since he seems to have found it directly from his interpolation formula as we remarked above.[64] It is not quite certain that Gregory knew the expansion we now attribute to Taylor, but Turnbull makes it seem very plausible. Gregory never actually displayed the result, but in a letter of January 29, 1671 he gave the expansion of r logtan $(\theta/2 + \pi/4)$. He wrote

$$a + \frac{a^3}{6r^2} + \frac{a^5}{24r^4} + \frac{61a^7}{5040r^6}$$

where $a = r\theta$. He also recorded the values of $a, a^2, a^3, \ldots, a^{16}$ on another part of the page for $\theta = 18°$.[65] Then on February 15, 1671 he wrote a letter giving a number of expansions: "As for Mr. Newton's universal method, I imagine I have some knowledge of it, both as to geometrick & mechanick curves, however I thank you for the series ye sent me, and send you these following in requital." He then went on to set r to be a radius, a the arc of a circle of this radius cut off by the angle θ, t/r the tangent of θ and s/r the secant of θ. Then Gregory wrote the expansions

$$a = t - \frac{t^3}{3r^2} + \frac{t^5}{5r^4} - \frac{t^7}{7r^6} + \frac{t^9}{9r^8},$$

$$t = a + \frac{a^3}{3r^2} + \frac{2a^5}{15r^4} + \frac{17a^8}{315r^6} + \frac{3233a^9}{181440r^8},$$

$$s = r + \frac{a^2}{2r} + \frac{5a^4}{24r^3} + \frac{61a^6}{720r^5} + \frac{277a^8}{8064r^7}.$$

Moreover, Gregory now set t to be $(1/r)\cdot$logtan $(\theta/2 + \pi/4)$ and s to be $(1/r)$ log$\cdot$sec θ. Then he had

$$s = \frac{a^2}{2r} + \frac{a^4}{12r^3} + \frac{a^6}{45r^5} + \frac{17a^8}{2520r^7} + \frac{3233a^{10}}{1814400r^9},$$

$$t = e + \frac{e^3}{6r^2} + \frac{e^5}{24r^4} + \frac{61e^7}{5040r^6} + \frac{277e^9}{72576r^8},$$

where $e = 2a - r\pi/2$.[66] He then goes on to say: "Ye shall here tak notice that the radius artificialis = 0, and that when ye find $q > 2a$, or the artificial secant of 45° to be greater than the given secant, to alter the signs and go on in the work according to the ordinary precepts of Algebra."

Looking back in admiration at the genius of Gregory, working as he did in a remarkable period of exceptional progress by extraordinary scholars, one is tempted to speculate about what he might have done if he had lived longer. Maclaurin succeeded him to the chair of mathematics at Edinburgh.

---

[64] Whiteside, *Patterns*, pp. 260–261.

[65] Gregory, *GTV*, pp. 350–359.

[66] Gregory, *GTV*, pp. 170–171 and 356–368. I have omitted two other series given in his letter. The reference to Newton's universal method is noteworthy.

1st.	$+x$	$\times$ 1.	1.	1.	1.	1.	1.	1.	1.	1.	1.	1.
2^{d}.	$-\frac{x^3}{3}$	$\times$ 0.	0 + 1 = 1.	1 + 1 = 2.	2 + 1 = 3.	3 + 1 = 4.	4 + 1 = 5.	6.	7.	8.	9.	10.
3^{d}.	$+\frac{x^5}{5}$	$\times$ 0.	0 + 0 = 0.	0 + 1 = 1.	1 + 2 = 3.	3 + 3 = 6.	6 + 4 = 10.	15.	21.	28.	36.	45.
4th.	$-\frac{x^7}{7}$	$\times$ 0.	0 + 0 = 0.	0 + 0 = 0.	0 + 1 = 1.	1 + 3 = 4.	4 + 6 = 10.	20.	35.	56.	84.	120.
5.	$+\frac{x^9}{9}$	$\times$ 0.	0 + 0 = 0.	0 + 0 = 0.	0 + 0 = 0.	0 + 1 = 1.	1 + 4 = 5.	15.	35.	70.	126.	210.
6.	$-\frac{x^{11}}{11}$	$\times$ 0.	0 + 0 = 0.	0 + 0 = 0.	0 + 0 = 0.	0 + 0 = 0.	0 + 1 = 1.	6.	21.	56.	126.	252.
		1st. *.	2^{d}. *.	3^{d}. *.	4th. *.	5^{t}. *.	6^{t}.	1.	7.	28.	84.	210.
								*. 7th.	1.	8.	36.	120.
									*. 8th.	1.	9.	45.
										*. 9th.	1.	10.
											*. 10th.	1.
												*. 11th.

Now if the meane termes in these progressions can bee calculated y^e first of y^m gives y^e area *aeqp*. Which is thus done

1st.	$+x$	$\times$ 1.	1.	1.	1.	1.	1.	1.	1.	1.	1.	1.	1.	1.	
2d.	$-\frac{x^3}{3}$	$\times$ 0.	$\frac{1}{2}$.	1.	$\frac{3}{2}$.	2.	$\frac{5}{2}$.	3.	$\frac{7}{2}$.	4.	$\frac{9}{2}$.	5.	$\frac{11}{2}$.	6.	
3d.	$+\frac{1}{5}x^5$	$\times$ 0.	$-\frac{1}{8}$.	0.	$\frac{3}{8}$.	1.	$\frac{15}{8}$.	3.	$\frac{35}{8}$.	6.	$\frac{63}{8}$.	10.	$\frac{99}{8}$.	15.	
4.	$-\frac{1}{7}x^7$	$\times$ 0.	$+\frac{1}{16}$.	0.	$\frac{1}{-16}$.	0.	$\frac{5}{16}$.	1.	$\frac{35}{16}$.	4.	$\frac{105}{16}$.	10.	$\frac{231}{16}$.	20.	
5.	$+\frac{1}{9}x^9$	$\times$ 0.	$-\frac{3}{128}$.	0.	$\frac{3}{128}$.	0.	$\frac{-5}{128}$.	0.	$\frac{35}{128}$.	1.	$\frac{315}{128}$.	5.	$\frac{1155}{128}$.	15.	
6.	$-\frac{1}{11}x^{11}$	$\times$ 0.	$\frac{7}{256}$.	0.	$\frac{-3}{256}$.	0.	$\frac{3}{256}$.	0.	$\frac{-7}{256}$.	0.	$\frac{63}{256}$.	1.	$\frac{693}{256}$.	6.	
7.	$\frac{1}{13}x^{13}$	$\times$ 0.	$\frac{-21}{1024}$.	0.	$\frac{7}{1024}$.	0.	$\frac{-5}{1024}$.	0.	$\frac{7}{1024}$.	0.	$\frac{-21}{1024}$.	0.	$\frac{231}{1024}$.	1.	$\frac{3003}{1024}$.[67]

[67] Newton, *Papers*, Vol. I, pp. 106–107. His last tabulation is for the integral evaluated at $\alpha = 0, 1/2, 1, 3/2, \ldots, 12/2$. The material is in his "Annotations from Wallis."

Newton was influential in this appointment, and others. His kindness and generosity to his friends is exemplified in the following quotation from his letters. "I am very glad to hear that you have a prospect of being joined to Mr. *James Gregory* in the professorship of the Mathematics at Edinburgh, not only because you are my friend, but principally because of your abilities, your being acquainted as well with the new improvements of mathematics, as with the former state of those Sciences, I heartely wish you a good success, and shall be very glad of hearing of your being elected. I am, with all sincerity, your faithful and most humble servant." The second is to the Lord Provost of Edinburgh and was written privately without Maclaurin's knowledge: "I am glad to understand that Mr. *Maclaurin* is in good repute amongst you for his skill in mathematics, for I think he deserves it very well, and to satisfy you that I do not flatter him, and also to encourage him to accept the place of assisting Mr. Gregory, in order to succeed him, I am ready (if you please to give me leave) to contribute twenty pounds *per annum* towards a provision for him, till Mr. *Gregory's* place become void, if I live so long, and I will pay it to his order in *London*."[68] Gregory was professor of mathematics at St. Andrews until 1674, when he went to the University of Edinburgh as the first holder of the chair of mathematics, a post he held for a little more than a year. He died in October 1675.

2.6. Maclaurin on the Euler–Maclaurin Formula

One of Maclaurin's best-known and important results in the field under discussion is the so-called Euler–Maclaurin formula for numerical integration.[69] Maclaurin remarks in a footnote: "... I take this opportunity to mention, that having occasionally shown in 1737, the 292, 293, pages of this treatise (after they were printed) to Mr. STIRLING, he took notice that a theorem similar to the first of these described in art. 352 had been communicated to him by Mr. EULER."[70] Actually, Maclaurin stated not one but four related theorems on this subject in Book I. They are given in Articles 352 and 353 and proved in Book II, Articles 828–832. We shall see that the four results arise from his way of considering cases. First of all, he distinguished between

[68] Maclaurin, *Account*. The letters are on pp. iv, v. The book was prepared by Maclaurin after Newton's death at the request of Newton's nephew, John Conduitt.

[69] Maclaurin, *Fluxions*. This work is clearly and elegantly written. The result itself is stated in Chap. X of Book I, pp. 293ff., but is proved and illustrated by many quite interesting examples in Chap. V of Book II, pp. 672ff. The so-called Maclaurin's expansion occurs in Book II, Art. 751, pp. 610–612. He says there: "This theorem was given by Dr. Taylor, *method. increm.*" It is a fundamental tool for him and perhaps this is why his name became attached to a simple corollary of Taylor's theorem.

[70] Maclaurin, *Fluxions*, Book II, p. 691*n*. Euler's publication appeared in Euler 1, XXII [1738] which we discuss in Chap. III. The conclusion seems clear: they both discovered the theorem independently.

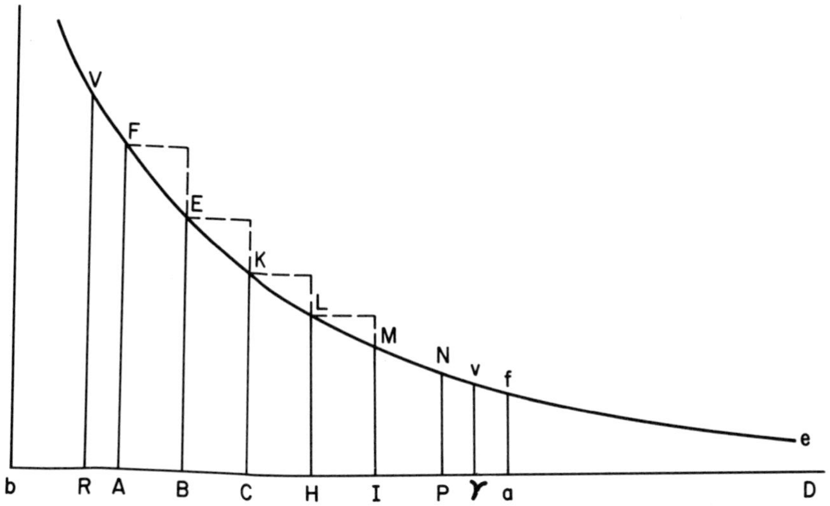

Figure 2.7. $RA = \frac{1}{2}AB$, $AB = BC = \cdots = IP = Pa$, $Pr = \frac{1}{2}AB$.

the case where he wanted to express a finite sum, as shown below in (2.15), and the case where he wanted the limit of the sum. He also developed two related theorems, as we shall see. In the case of limits of sums he assumes his curve is asymptotic to the axis bD in Figure 2.7, is convex to this asymptote and approaches it sufficiently rapidly that the area under the curve from the ordinate AF to infinity exists. In fact, he shows in Article 350 that the sum of the ordinates $AF, BE, CK, \ldots$ has a limit if and only if the area to infinity under the curve exists. His argument is this: "For let the rectangles [cf. Fig. 2.7] FB, EC, KH, LI, MP be completed, and the area $APNF$ being continued over the same base, it is always less than the sum of all those rectangles, but greater than the sum of all the rectangles after the first. Therefore the area $APNF$ and the sum of those rectangles either both have limits, or both have none."[71] He goes on in Article 352 to say: ". . . but when the former limit [the area] is known, we may by it approximate to the value of the latter [the sum]."[72]

Maclaurin now states his first theorem: "Let the limit of the area $APNF$ be expressed by A, the ordinate AF by α, the first fluxion of AF (the fluxion of the base being measured by AB, or unit) by b, the third, fifth, and the subsequent fluxions of AF taken alternately, and always positively, by d, f, &c. Then the sum of the progression represented by AF, BE, CK, HL, &c. shall be found nearly by computing

$$A + \frac{1}{2}a + \frac{1}{12}b - \frac{1}{720}d + \frac{1}{30240}f + \text{\&c.''}$$[73]

[71] Maclaurin, *Fluxions*, Book I, pp. 289–290.

[72] Maclaurin, *Fluxions*, Book I, pp. 291–292.

[73] Maclaurin, *Fluxions*, Book I, p. 292. He had no reasonable notation for an integral. This was true for all the British mathematicians of the era, as we remarked on p. 57n.

In our terminology he stated the result expressed in the relation

$$\sum_{\alpha=0}^{\infty} F(a+\alpha) = \int_a^{\infty} F(x)\,dx + \frac{1}{2}F(a) + \frac{1}{12}F'(a) - \frac{1}{720}F'''(a) + \frac{1}{30240}F^{\mathrm{v}}(a)\cdots \quad . \tag{2.13}$$

In Article 353 he solves this relation for the integral and thereby finds an approximation for the integral when he knows the value of the sum.

The other theorem in Article 352 is this: "Or, if *AR* be taken towards *b* equal to one half of *AB*, the ordinate at *R* meet the curve in *V*, the limit of the area *RPNV* (or its value when *RP* is supposed to be produced infinitely) be now expressed by *A*, the first, third, fifth, and subsequent fluxions of *RV* taken alternately, and always positively, by *b, d, f,* &c. then the sum of the terms *AF, BE, CK, HL,* &c. may be found by computing

$$A - \frac{1}{24}b + \frac{7}{5760}d - \frac{11}{967550}f + \text{\&c.''}$$ [74]

This result may be expressed as

$$\sum_{\alpha=0}^{\infty} F(a+\alpha) = \int_{a-1/2}^{\infty} F(x)\,dx + \frac{1}{24}F'\left(a-\frac{1}{2}\right) - \frac{7}{5760}F'''\left(a-\frac{1}{2}\right) + \frac{31}{967680}F^{\mathrm{v}}\left(a-\frac{1}{2}\right) + \cdots \quad . \tag{2.14}$$

He then stated what amount to two more results: "When it is not the limit of the progression that is required, but the sum of any number of terms of which *AF* is the first and *af* the last; then we may approximate to this sum by the first series if we suppose *A* to represent now the curvilineal area *AafF*, *a* the difference of *AF* and *af*, *b* the difference of their first fluxions, *d* the difference of their third fluxions; and so on. Or, if from *a* towards *A* we take *ar* equal to one half of *AB*, and the ordinate *rv* meet the curve in *v*, we may make use of the second series, provided *A* represent in it the area *RrvV*, *b* represent the difference of the fluxions of *RV* and *rv*, *d* the difference of

[74] Maclaurin, *Fluxions*, Book I, p. 292. There is a typographical error in this result. It should read

$$A + \frac{1}{24}b - \frac{7}{5760}d + \frac{31}{967680}f + \text{\&c.}$$

I have examined two copies of the book and find the last denominator printed incorrectly but both corrected in ink to 967680.

their third fluxions, and so on."[75] Thus we now have

$$\sum_{\alpha=0}^{\beta-1} F(a+\alpha) = \int_a^{a+\beta} F(x)\,dx + \frac{1}{2}[F(a) - F(a+\beta)] + \frac{1}{12}[F'(a) - F'(a+\beta)]$$
$$- \frac{1}{720}[F'''(a) - F'''(a+\beta)] + \frac{1}{30240}[F^{\text{v}}(a) - F^{\text{v}}(a+\beta)]$$
$$+ \cdots; \tag{2.15}$$

and

$$\sum_{\alpha=0}^{\beta} F(a+\alpha) = \int_{a-1/2}^{a-1/2+\beta} F(x)\,dx + \frac{1}{24}\left[F'\left(a-\frac{1}{2}\right) - F'\left(a-\frac{1}{2}+\beta\right)\right]$$
$$- \frac{7}{5760}\left[F'''\left(a-\frac{1}{2}\right) - F'''\left(a-\frac{1}{2}+\beta\right)\right]$$
$$+ \frac{31}{967680}\left[F^{\text{v}}\left(a-\frac{1}{2}\right) - F^{\text{v}}\left(a-\frac{1}{2}+\beta\right)\right] + \cdots . \tag{2.16}$$

Notice the first sum excludes the term $F(\alpha + \beta)$ while the second includes it.

Maclaurin gave his proofs in his second volume as well as a number of highly interesting examples, which we shall now examine. It is of incidental interest that a number of mathematicians of this era were also hard at work estimating the sums of various series. These include de Moivre, Brook Taylor, François Nicole (1683–1758) as well as Stirling in his *Methodus Differentialis*.[76] Maclaurin's proofs illustrate the mathematical notations and techniques of the era. They are based on his expansion and "serve for the resolution of many problems that are usually referred to what is called Sir Isaac Newton's *differential method*." For his first one [cf. Fig. 2.7] he said: "828. Suppose the base $AP = z$, the ordinate $PM = y$, and the base being supposed to flow uniformly let $\dot{z} = 1$. Let the first ordinate AF be represented by a, $AB = 1$, and the area $ABEF = A$. As A is the area generated by the ordinate y, so let B, C, D, E, F, &c. represent the areas upon the same base AB generated by the respective ordinates

$$\dot{y}, \quad \ddot{y}, \quad \dddot{y}, \quad \ddddot{y}, \quad \text{\&c.}$$

Then

$$AF = a = A - \frac{B}{2} + \frac{C}{12} - \frac{E}{720} + \frac{G}{30240} - \text{\&c.}$$

For by art. 752.

$$A = a + \frac{\dot{a}}{2} + \frac{\ddot{a}}{6} + \frac{\dddot{a}}{24} + \frac{\ddddot{a}}{120} + \text{\&c.}$$

---

[75] Maclaurin, *Fluxions*, Book I, pp. 292–293.

[76] Maclaurin's first volume was largely printed in 1737. Later on p. 97 we give a discussion of Stirling's work. Nicole also wrote extensively on numerical analysis. See Nicole [1717], [1723], [1724], and [1727].

whence we have the equation (Q)

$$a = A - \frac{\dot{a}}{2} - \frac{\ddot{a}}{6} - \frac{\dddot{a}}{24} - \frac{\ddddot{a}}{120} - \&\text{c.}$$

In like manner

$$\dot{a} = B - \frac{\ddot{a}}{2} - \frac{\dddot{a}}{6} - \frac{\ddddot{a}}{24} - \&\text{c.}$$

$$\ddot{a} = C - \frac{\dddot{a}}{2} - \frac{\ddddot{a}}{2} - \&\text{c.}$$

$$\dddot{a} = D - \frac{\ddddot{a}}{2} - \&\text{c.}$$

$$\ddddot{a} = E - \&\text{c.}$$

By which latter equations if we exterminate

$$\dot{a}, \quad \ddot{a}, \quad \dddot{a}, \quad \ddddot{a}, \quad \&\text{c.}$$

from the value of a in the equation Q, we shall find that

$$a = A - \frac{B}{2} + \frac{C}{12} - \frac{E}{720} + \&\text{c.}$$

The coefficients are continued thus: Let k, l, m, n, &c. denote the respective coefficients of

$$\dot{a}, \quad \ddot{a}, \quad \dddot{a}, \quad \&\text{c.}$$

in the equation Q; that is, let

$$k = \frac{1}{2}, \quad l = \frac{1}{6}, \quad m = \frac{1}{24}, \quad n = \frac{1}{120}, \ \&\text{c.}$$

suppose

$$K = k = \frac{1}{2}, \quad L = kK - l = \frac{1}{12}, \quad M = kL - lK + m = 0,$$

$$N = kM - lL + mK - n = -\frac{1}{720},$$

and so on; then $a = A - KB + LC - MD + NE -$ &c. where the coefficients of the alternate areas D, F, H, &c. vanish." The proof is then completed by summing the successive ordinates and the areas A, B, C, &c.[77]

[77] Maclaurin, *Fluxions*, Book II, pp. 672–673. Note Maclaurin's definitions of the coefficients $K, L, M, N, \ldots$; they are closely related to the Bernoulli numbers, which we shall discuss on p. 96.

When $\dot{z}$ is not taken to be 1, he finds a form of the Maclaurin–Euler result, which he writes

$$S = \frac{A}{z} - \frac{\alpha}{2} + \frac{z\beta}{12\dot{z}} - \frac{z^3\delta}{720\dot{z}^3} + \frac{z^5\zeta}{30240\dot{z}^5} - \frac{z^7\theta}{1209600\dot{z}^7} + \&\text{c.} \qquad (2.17)$$

In this result S is the "sum of the equidistant ordinates AF, BE, CK, &c. exclusive of the last ordinate $af, \ldots$, the total area $AFfa$ upon the base Aa by A, the excess of af above AF by α, the respective excesses of their first, third, fifth fluxions, &c. by β, δ, ζ, &c."

To establish his second relation (2.16) he first proves a lemma. "831. The ordinate AF being still represented by a, let AR and Ar be taken on opposite sides of the point A equal to each other, RV and rv the ordinates at R and r terminate the area $RVvr$; let y represent any ordinate as PM of the figure, and the base being supposed to flow uniformly let A, C, E, &c. represent the areas upon the base Rr that are generated by the respective ordinates

$$y, \quad \ddot{y}, \quad \ddot{\ddot{y}}, \quad \&\text{c.}$$

Then supposing $AR = z$, the middle ordinate

$$AF(=a) = \frac{A}{2z} - \frac{zC}{12\dot{z}^2} + \frac{7z^3E}{720\dot{z}^4} - \frac{31z^5G}{30240\dot{z}^6} + \&\text{c.}\text{''}^{78}$$

His proof follows from this expansion of y about AF by integrating the series from $-z$ to $+z$; he does the same thing for

$$\ddot{y}, \ddot{\ddot{y}}, \ldots .$$

Then as before he "exterminates" these derivatives in favor of the relevant areas. With the help of this result he could now show the validity of the relation (2.16) above without difficulty. (The result is given by Whittaker, *WR*, p. 140, as an exercise for the reader.)

Maclaurin's examples are not only interesting, but they also help us to see the kinds of problems the mathematicians of the period were investigating. One problem, which was also the basis for James Bernoulli's numbers, is the following: Maclaurin in Article 833 considers the arithmetical progression m, $m + e, m + 2e, \ldots, n = m + \beta e$ and any integer $r \neq -1$. He then asks for the value of the sum

$$S = \sum_{\alpha=0}^{\beta} (m + \alpha e)^r.$$

To find the value of S he integrates the function $y = x^r$ and finds for his area $A = AFfa$ in (2.17) the value $(n^{r+1} - m^{r+1})/(r + 1)$. Moreover, $z = e$, $\dot{z} = 1$, $af - AF = n^r - m^r$, etc. This gives

$$S = \frac{n^{r+1} - m^{r+1}}{\overline{r+1} \times e} + \frac{n^r + m^r}{2} + \frac{re}{12} \times \overline{n^{r-1} - m^{r-1}} - \&\text{c.}$$

[78] Maclaurin, *Fluxions*, Book II, p. 674.

He sets $e = 1$ and $m = 0$ and obtains the relation

$$\sum_{\alpha=0}^{n} \alpha^r = \frac{n^{r+1}}{r+1} + \frac{n^r}{2} + \frac{rn^{r-1}}{12} - \frac{r \times \overline{r-1} \times \overline{r-2} \times n^{r-3}}{720} + \text{\&c.},$$

"this series being continued to as many terms as there are units in

$$2 + \frac{r-1}{2}$$

only, where r is an odd number..."[79] He then indicated the values of S for $r = 1, 2$, and 3: "Thus if $r = 1$,

$$S = \frac{n^2}{2} + \frac{n}{2};$$

if $r = 2$,

$$S = \frac{n^3}{3} + \frac{n^2}{2} + \frac{n}{6},$$

and if $r = 3$,

$$S = \frac{n^4}{4} + \frac{n^3}{2} + \frac{n^2}{4}.$$

This is the theorem given by Mr. JAMES BERNOUILLI. *Ars conjectandi, p.* 97. When r is a fraction or negative number, the sum of the powers of the same numbers (by supposing $m = 1$) is

$$\frac{n^{r+1}-1}{r+1} + \frac{n^r+1}{2} + \frac{rn^{r-1}-r}{12} - \frac{r \times \overline{r-1} \times \overline{r-2}}{720} \times \overline{n^{r-3}-1} + \text{\&c.}\text{"}[80]$$

He points out that the case where r is negative and $\neq -1$ had earlier been worked out "from different principles by Mr. DE MOIVRE."

In Article 834 Maclaurin considers the sum of the rth powers of the progression $m + e, m + 3e, m + 5e, \ldots, n - e$, using his second relation (2.16). When he puts $e = m = \frac{1}{2}$ in his resulting formula he finds

$$S = \frac{n^{r+1}}{r+1} - \frac{rn^{r-1}}{24} + \frac{7r \times \overline{r-1} \times \overline{r-2} \times n^{r-3}}{5760}$$
$$- \frac{31r \times \overline{r-1} \times \overline{r-2} \times \overline{r-3} \times \overline{r-4}}{967680} \times n^{r-5}$$
$$+ \text{\&c.} - \frac{1}{\overline{r+1} \times 2^{r+1}} + \frac{r}{24 \times 2^{r-1}} - \frac{7r \times \overline{r-1} \times \overline{r-2}}{5760 \times 2^{r-3}} + \text{\&c.}$$

In case r is negative he sets $s = -r$ and assumes $s > 1$. "To compute the

---

[79] Maclaurin, *Fluxions*, Book II, pp. 676–677. Note the sum S goes to β; thus the term n^r has been added to both sides of the relation (2.15).

[80] Maclaurin, *Fluxions*, Book II, p. 677. Bernoulli's results appear in *Ars Conjectandi*, Bernoulli, Ja, *AC*.

sum of the progression

$$1 + \frac{1}{4} + \frac{1}{9} + \frac{1}{16} + \frac{1}{25} + \&c.$$

find the sum of the terms at the beginning of the series as far as

$$\frac{1}{\overline{m + \frac{1}{2}}^2}$$

exclusively, then compute the sum of the subsequent terms by this theorem. Thus if we add the three first only

$$1 + \frac{1}{4} + \frac{1}{9} = 1.36111 \ \&c.$$

suppose

$$m + \frac{1}{2} = 4, \quad \text{or} \quad m = \frac{7}{2},$$

and the sum of the following terms will be

$$\frac{2}{7} \times \overline{1 - \frac{1}{3 \times 49} + \frac{1}{15 \times 343} - \&c.}$$

three terms of which series only collected and added to the former number 1.36111 &c. give 1.64493 &c. for the sum of the series required, true to the fifth decimal."[81]

Having these results, Maclaurin now turned to the case $r = -1$: he summed the relevant logarithms and thereby gave what we today call Stirling's approximation to the factorial.[82] Maclaurin's proof in Article 842 is quite direct. He first shows by his relation (2.15) that the sum of the logarithms of $1, 2, 3, \ldots, n - 1$ is

$$\begin{aligned}\sum_{\alpha=1}^{n-1} \log \alpha &= \overline{n - \frac{1}{2}} \times \log n - n + \frac{1}{12n} - \frac{1}{360n^3} + \frac{1}{1260n^5} - \&c. \\ &\quad + \left(1 - \frac{1}{12} + \frac{1}{360} + \frac{1}{1260} + \&c.\right) \\ &= \overline{n - \frac{1}{2}} \times \log n - n + \frac{1}{12n} - \frac{1}{360n^3} + \frac{1}{1260n^5} - \&c. + \frac{\log 2\pi}{2}.\end{aligned}$$ [83]

Thus he has the relation

$$(n-1)! = n^{n-1/2}\sqrt{2\pi}\cdot\exp\left[-n + \frac{1}{12n} - \frac{1}{360n^3} + \cdots\right],$$

which goes over into the usual one upon multiplying both sides by n.

---

[81] Maclaurin, *Fluxions*, Book II, p. 678. (Actually 1.64483!)

[82] Stirling [1730], and [1739].

[83] That $\frac{1}{2}\log 2\pi = 1 - 1/12 + 1/360 - 1/1260 + \cdots$ is due to Stirling. Maclaurin, *Fluxions*, Book II, pp. 683–684. He used the symbol c for 2π.

Maclaurin was able to use his relations (2.15), (2.16), (2.17) for many things. One quite interesting result for us is contained in his Article 848 where he derives the Newton–Cotes Numerical integration formulas. He now divides the base $Aa = R$ in Figure 2.7 into n equal parts, calls the area $AFfa =$ Q, the sum of AF and $af = A$, the sum of all intermediate ordinates $BE + CK + \cdots = B$, and uses $\alpha, \beta, \delta, \zeta, \ldots$ as before. In the relation (2.17) set $z = R/n$ and $\dot{z} = 1$. Then he immediately has

$$
\begin{aligned}
S + \frac{\alpha}{2} &= AF + B + \frac{\alpha}{2} \\
&= \frac{A}{2} + B \\
&= \frac{nQ}{R} + \frac{R\beta}{12n} - \frac{R^3\delta}{720n^3} + \frac{R^5\zeta}{30240n^5} - \&\text{c}. \qquad (2.18)
\end{aligned}
$$

Next we wishes to eliminate β from this relation. He does so by setting $n = 1$ and finds

$$\frac{AF + af}{2} = \frac{A}{2} = \frac{Q}{R} + \frac{R\beta}{12} - \frac{R^3\delta}{720} + \frac{R^5\zeta}{30240} - \&\text{c}.$$

He solves this for $R\beta/12$ and substitutes it into (2.18). After a little arithmetical calculation he finds

$$AFfa = Q = \overline{\frac{A}{2n+2} + \frac{nB}{nn-1}} \times R - \frac{R^4\delta}{720nn} + \frac{R^6\zeta}{30240} \times \frac{nn+1}{n^4} - \&\text{c}.$$

In modern terms this is

$$
\begin{aligned}
\frac{1}{a - A}\int_A^a F(x)\,dx = {}& \frac{1}{(n+1)}\frac{[f(a) + F(A)]}{2} \\
&+ \frac{n}{n^2 - 1}\sum_{\alpha=1}^{n-1} F\left[A + \frac{(a - A)}{n}\alpha\right] \\
&- \frac{(a - A)^3}{720n^2}[F'''(a) - F'''(A)] \\
&+ \frac{(a - A)^5(n^2 + 1)}{30240n^4}[F^{\text{v}}(a) - F^{\text{v}}(A)] - \cdots .
\end{aligned}
$$

Maclaurin considers explicitly the cases $n = 2$ and 3 and says: "By neglecting δ, ζ, θ, &c. we shall have two of the theorems given by Sir Isaac Newton and others for computing the area from equidistant ordinates, the latter of which

$$\left(\text{viz. } AFfa = \frac{A + 3B}{8} \times R\right)$$

is much recommended by Mr. COTES." The case $n = 2$ gives — let $a - A = nh$ —

$$\int_A^a F(x)\,dx = \frac{h}{3}[F(A) + F(A + 2h)] + \frac{4h}{3}F(A + h) + \cdots$$

$$= \frac{h}{3}[F(A) + 4F(A + h) + F(A + 2h)] + \cdots,$$

which is Simpson's rule, as we mentioned earlier. The case $n = 3$ gives

$$\int_A^a F(x)\,dx = \frac{3h}{8}[F(A) + F(A + 3h)] + \frac{9h}{8}[F(A + h) + F(A + 2h)] + \cdots$$

$$= \frac{3h}{8}[F(A) + 3F(A + h) + 3F(A + 2h) + F(A + 3h)] + \cdots,$$

which is often called the *Three-eighth rule*. It is curious that this is the one Cotes favored.

Maclaurin goes on to devise still other algorithms of this sort: "849. By exterminating δ, ζ, θ, &c. successively, other theorems will be found by which the area will be more and more accurately determined from the ordinates. Let there be five ordinates, A the sum of the first and last, B the sum of the second and fourth, and C the middle ordinate; then the area

$$AFfa = \frac{7A + 32B + 12C}{90} \times R - \frac{31R^6\zeta}{6 \times 16 \times 16 \times 30240} + \&c.$$

For by the rule of three ordinates

$$\frac{Q}{R} = \frac{A + 4C}{6} - \frac{R^3\delta}{4 \times 720} + \frac{5R^5\zeta}{16 \times 30240}.$$

By dividing the base into two equal parts and computing from the same rule the area that stands upon each part, and adding these areas together,

$$\frac{2Q}{R} = \frac{A + 4B + 2C}{6} - \frac{R^3\delta}{32 \times 720} + \frac{5R^5\zeta}{2 \times 16 \times 16 \times 30240} - \&c.$$

then by exterminating δ by these two equations, the proposition appears."[84] He remarked further that still higher order rules can be found ". . . and some judgment formed of the accuracy of the several rules, by comparing the quantities that are neglected in them."[85] Maclaurin then proceeded to show how his formulas could be used to interpolate as well as calculate sums and areas.

Before proceeding, we should remark on two curious things: first, Maclaurin does not seem to refer in his book to the integration formula of

[84] Maclaurin, *Fluxions*, Book II, p. 688. There is an error in Maclaurin's theorem. The factor 31 in the coefficient of ζ should be 3.

[85] Maclaurin, *Fluxions*, Book II, pp. 688ff.

his predecessor Gregory, which is closely related to his; second, he does not use his second relation (2.16) to develop integration formulas paralleling those of Newton–Cotes. If he had done this, he would have developed both the so-called closed and open types of numerical quadrature formulas.[86]

The formulas of Gregory and Euler–Maclaurin differ only in that the former are couched in terms of differences and the latter in terms of derivatives. If we substitute the appropriate formulas relating differences and derivatives in one we easily obtain the other.[87]

Jacob I (James or Jacques) Bernoulli considered in his *Ars Conjectandi* a variety of topics including a study of an array of the so-called figurate numbers. They form in effect a Pascal triangle.

1	0	0	0	0	0
1	1	0	0	0	0
1	2	1	0	0	0
1	3	3	1	0	0
1	4	6	4	1	0
1	5	10	10	5	1.

He considered a number of properties of the array and by means of them ingeniously determined the sums of powers of the consecutive integers.[88] He knew the sum of the first column of n numbers was n and of the second column was $n(n-1)/2$ and from this he inferred (to use his notation) that

$$\int n - \int 1 \propto \frac{nn-n}{2}, \quad \& \quad \int n \propto \frac{nn-n}{2} + \int 1 \propto \frac{nn-n}{2} + \frac{1}{2}nn + \frac{1}{2}n,$$

where he used an integral sign for a sum. Again he knew the sum of the third column to be $n(n-1)(n-2)/3!$ and the general term to be $(n-1)(n-2)/2!$. Thus

$$\int \overline{\frac{nn-3n+2}{2}} \quad \text{or} \quad \int \frac{1}{2}nn - \int \frac{3}{2}n + \int 1 \propto \frac{n3-3nn+2n}{6}.$$

From this he inferred that

$$\int \frac{1}{2}nn \propto \frac{n3-3nn+2n}{6} + \int \frac{3}{2}n - \int 1 \propto \frac{n3-3nn+2n}{6} + \frac{3nn+3n}{4} - n$$
$$\propto \frac{1}{6}n3 + \frac{1}{4}nn + \frac{1}{12}n.$$

He worked out in detail the sums $\sum n^c$ for $c = 1, 2, \ldots, 10$ and wrote out

[86] Steffensen [1950], pp. 158ff.

[87] Whittaker, *WR*, pp. 143–144.

[88] Ja. Bernoulli, *AC*, pp. 86–99. Here he used the symbol $\propto$ for equality.

the formula for the general c as

$$\int n^c \infty \frac{1}{c+1} n^c + 1 + \frac{1}{2} n^c + \frac{c}{2} A n^{c-1} + \frac{c \cdot c - 1 \cdot c - 2}{2 \cdot 3 \cdot 4} B n^{c-3}$$

$$+ \frac{c \cdot c - 1 \cdot c - 2 \cdot c - 3 \cdot c - 4}{2 \cdot 3 \cdot 4 \cdot 5 \cdot 6} C n^{c-5}$$

$$+ \frac{c \cdot c - 1 \cdot c - 2 \cdot c - 3 \cdot c - 4 \cdot c - 5 \cdot c - 6}{2 \cdot 3 \cdot 4 \cdot 5 \cdot 6 \cdot 7 \cdot 8} D n^{c-7} \ldots \&.$$

He says that capital letters $A, B, C, D, \ldots$ are the last coefficients (the coefficients of n^1) in the formulas for $\int nn, \int n^4, \int n^6, \int n^8$, &c. Thus, e.g., $A = 1/6$, $B = -1/30$, $C = 1/42$, $D = -1/30, \ldots$. Below is a reproduction of his array to illustrate his remarks. Note that the sum of the coefficients in each line is 1. This enables him to determine the $A, B, C, \ldots$ recursively; thus, he points out that $1/9 + 1/2 + 8A/2 + 8 \cdot 7 \cdot 6B/4! + 8 \cdot 7 \cdot 6 \cdot 5 \cdot 4C/6! + 8!\, D/8! = 1$ and so $1/9 + 1/2 + 2/3 - 7/15 + 2/9 + D = 1$.

Summæ Potestatum.

$$\int n \infty \frac{1}{2} nn + \frac{1}{2} n.$$

$$\int nn \infty \frac{1}{3} n^3 + \frac{1}{2} nn + \frac{1}{6} n.$$

$$\int n^3 \infty \frac{1}{4} n^4 + \frac{1}{2} n^3 + \frac{1}{4} nn.$$

$$\int n^4 \infty \frac{1}{5} n^5 + \frac{1}{2} n^4 + \frac{1}{3} n^3 * - \frac{1}{30} n.$$

$$\int n^5 \infty \frac{1}{6} n^6 + \frac{1}{2} n^5 + \frac{5}{12} n^4 * - \frac{1}{12} nn.$$

$$\int n^6 \infty \frac{1}{7} n^7 + \frac{1}{2} n^6 + \frac{1}{2} n^5 * - \frac{1}{6} n^3 * + \frac{1}{42} n.$$

$$\int n^7 \infty \frac{1}{8} n^8 + \frac{1}{2} n^7 + \frac{7}{12} n^6 * - \frac{7}{24} n^4 * + \frac{1}{12} nn.$$

$$\int n^8 \infty \frac{1}{9} n^9 + \frac{1}{2} n^8 + \frac{2}{3} n^7 * - \frac{7}{15} n^5 * + \frac{2}{9} n^3 * - \frac{1}{30} n.$$

$$\int n^9 \infty \frac{1}{10} n^{10} + \frac{1}{2} n^9 + \frac{3}{4} n^8 * - \frac{7}{10} n^6 * + \frac{1}{2} n^4 * - \frac{1}{12} nn.$$

$$\int n^{10} \infty \frac{1}{11} n^{11} + \frac{1}{2} n^{10} + \frac{5}{6} n^9 * - 1\, n^7 * + 1\, n^5 * - \frac{1}{2} n^3 * + \frac{5}{66} n.$$

Using this array, Bernoulli says that in a short while (*intra semi-quadrantem horae*) one can find the sum of the tenth powers of the first 1000 numbers.

He gives it as

$$91,409,924,241,424,243,424,241,924,242,500.$$

It is of course easy to find this from his formula above.

In his discussion of the figurate numbers he refers to Johann Faulhaber, Mercator, Wallis and several others as having studied these numbers. To these he could have added many names, including Vieta and Briggs.

This work led to further developments of what we now call Bernoulli's numbers and polynomials, which are worth mentioning here. Nörlund and Steffensen define the Bernoulli polynomials with the help of the relations

$$\Delta B_n(x) = nx^{n-1} \qquad (n = 0, 1, \ldots),$$

$$\frac{dB_n(x)}{dx} = nB_{n-1}(x) \qquad (n = 1, 2, \ldots).^{89}$$

(C. Jordan's definition differs from this one by a factor of $(n-1)!$ Let us use the one above.) It is easy to see that B_n is a polynomial of degree n, and by a little calculation to find

$$\sum_{\alpha=0}^{n-1} \binom{n}{\alpha} B_\alpha(x) = nx^{n-1}.$$

This gives the polynomials: $B_0(x) = 1$, $B_1(x) = x - 1/2$, $B_2(x) = x^2 - x + 1/6$, $B_3(x) = x^3 - 3x^2/2 + x/2, \ldots$. The values $B'_n = B_n(0)$ are called Bernoulli's numbers. Thus $B'_0 = 1$, $B'_1 = -1/2$, $B'_2 = 1/6$, $B'_4 = -1/30$, $B'_6 = 1/42$, $B'_8 = -1/30$, $B'_{10} = 5/66, \ldots$. ($B'_{2\alpha+1} = 0$ for $\alpha \geq 1$.)

Let us now define in Bernoulli's table of sums of powers a family of functions $b_\alpha(n)$ with the help of his relations:

$$b_1(n+1) = \int n = \frac{n^2}{2} + \frac{n}{2},$$

$$b_2(n+1) = \int n^2 = \frac{n^3}{3} + \frac{n^2}{2} + \frac{n}{6},$$

$$b_3(n+1) = \int n^3 = \frac{n^4}{4} + \frac{n^3}{2} + \frac{n^2}{4},$$

$$b_4(n+1) = \int n^4 = \frac{n^5}{5} + \frac{n^4}{2} + \frac{n^3}{3} + \frac{n}{30}.$$

We notice that the derivatives of these functions $b_\alpha(n)$ give

$$\frac{db_\alpha(x)}{dx} = B_\alpha(x) \qquad (\alpha = 1, 2, \ldots).$$

---

[89] Steffensen [1950], pp. 119ff.

One of the most interesting relations concerning Bernoulli polynomials is

$$S(x) = \sum_{\alpha=0}^{x-1} \alpha^{n-1} = \frac{1}{n}[B_n(x) - B'_n],$$

where x is integral. To see this let us write

$$\alpha^{n-1} = \frac{1}{n}\Delta B_n(\alpha)$$

and sum both sides from $\alpha = 0$ to $x - 1$. Then clearly

$$S(x) = \frac{1}{n}[B_n(x) - B_n(0)] = \frac{1}{n}[B_n(x) - B'_n].$$

Another way this may be written is

$$\sum_{\alpha=0}^{x-1} \alpha^{n-1} = \int_0^x B_{n-1}(t)\,dt = \frac{1}{n}\int_0^x \frac{dB_n(t)}{dt}\,dt = \frac{1}{n}[B_n(x) - B'_n].$$

Euler called the $|B'_{2\alpha}|$ the Bernoullian numbers and wrote them usually as capital Gothic letters. He used them *inter alia* to express the sum of the x^n for a general n.

2.7. Stirling

James Stirling (1692–1770) was a Scot who studied first in Glasgow and then in Balliol College, Oxford. When he was expelled from Oxford because of his Jacobite sympathies he went to Venice, where he lived for ten years. On his return to England he enjoyed Newton's friendship, and his work was greatly influenced by Newton's ideas. Among his most important works is the *Methodus Differentialis* of 1730 mentioned earlier. This was translated into English by a "Francis Holliday, Master of the Grammar Free-School at Haughton-Park near Retford, Nottinghamshire with the Author's approbation."[90] In this work Stirling went very far towards developing the calculus of finite differences. He developed among other things, the discrete analog of the antiderivative or indefinite integral, as we shall see. Stirling's first result is his:

"**Proposition** I. *If the terms of any series be formed by writing the numbers* 1, 2, 3, 4, 5, *&c. for z, in the quantity* $A + Bz + Cz\cdot\overline{z-1} + Dz\cdot\overline{z-1}\cdot\overline{z-2} + Ez\cdot\overline{z-1}\cdot\overline{z-2}\cdot\overline{z-3}$ + *&c. then the sum of the terms from the beginning whose number is z, will be* $Az + \overline{z+1}$ *into*

$$\frac{1}{2}Bz + \frac{1}{3}Cz\cdot\overline{z-1} + \frac{1}{4}Dz\cdot\overline{z-1}\cdot\overline{z-2} + \frac{1}{5}Ez\cdot\overline{z-1}\cdot\overline{z-2}\cdot\overline{z-3} + \&c.\text{''}$$[91]

[90] Stirling [1739].

[91] Stirling [1739], p. 17.

What is meant here is this: the function

$$f(z) = A + Bz + Cz(z-1) + \cdots \qquad (z = 1, 2, 3, \ldots)$$

is to be summed from 1 to z, i.e., the expression

$$y(z+1) = \sum_{\alpha=1}^{z} f(\alpha) = Az + \frac{1}{2}B(z+1)z + \frac{1}{3}C(z+1)z(z-1) + \cdots$$

is to be formed. It is quite clear that Stirling has expressed his function $f(z)$ in a Harriot–Briggs or Newton expansion and has then summed the resulting expression term by term. His proof is this: form

$$\Delta y(z) = y(z+1) - y(z) = A + Bz + Cz\cdot(z-1) + \cdots = f(z).$$

In other words the sum y is a function whose first difference is f. "Whence, on the contrary, if this value of the term $[f(z)]$ be given as in the proposition, the sum will be that which is assigned. Moreover this sum is nothing when z is nothing, and consequently the theorem is manifest. Q.E.D."[92]

In what follows I have made use of material in Jordan [1939], Chapter IV, pp. 142–229. Here is a very complete discussion of Stirling's numbers which, along with the notion of the factorial, are of the utmost importance in the theory of finite differences. Jordan points out that this ". . . however has not been fully recognized; the numbers have been neglected and are seldom used."

In modern terms Stirling was concerned with solving the difference equation

$$y(x+1) - y(x) = \Delta y(x) = f(x) \tag{2.19}$$

for a variety of functions f. Let us discuss the problem before examining Stirling's ideas. If the equation in (2.19) is solvable for each f in some suitable space, then the operator Δ has an inverse, and we may write

$$y(x) + k = \Delta^{-1}f(x). \tag{2.20}$$

Note that k is an arbitrary constant if x takes on discrete values such as $x = 1, 2, \ldots$, or is a periodic function of period 1 if it varies over an interval $a \leq x \leq a + 1$, as we see by operating with Δ on both sides of (2.20). To simplify what we do we shall suppose that x takes on only discrete values.

Using the operator $Eg(x) = g(x+1)$, we may write Δ as $\Delta = E - I = -(I - E)$ where I is the identity operator. Then symbolically we have

$$\Delta^{-1} = -(I + E + E^2 + \cdots)$$

and hence

$$\Delta^{-1}f(x) = -\sum_{\alpha=0}^{\infty} f(x+\alpha) = -\sum_{\alpha=x}^{\infty} f(\alpha) = \sum_{\alpha=0}^{x-1} f(\alpha) - \sum_{\alpha=0}^{\infty} f(\alpha), \tag{2.21}$$

[92] Stirling [1749], p. 18. We will see below on p. 99 how the last remark fixes the arbitrary constant that arises from taking indefinite sums. We will also see why I set $y(z+1)$ equal to a sum from α to z instead of from α to $z+1$.

provided the relevant series are convergent. We may now introduce the notion of an antidifference or indefinite sum, Δ^{-1}, as the finite difference analog of the integral in the differential and integral calculi. To make the analogy as close as possible it is convenient to introduce the summation operator

$$\mathop{\mathrm{S}}_{a}^{b} f(x) = \sum_{\alpha=a}^{b-1} f(\alpha) \qquad (a < b \text{ integers}). \tag{2.22}$$

(Notice that the sum goes from a through $b - 1$ but excludes b. This is done to simplify notations, as we shall see, and preserve the analogy between sums and integrals.) Recall here what we said above about Stirling's Proposition I. In terms of this definition the relation (2.21) may be written as

$$\Delta^{-1}f(x) = \mathop{\mathrm{S}}_{0}^{x} f(\alpha) - \mathop{\mathrm{S}}_{0}^{\infty} f(\alpha)$$

and so

$$y(x) + k = \mathop{\mathrm{S}}_{0}^{x} f(\alpha) - \mathop{\mathrm{S}}_{0}^{\infty} f(\alpha)$$

or

$$y(x) = c + \mathop{\mathrm{S}}_{0}^{x} f(\alpha),$$

where c is an arbitrary constant. (Note the analogy between this and the corresponding integral formula in our usual calculus.)

It is not difficult to avoid the problem of the convergence of the series we just discussed by proceeding in this way:

$$\begin{aligned}
&y(x) - y(x-1) = f(x-1),\\
&y(x-1) - y(x-2) = f(x-2),\\
&\vdots\\
&y(a+1) - y(a) = f(a),
\end{aligned}$$

and hence we have

$$y(x) - y(a) = \sum_{\alpha=a}^{x-1} f(\alpha) = \mathop{\mathrm{S}}_{a}^{x} f(\alpha),$$

$$y(x) = y(a) + \mathop{\mathrm{S}}_{a}^{x} f(\alpha); \tag{2.23}$$

here $y(a)$ is the constant c above. Moreover, we see that

$$\mathop{\mathrm{S}}_{a}^{b} f(\alpha) = -\mathop{\mathrm{S}}_{b}^{\infty} f(\alpha) + \mathop{\mathrm{S}}_{a}^{\infty} f(\alpha) = \Delta^{-1}f(b) - \Delta^{-1}f(a). \tag{2.24}$$

These various relations show the connections between the operations Δ^{-1} and S.

Stirling was much concerned in his *Methodus* with evaluating expressions such as (2.21) for a variety of functions f. As we saw he essentially understood how S and Δ^{-1} are related and exploited the connection quite ingeniously to sum series by solving difference equations, which he called differential equations.

Let us digress briefly to discuss the behavior of those functions $(x)_n = x(x-1)\cdots(x-n+1)$ which are the analogs of the powers x^n in the differential calculus and which we mentioned just before the relation (2.8). Stirling made great use of these functions but did not give them an explicit name. It remained for Alexandre Théophile Vandermonde (1735–1796) in 1772 to introduce the notation $[x]^n$ for the function we now usually write as $(x)_n$.[93] Clearly if m, n are positive integers with $n > m$

$$(x)_n = (x)_m \cdot (x-m)_{n-m}. \tag{2.25}$$

If we permit this relation to hold for m, n any integers, then it serves as a definition for $(x)_0$ and $(x)_{-n}$. For example set $m = 0$; then we have $(x)_n = (x)_0 \cdot (x)_n$ or $(x)_0 = 1$. Similarly if we set $n = 0$, $(x)_0 = 1 = (x)_m \cdot (x-m)_{-m}$. I.e.,

$$(x)_{-n} = \frac{1}{(x+n)_n} = \frac{1}{(x+1)(x+2)\cdots(x+n)}.$$

Now let us form

$$\Delta(x)_n = (x+1)_n - (x)_n = n(x)_{n-1},$$
$$\Delta(x)_{-n} = (x+1)_{-n} - (x)_{-n} = -n(x)_{-n-1}.$$

We may combine our formulas by having recourse to the well-known Gamma function. Clearly $(x)_n = \Gamma(x+1)/\Gamma(x-n+1)$ for arbitrary n. Then

$$\Delta(x)_n = \frac{(x+1)\Gamma(x+1)}{(x-n+1)\Gamma(x-n+1)} - \frac{\Gamma(x+1)}{\Gamma(x-n+1)} = n(x)_{n-1},$$

and we have at once

$$\Delta^{-1}(x)_m = \frac{(x)_{m+1}}{(m+1)} + k, \qquad \Delta^{-1}(x)_{-m} = \frac{(x)_{-m+1}}{(1-m)} + l,$$

where k and l are arbitrary constants. This tells us that the sum of the $(x)_m$ is given by

$$\mathop{\mathsf{S}}_0^x (\alpha)_m = \Delta^{-1}(x)_m - \Delta^{-1}(0)_m = \frac{1}{m+1}(x)_{m+1} \qquad (m = 0, 1, \ldots), \tag{2.26}$$

and

$$\mathop{\mathsf{S}}_{x-1}^{\infty} (\alpha)_{-m} = \Delta^{-1}(\infty)_{-m} - \Delta^{-1}(x-1)_{-m} = \frac{1}{m-1}(x-1)_{-m+1}$$
$$(m = 2, 3, \ldots);$$

---

[93] Vandermonde [1774]. He considered positive and negative n as well as x and n non-integral. He showed that $2[1/2]^{1/2} = \sqrt{\omega}$ where ω was his notation for π.

or, stated otherwise,

$$\sum_{\alpha=x-1}^{\infty} \frac{1}{(\alpha+1)(\alpha+2)\cdots(\alpha+m)} = \sum_{\alpha=x}^{\infty} \frac{1}{\alpha(\alpha+1)\cdots(\alpha+m-1)}$$

$$= \frac{1}{m-1}\frac{1}{x(x+1)\cdots(x+m-2)}. \quad (2.26')$$

This was basic in Stirling's *Methodus*.

Very early in the book he calculated what are now called Stirling's numbers of the first and second kind.[94] He introduced the numbers $\mathfrak{S}_n^m$ in order to express x^n as a series in $(x)_1, (x)_2, \ldots, (x)_n$:

$$x^n = \sum_{\alpha=1}^{n} \mathfrak{S}_n^\alpha (x)_\alpha = \mathop{\mathrm{S}}_{1}^{n+1} \mathfrak{S}_n^\alpha (x)_\alpha.$$

Let us evaluate these quantities $\mathfrak{S}_n^m$. By means of the Harriot–Briggs formula we can write x^n as a series in the functions $(x)_m$

$$x^n = \sum_{m=1}^{n} \frac{1}{m!}(x)_m \Delta^m x^n|^{x=0}.$$

Then by definition

$$\mathfrak{S}_n^m = \frac{1}{m!}\Delta^m x^n|^{x=0}. \quad (2.27)$$

These are Stirling's numbers of the second kind. His numbers of the first kind S_n^m are defined analogously as

$$S_n^m = \frac{1}{m!}\frac{d^m x^n}{dx^m}|^{x=0}. \quad (2.27')$$

To evaluate these numbers let us, following Jordan, symbolically expand $\Delta^m = (-1)^m(I - E)^m$ into a series:

$$\Delta^m = (-1)^m \sum_{\alpha=0}^{m} (-1)^\alpha \binom{m}{\alpha} E^\alpha.$$

If we then apply these operators to x^n, we find

$$\left[\Delta^m \frac{x^n}{m!}\right]_{x=0} = \mathfrak{S}_n^m = \frac{(-1)^m}{m!} \sum_{\alpha=0}^{m} (-1)^\alpha \binom{m}{\alpha} \alpha^n. \quad (2.28)$$

From this we may form the table below given by Stirling.

---

[94] His so-called numbers of the second kind are contained in his first table, Stirling [1749], p. 7; those of the first kind are in his second table on p. 10. These numbers form infinite matrices whose finite sections are quite interesting.

1	1	1	1	1	1	1	1	1	&c.
	1	3	7	15	31	63	127	255	&c.
		1	6	25	90	301	966	3025	&c.
			1	10	65	350	1701	7770	&c.
				1	15	140	1050	6951	&c.
					1	21	266	2646	&c.
						1	28	461	&c.
							1	36	&c.
								1	&c.
									&c.

In his table each column $\mathfrak{S}_n^m$ ($m = 1, 2, \ldots, n$) corresponds to a given n; thus column four: 1, 7, 6, 1 gives

$$x^4 = (x)_1 + 7(x)_2 + 6(x)_3 + (x)_4.$$

It is easy to see with the help of (2.27) or (2.28) that $\mathfrak{S}_0^0 = 1 = \mathfrak{S}_n^1$ and that $\mathfrak{S}_0^m = 0$ for $m \neq 0$, and in fact that $\mathfrak{S}_m^n = 0$ for $m > n$. The others can of course be formed from our relation (2.27) but there is a very simple recurrence relation which is the easiest way to form a table of the $\mathfrak{S}_n^m$. The relation is

$$\mathfrak{S}_{n+1}^m = \mathfrak{S}_n^{m-1} + m\mathfrak{S}_n^m. \tag{2.29}$$

Given $\mathfrak{S}_0^0 = 1$ and $\mathfrak{S}_0^m = 0$ if $m \neq 0$ we can find the others quickly and systematically from (2.29).[95] The S_n^m satisfy a similar recurrence relation to (2.29),

$$S_{n+1}^m = S_n^{m-1} - nS_n^m, \tag{2.29'}$$

which we derive shortly.

Stirling's procedure began with a function $f(x)$, the xth term in a sequence whose sum he wished to evaluate; he expressed it in the form

$$f(x) = \sum_{n=0}^{\infty} \frac{f^{(n)}(0)}{n!} \sum_{m=1}^{n} \mathfrak{S}_n^m \cdot (x)_m.$$

He was then able to apply the operator S term by term to the $(x)_m$ to find the sum of the series. Clearly by (2.26) the result is

$$\mathsf{S}_0^x f(x) = \sum_{n=0}^{\infty} \frac{f^{(n)}(0)}{n!} \sum_{m=1}^{n} \frac{1}{m+1} \mathfrak{S}_n^m \cdot (x)_{m+1}.$$

Thus, e.g., he chose $f(z) = (2z - 1)^2$ and wrote it as $(2z - 1)^2 = 1 + 4z(z - 1)$. Then in his Proposition I above he had $A = 1$, $B = 0$, $C = 4$,

[95] Jordan [1921], p. 169. The derivation follows directly from the fact that $\Delta^m x^{n+1} = \Delta^m x \cdot x^n = (x + m)\Delta x^n + m\Delta^{m-1}x^n$, which in turn is an immediate consequence of the fact that $\Delta^m(uv) = v(x + m)\Delta^m u + \binom{m}{1}\Delta v(x + m - 1)\Delta^{m-1}u + \cdots + u\Delta^m v(x)$ applied to the case $u = x^n$, $v = x$.

$0 = D = E = \cdots$, and his conclusion was that

$$\sum_{\alpha=1}^{z} (2\alpha - 1)^2 = z + \frac{4}{3}(z + 1)z(z - 1) = \frac{1}{3}(4z^3 - z).$$ [96]

Another example he gave was to find the sum of the "cubes 1, 8, 27, 64, 125, 216, &c." Now $f(z) = z^3$; and $z^3 = z + 3z\cdot\overline{z - 1} + z\cdot\overline{z - 1}\cdot\overline{z - 2}$. We have $A = 0$, $B = 1$, $C = 3$, $D = 1$, $E = F = \cdots = 0$. Then the sum is

$$\overline{z + 1} \text{ into } \frac{1}{2}z + z\cdot\overline{z - 1} + \frac{1}{4}z\cdot\overline{z - 1}\cdot\overline{z - 2} \quad\text{or}\quad \frac{zz}{4} \times \overline{z + 1}^2.$$

"And hence it appears that the sums of these cubes are the squares of the numbers 1, 3, 6, 10, 15, &c. namely of triangular numbers."[97]

Stirling remarked next on the ease of using difference tables for summation. He showed this in the example of a sequence $1, -1, 0, 8, 27, 61, 114, 190, \ldots$ whose sum he desired. By differencing he found the third differences to be constant and thus the general term to be $1 - 2z + 3z(z - 1)/2 + 4z(z - 1)(z - 2)/6$; "consequently the sum will be

$$\frac{z}{1} - 2 \times \frac{z}{1} \times \frac{z - 1}{2} + 3 \times \frac{z}{1} \times \frac{z - 1}{2} \times \frac{z - 2}{3} + 4 \times \frac{z}{1} \times \frac{z - 1}{2} \times \frac{z - 2}{3} \times \frac{z - 3}{4},$$

which, reduced into order, becomes

$$\frac{\overline{z - 3}\cdot\overline{z - 2}\cdot z\cdot\overline{z + 2}}{6}.$$

And the series is formed by writing 0, 1, 2, 3, 4, &c. in the quantity

$$\frac{4z^3 - 3zz - 13z + 6}{6}."$$ [98]

The next topic Stirling took up was contained in his:

"**Proposition II.** *If the terms of any series whatsoever be formed by writing any numbers differing by unit in the quantity,*

$$\frac{A}{z\cdot\overline{z + 1}} + \frac{B}{z\cdot\overline{z + 1}\cdot\overline{z + 2}} + \frac{C}{z\cdot\overline{z + 1}\cdot\overline{z + 2}\cdot\overline{z + 3}} + \frac{D}{z\cdot\overline{z + 1}\cdot\overline{z + 2}\cdot\overline{z + 3}\cdot\overline{z + 4}} + \&c.$$

the sum of all the terms in infinitum, *beginning at any given term, will be*

$$\frac{A}{z} + \frac{B}{2z\cdot\overline{z + 1}} + \frac{C}{3z\cdot\overline{z + 1}\cdot\overline{z + 2}} + \frac{D}{4z\cdot\overline{z + 1}\cdot\overline{z + 2}\cdot\overline{z + 3}} + \&c."$$ [99]

---

[96] Stirling [1739], pp. 18–19.
[97] Stirling [1739], p. 19.
[98] Stirling [1749], pp. 19–20. Notice that here Stirling counted from 0 instead of 1 as in Prop. I.
[99] Stirling [1749], p. 20.

The proof is made with the help of the indefinite sum relation (2.26′). His prescription is given in Corollary I where he says: "If the term be

$$\frac{p}{z\cdot z+1\cdot z+2\cdot z+3\cdot z+4,\ \&c.}$$

reject the last factor, then divide the remainder by the number of factors which remain, and you will have sum of the terms. Suppose the term be

$$\frac{A}{z\cdot z+1};$$

reject the last factor $z + 1$, and there will remain

$$\frac{A}{z},$$

and when there is only one remaining factor z, then

$$\frac{A}{z}$$

will be the sum of all the terms."[100] He goes on to say: "Suppose the term be

$$\frac{A}{z},$$

reject the factor z, and because nothing remains, divide A by 0, and you will have for the sum a quantity infinitely great, of which Dr. *Brook Taylor* was the first, that I know of, that handled this in his *Methodus Incrementorum*; likewise M. *Nichol* hath largely and elegantly treated of this in the *Memoires de l'Academie Royale de Sciences*."[101]

To deal with reciprocals of x^n Stirling prepared another table, which gave him the absolute values of the Stirling numbers of the first kind S_n^m. His table is reproduced below.[102]

"*Draw the terms of this progression n,* $1 + n$, $2 + n$, $3 + n$, &c. *continually into one another, and let the products be disposed as in the following table,*

1									
1	1								
2	3	1							
6	11	6	1						
24	50	35	10	1					
120	274	225	85	15	1				
720	1764	1624	735	175	21	1			
5040	13068	13132	6769	1960	322	28	1		
40320	109584	105056	67284	22449	4536	546	36	1	
&c.	&c.	&c.	&c.	&c.	&c.	&c.	&c.	&c.	&c.

[100] Stirling [1749], pp. 20–21.

[101] Stirling [1749], p. 21. The papers of Nicole appear in the volumes for the years 1719, 1723, 1724, and 1727. Cf. Bibliography.

[102] Stirling [1739], p. 10. Cf. also, Jordan [1939], pp. 142ff. We discuss these numbers shortly.

according to the order of the powers of n, only preserving the co-efficients, and *there will come out this* TABLE."

In this array each column m gives the values of $|S_n^m|$ $(n = 1, 2, \ldots)$. (There are a variety of notations for these numbers but we will use Jordan's.) To illustrate how Stirling used them, he wrote

$$\frac{1}{z^3} = \frac{1}{z \cdot z + 1 \cdot z + 2} + \frac{3}{z \cdot z + 1 \cdot z + 2 \cdot z + 3} + \frac{11}{z \cdot z + 1 \cdot z + 2 \cdot z + 3 \cdot z + 4} + \&c.$$

We see that the numerators are the $|S_n^2|$ $(n = 1, 2, \ldots)$.

Stirling needed these numbers so that he could take any expression of the form

$$\frac{A}{z^2} + \frac{B}{z^3} + \frac{C}{z^4} + \cdots$$

and represent it as

$$\frac{a}{z(z+1)} + \frac{b}{z(z+1)(z+2)} + \frac{c}{z(z+1)(z+2)(z+3)} + \cdots$$

where

$$a = A, \qquad b = A + B, \qquad c = 2A + 3B + C, \qquad d = 6A + 11B + 6C + D,$$
$$e = 24A + 50B + 35C + 10D + E,$$
$$f = 120A + 274B + 225C + 85D + 15E + F, \qquad \&c.^{103}$$

Notice that the coefficients of $a, b, c, d, \ldots$ in Stirling's formulas are formed from the rows of his matrix $|S_n^m|$ given above. Thus if we write

$$\sum_{\alpha=2}^{\infty} A_\alpha z^{-\alpha} = z^{-1} \sum_{\beta=1}^{\infty} a_\beta \cdot (z)_{-\beta}, \tag{2.30}$$

and if no convergence problems arise, Stirling asserts that

$$a_\beta = \sum_{\alpha=1}^{\beta} |S_\beta^\alpha| A_{\alpha+1} \qquad (\beta = 1, 2, \ldots), \tag{2.31}$$

$$z^{-\alpha} = \sum_{\beta=1}^{\infty} |S_\beta^\alpha| (z)_{-\beta} \qquad (\alpha = 1, 2, \ldots).$$

(We derive these shortly.)

Now the sum relation (2.26′) enabled Stirling to sum terms of the form $(z)_{-n}$. Let us look at a typical term, say $1/z(z+1)(z+2)$. If we set $m = 3$ in (2.26′) we have

$$\sum_{\alpha=x}^{\infty} \frac{1}{\alpha(\alpha+1)(\alpha+2)} = \frac{1}{2x(x+1)},$$

which is in accord with his rule stated earlier.

[103] Stirling [1749], pp. 10–11.

Let us now see why the S_n^m satisfy the relations (2.31). They are usually defined quite otherwise, as we saw in (2.27′). Jordan's definition is contained in the relation

$$(x)_n = \sum_{m=1}^{n} S_n^m \cdot x^m. \tag{2.32}$$

Thus $S_n^m = 0$ for $m > n$, and a comparison of (2.32) with

$$x^m = \sum_{\alpha=1}^{m} \mathfrak{S}_m^\alpha \cdot (x)_\alpha \tag{2.33}$$

shows that the matrices $S = (S_n^m)$, $\mathfrak{S} = (\mathfrak{S}_n^m)$ $(m, n = 1, 2, \ldots, p)$ for any positive integer p are inverses of each other. I.e.,

$$\sum_{m=1}^{p} S_n^m \mathfrak{S}_m^r = \delta_{nr} \qquad (n, r = 1, 2, \ldots, p).$$

We could then, in principle, calculate the S_n^m out of the $\mathfrak{S}_n^m$. Alternatively we can write $(x)_n$ in a Maclaurin's series in the form

$$(x)_n = \sum_{\alpha=1}^{n} \left(\frac{1}{\alpha!}\frac{d^\alpha (x)_n}{dx^\alpha}\right)\Bigg|^{x=0} x^\alpha;$$

and notice that the S_n^m are given by (2.27′), i.e.,

$$S_n^m = \frac{1}{m!}\frac{d^m (x)_n}{dx^m}\Bigg|^{x=0}$$

This provides another mechanism to evaluate the S_n^m. Finally we can find a recurrence relation for the S_n^m. We know by Leibniz's formula for derivatives of a product that

$$d^m(x)_{n+1}/dx^m = d^m[(x-n)(x)_n]/dx^m = (x-n)d^m(x)_n/dx^m + md^{m-1}(x)_n/dx^{m-1},$$

and hence, at $x = 0$, we find the recurrence relation (2.29′).

To see how Stirling's relations (2.31) can be derived we note first that (2.32) permits us to form the derivatives of $(x)_n$ easily. Let us express a function F by means of a Newton or (Harriot–Briggs) series in the form

$$F(x) = \sum_{\alpha=0} \frac{(x-z)_\alpha}{\alpha!} \Delta^\alpha F(z)$$

and observe that

$$\frac{d^\beta F(x)}{dx^\beta} = \sum_{\alpha=\beta} \frac{\Delta_z^\alpha}{\alpha!}\frac{d^\beta (x-z)_\alpha}{dx^\beta},$$

where $\Delta_z^\alpha = \Delta^\alpha F(x)$ at $x = z$. But we have by (2.32)

$$(x-z)_\alpha = \sum_{m=1}^{\alpha} S_\alpha^m (x-z)^m;$$

and so

$$\frac{d^\beta (x - z)_\alpha}{dx^\beta} = \sum_{m=\beta}^{\alpha} S_\alpha^m (m)_\beta (x - z)^{m-\beta}.$$

For $x = z$ this implies that

$$\left. \frac{d^\beta (x - z)_\alpha}{dx^\beta} \right|^{x=z} = S_\alpha^\beta \cdot \beta!,$$

and hence that

$$\left. \frac{d^\beta F(x)}{dx^\beta} \right|^{x=z} = \sum_{\gamma=\beta} \frac{\beta!}{\gamma!} \Delta^\gamma F(x) S_\gamma{}^\beta.$$

Applying this result to the case $F(x) = 1/x$, we have

$$\left. \frac{d^\beta F(x)}{dx^\beta} \right|^{x=z} = \frac{(-1)^\beta \beta!}{z^{\beta+1}}, \qquad \Delta^\gamma F(z) = \frac{(-1)^\gamma \gamma!}{(z + \gamma)_{\gamma+1}};$$

and therefore

$$\frac{(-1)^\beta \beta!}{z^{\beta+1}} = \sum_{\gamma=\beta} \frac{(-1)^\gamma \beta!}{(z + \gamma)_{\gamma+1}} S_\gamma^\beta = \sum_{\gamma=\beta} (-1)^\gamma \beta!\, z^{-1} (z)_{-\gamma} S_\gamma^\beta,$$

since $(z + \gamma)_{\gamma+1} = z(z + \gamma)_\gamma = z/(z)_{-\gamma}$. (Recall that by definition $(z)_{-\gamma} = 1/(z + \gamma)_\gamma$.)

This relation is clearly expressible as

$$\begin{aligned} z^{-\beta} &= \sum_{\gamma=\beta} (-1)^{\beta+\gamma} S_\gamma^\beta (z)_{-\gamma} \\ &= \sum_{\gamma=\beta} |S_\gamma^\beta| (z)_{-\gamma} \\ &= \sum_{\gamma=0} |S_\gamma^\beta| (z)_{-\gamma}, \end{aligned}$$

since $S_\gamma^\beta = 0$ for $\gamma < \beta$ and since $|S_\alpha^\beta| = (-1)^{\alpha+\beta} S_\alpha^\beta$. This is the second of Stirling's relations (2.31). If in this relation we multiply both sides by z^{-1} and replace $\beta + 1$ by δ, then we have

$$z^{-\delta} = \sum_{\gamma=0} |S_\gamma^{\delta-1}| (z - 1)_{-\gamma-1}.$$

From this we find with the help of (2.30) that

$$\sum_{\delta=2} A_\delta z^{-\delta} = \sum_{\gamma=0} (z - 1)_{-\gamma-1} \sum_{\delta=2} A_\delta |S_\gamma^{\delta-1}| = z^{-1} \sum_{\gamma=1} a_\gamma (z)_{-\gamma}.$$

Equating comparable coefficients and noting that $z^{-1}(z)_{-\gamma} = (z - 1)_{-\gamma-1}$, we find

$$a_\beta = \sum_{\alpha=2} |S_\beta^{\alpha-1}| A_\alpha = \sum_{\alpha=1} |S_\beta^\alpha| A_{\alpha+1}$$

which is Stirling's first relation (2.31).

Consider with Stirling summing the series whose general term is $1/(z^2 + nz)$, where z ranges over the positive integers. He writes it as

$$\frac{1}{z^2(1 + n/z)} = \frac{1}{z^2} - \frac{n}{z^3} + \frac{n^2}{z^4} \pm \cdots$$

and notes in effect that $A_\alpha = (-1)^\alpha n^{\alpha-2}$ in the relation (2.30). Thus $a_1 = 1$, $a_2 = 1 - n$, $a_3 = 2 - 3n + n^2$, $a_4 = 6 - 11n + 6n^2 - n^3$ and

$$\frac{1}{z^2 + nz} = \frac{1}{z(z + 1)} + \frac{1 - n}{z(z + 1)(z + 2)} + \frac{2 - 3n + n^2}{z(z + 1)(z + 2)(z + 3)} + \cdots.$$ [104]

Next he sets $n = 3$, and notes that the expression above now collapses to

$$\frac{1}{z(z + 3)} = \frac{1}{z(z + 1)} - \frac{2}{z(z + 1)(z + 2)} + \frac{2}{z(z + 1)(z + 2)(z + 3)},$$

and that the sum from z to ∞ is therefore by his Proposition II above

$$\frac{1}{z} - \frac{1}{z(z + 1)} + \frac{2}{3z(z + 1)(z + 2)} = \frac{3z^2 + 6z + 2}{3z^3 + 9z^2 + 6z}.$$

He then takes up the series

$$\frac{1}{1\cdot 4\cdot 7} + \frac{1}{4\cdot 7\cdot 10} + \frac{1}{7\cdot 10\cdot 13} + \cdots + \frac{1}{27z(z + 1)(z + 2)} + \cdots,$$

where $z = 1/3, 4/3, 7/3, \ldots$. Clearly, by his Proposition II, its sum is $1/54z(z + 1)$, and when he sets $z = 1/3$ the sum of the entire series is $1/24$. His next example is the series

$$\frac{1}{1\cdot 4} + \frac{1}{2\cdot 5} + \frac{1}{3\cdot 6} + \cdots + \frac{1}{z(z + 3)} + \cdots \quad .$$

By what we saw above, its sum from $z = 1$ to ∞ is $11/18$.

Instead of multiplying examples, we may consider his Corollary to Proposition II. It states that "any term

$$\frac{A}{z\cdot z + a\cdot z + b\cdot z + c, \text{ \&c.}}$$

may always be resolved into two summable terms, or perhaps more, and finite in number, when a, b, c, &c. are whole numbers; therefore, in this case, the series will be summable."[105] Of course it is not always the case that the general term is so expressible. As a counterexample on p. 23 he takes up Brouncker's expansion of log 2:

$$\frac{1}{1\cdot 2} + \frac{1}{3\cdot 4} + \frac{1}{5\cdot 6} + \frac{1}{7\cdot 8} + \frac{1}{9\cdot 10} + \text{\&c.},$$

[104] Stirling [1749], p. 11.

[105] Stirling [1742], p. 21.

where

$$z = \tfrac{1}{2}, 1\tfrac{1}{2}, 2\tfrac{1}{2}, 3\tfrac{1}{2}, \qquad \&c.$$

The general term $\frac{1}{4}z(z + \frac{1}{2})$, "reduced to a summable form, is

$$\frac{1}{4z\cdot z+1} + \frac{1}{8z\cdot z+1\cdot z+2} + \frac{1\cdot 3}{16z\cdot z+1\cdot z+2\cdot z+3} + \frac{1\cdot 3\cdot 5}{32z\cdot z+1\cdot z+2\cdot z+3\cdot z+4} + \&c.$$

But because the difference of the Factors, in the expression assigning the terms, is a broken number, it runs into an infinite series; which shews that the series is not summable; but by going back from the term to the sum, it will be

$$\frac{1}{4z} + \frac{1}{16z\cdot z+1} + \frac{1\cdot 3}{48z\cdot z+1\cdot z+2} + \frac{1\cdot 3\cdot 5}{128z\cdot z+1\cdot z+2\cdot z+3} + \&c.$$

which is a series converging so much the swifter, by how much greater the quantity z is." In this series he substituted $z = 27/2$ which gave him for the sum from the 14th term onwards .018861219. He then added up manually the first thirteen terms and found .674285961 and thus found $\log 2 =$ .693147180.

Using Brouncker's series, Stirling then gives a most interesting and important discussion to show the true importance of formulas to sum series. He says: "But that it is impossible in practice to obtain the sums of these series by a mere collection of terms, will be manifest from the following computation, where you have the sum of a hundred, thousand, ten thousand, and so on to ten millions, hundred millions of terms.

The sum of	100	Terms	.690653446
	1000		.692897242
	10000		.693122181
	100000		.693144680
	1000000		.693146930
	10000000		.693147155
	100000000		.693147178
	1000000000		.693147180

From this calculus it appears that a hundred terms gives the sum accurate to two figures; and if we gradually sum up ten times the number of terms, it is an uncertainty whether another figure can be gain'd: therefore if any one would find an accurate value of this series to nine places of figures (which requires no art but only addition) they would require one thousand million of terms; and this series converges much swifter than many others, whose values are finite quantities."[106] Next Stirling considers summing power series in Proposition III. He states:

"**Proposition III.** *If the terms of any series be formed by writing any numbers differing by unity, for z in the quantity* x^{z+n} *into*

$$\frac{a}{z}+\frac{b}{z\cdot z+1}+\frac{c}{z\cdot z+1\cdot z+2}+\frac{d}{z\cdot z+1\cdot z+2\cdot z+3}\ \&c.$$

the sum will be equal to x^{z+n} *into*

$$\frac{a}{1-x\cdot z}+\frac{b-Ax}{1-x\cdot z\cdot z+1}+\frac{c-2Bx}{1-x\cdot z\cdot z+1\cdot z+2}+\frac{d-3Cx}{1-x\cdot z\cdot z+1\cdot z+2\cdot z+3},\ \&c.$$

The quantities A, B, C, D, &c. denote the coefficients of the terms preceding those in which they are found; namely,

$$A=\frac{a}{1-x},\qquad B=\frac{b-Ax}{1-x},\qquad C=\frac{c-2Bx}{1-x},\qquad \&c.$$

But I except that case in which x is equal to unity..."[107]

The proof consists of forming the first difference of the sum and comparing the resulting quantity to the general term. As one example of this proposition Stirling gives a quite interesting and important example: to find $\pi/4$ by Leibniz's formula $1-1/3+1/5-1/7+\cdots$. He starts with the series $1+t/3+t^2/5+\cdots$ whose general term is expressible as $t^{z-1/2}$ into $1/2z$ where z takes on the values $1/2, 3/2, 5/2, \ldots$. In this case $x=t$, $n=-1/2$, $a=1/2$, $0=b=c\ldots$. Then

$$\sum_{\alpha=z}^{\infty}\frac{t^{\alpha}}{2\alpha+1}=\frac{t^{z-1/2}}{1-t}\times\left[\frac{1/2}{z}+\frac{t/2}{(t-1)z(z+1)}+\frac{t^2}{(t-1)^2z(z+1)(z+2)}+\cdots\right];$$

and for $t=-1$ it becomes

$$(-1)^{z+1/2}[1/4z+1/8z(z+1)+1/8z(z+1)(z+2)+\cdots].$$

Stirling summed by addition the first twelve terms of the series $(1-1/3)+(1/5-1/7)+\cdots+(1/21-1/23)=2[1/1\cdot 3+1/5\cdot 7+\cdots+1/21\cdot 23]$ and found .7646006915. He then used his series formula for $z=25/2$, its thirteenth value. Using only ten terms in his formula, he found .0207974719, to which he added his previous total, finding that $\pi/4=.7853981634$, a result "which M. *Leibnitz* long ago greatly desired."

---

[106] Stirling [1749], pp. 24ff.
[107] Stirling [1749], pp. 26ff.

He comments that Newton's formula for $\pi/4$, $1 + 1/3 - 1/5 - 1/7 + 1/9 + 1/11 \pm \cdots$, could also be easily summed by the same method, provided it is first expressed as the sum of the two separate series $1 - 1/5 + 1/9 - 1/13 \pm \cdots$ and $1/3 - 1/7 + 1/11 - 1/15 \pm \cdots$, and then each is treated by the methods shown above.

There are considerably more things in Stirling's book than we can take up here, but I would be remiss were I not to mention Proposition XXVIII: "*To find the sum of any number of logarithms, whose numbers are in arithmetical progression.*" This obviously includes his famous approximation to the factorial as a special case.[108] Stirling calls his progression $x + n$, $x + 3n$, $\ldots, z - n$. Then he says (note that Stirling writes $z \log z$ as zl, z) that the sum of the logarithms of these numbers is the difference of the two series

$$\begin{aligned} &\frac{zl, z}{2n} - \frac{az}{2n} - \frac{an}{12z} + \frac{7an^3}{360z^3} - \frac{31an^5}{1260z^5} + \frac{127an^7}{1680z^7} - \frac{511an^9}{1188z^9} + \&\text{c.}, \\ &\frac{xl, x}{2n} - \frac{ax}{2n} - \frac{an}{12x} + \frac{7an^3}{360x^3} - \frac{31an^5}{1260x^5} + \frac{127an^7}{1680x^7} - \frac{511an^9}{1188x^9} + \&\text{c.}, \end{aligned} \tag{2.34}$$

where "$a = .43429.44819.03252$." "And these series are thus continued *in infinitum*; put

$$-\frac{1}{3\cdot 4} = A,$$

$$-\frac{1}{5\cdot 8} = A + 3B,$$

$$-\frac{1}{7\cdot 12} = A + 10B + 5C,$$

$$-\frac{1}{9\cdot 16} = A + 21B + 35C + 7D,$$

$$-\frac{1}{11\cdot 20} = A + 36B + 126C + 84D + 9E,$$

where the numbers which are multiplied into A, B, C, D, &c. in the different values, are the alternate uncias in the odd powers of a binomial. These being premised, the coefficient of the third term will be

$$-\frac{1}{12} = A,$$

that of the fourth

$$+\frac{7}{300} = B,$$

---

[108] Stirling [1739], pp. 123ff. Cf. the proof of Maclaurin above.

of the fifth

$$-\frac{31}{1263} = C, \text{ \&c.''}^{109}$$

Stirling's demonstration is made very simply. He says: "Let the variable quantity z be diminished by its decrement $2n$, or which is the same, substitute $z - 2n$ for z in the series.

$$\frac{zl, z}{2n} - \frac{az}{2n} - \frac{an}{12z} + \frac{7an^3}{360z^3} - \frac{31an^5}{1260z^5} + \text{\&c.}$$

and the successive value of the same will come out

$$\frac{\overline{z - 2nl}, \overline{z - 2n}}{2n} - \frac{a}{2n} \times \overline{z - 2n} - \frac{an}{12 \cdot \overline{z - 2n}}$$
$$+ \frac{7an^3}{360 \cdot \overline{z - 2n}|^3} - \frac{31an^5}{1260 \cdot \overline{z - 2n}|^5} + \text{\&c.}$$

Subtract this from the former value, the terms first being reduced by division to the same form, and there will remain

$$lz - \frac{an}{z} - \frac{ann}{2z^2} - \frac{an^3}{3z^3} - \frac{an^4}{4z^4} - \text{\&c.}$$

that is, the logarithm of $z - n$. Therefore, universally, the decrement of two successive values of the series is equal to the logarithm of $z - n$, which expresses in general any of the logarithms that were to be summed. Therefore the series will be the sum of the proposed logarithms, if the other series be taken from it. For the sums, as well as the areas, must sometimes be corrected, that the true ones may come out."[110]

Now he says: "But if you will have the sum of any number of the logarithms [base 10] of the natural numbers 1, 2, 3, 4, 5, &c. put $z - n$ to be the last of the numbers, n being $=\frac{1}{2}$; and three or four terms of this series

$$zl, z - az - \frac{a}{24z} + \frac{7a}{2880z^3} - \text{\&c.}$$

added to half the logarithm of the circumference of a circle whose radius is unity, that is, to 0.39908.99341.79, will give the sum sought; and that with less labour the more logarithms are to be summed. Thus if you put

$$z - \frac{1}{2} = 1000, \quad \text{or} \quad z = \frac{2001}{2},$$

the value of the series will be 2567.20555.42879, as before; which added to

---

[109] Stirling [1739], p. 124. The *uncias* are the binomial coefficients. Thus those corresponding to the fifth power are <u>1</u>, 5, <u>10</u>, 10, <u>5</u>, 1, and the underscored ones are Stirling's constants in the formula $A + 10B + 5C = -1/7 \cdot 12$ above.

[110] Stirling [1749], p. 125.

the constant logarithm makes 2567.60464.42221 for the sum of the logarithms of a thousand of the first numbers of this series 1, 2, 3, 4, 5, &c."[111]

The approximation to $n!$ that Stirling actually achieved is not quite the one we today credit him with. The modern one is that given by Maclaurin in the preceding section. Stirling's is the one we could have obtained from Maclaurin's series in Article 834 (cf. p. 91). In Stirling's formulas (2.34) above we set $n = \frac{1}{2}$, $x = \frac{1}{2}$, $z - n = N$. Then

$$\sum_{\alpha=1}^{N} \log \alpha = (N + \tfrac{1}{2}) \times \log (N + \tfrac{1}{2}) - (N + \tfrac{1}{2}) - \frac{1}{24(N + \frac{1}{2})}$$
$$+ \frac{7}{8 \times 360(N + \frac{1}{2})^3} \pm \cdots - \left(\frac{1}{2} \log \frac{1}{2} - \frac{1}{2} - \frac{1}{12} + \frac{7}{360} \pm \cdots\right),$$

and hence

$$N! = \frac{\sqrt{2\pi}(N + \frac{1}{2})^{N+1/2}}{\operatorname{Exp}\,[(N + \frac{1}{2})]} \cdot \operatorname{Exp}\left(-\frac{1}{24(N + \frac{1}{2})} + \frac{7}{8 \cdot 360(N + \frac{1}{2})^3} \pm \cdots\right).^{112}$$

It is not possible to describe here all the results in Stirling's opus, but there are a few on interpolation and its applications that are worth mentioning. In Proposition XX he considers the following basic problem: "*Let there be given a series of equidistant ordinates running on both ways* ad infinitum, *required to find a parabolic line which shall pass thro' the extremities of all of them.*"[113]

His solution in the case where there is one middle ordinate is the well-known Stirling series

$$f_x = f_0 + \sum_{\alpha=0}^{\infty} x(x^2 - 1^2)(x^2 - 2^2)\cdots(x^2 - \alpha^2)(A_\alpha + B_\alpha x).$$

where A_α, B_α are given as $A_\alpha = \mu\delta_0^{(2\alpha+1)}/(2\alpha + 1)!$, $B_\alpha = \delta_0^{(2\alpha+2)}/(2\alpha + 2)!$. Stirling makes several interesting observations on the problem. He says, e.g.: "*Newton* indeed describes a parabola through given points, others have considered the assignation of terms from given differences; but how it may be conceived, or under what form it may be exhibited, the problem is the same... But it is to be observed, that the form of the ordinate composed of the powers $A + Bz + Cz^2 + Dz^3 + \&c.$ which *Newton* assumes in demonstrating the foundation of his method, is badly designed for this purpose; for the value of every coefficient comes out into an infinite series:

[111] Stirling [1749], p. 125. Stirling actually had a multiplicative constant in the terms of his series to adjust things to the base 10 instead of the base e. He discusses his series for $\log \sqrt{2\pi}$ on pp. 110–114 and relates 2π to "$2 \times \frac{8}{9} \times \frac{24}{25} \times \frac{48}{49} \times \frac{80}{81} \times$ &c., *in infinitum*, which is equal to the circumference of a circle by Dr. Wallis's *Arithmetic of Infinites*."

[112] This form of the approximation is given in Feller, *Prob.*, Vol. 1, 1957, p. 64. He gives it as an alternative form in Exercises 26, 27.

[113] Stirling [1749], p. 95. He discusses the two cases: there are one or two middle ordinates. This is essentially the same discussion we saw in Newton.

but if any one will assume the terms here used, he will with very little trouble get the above conclusion."[114]

Among the last topics treated are some concerned with numerical integration. Consider:

Proposition XXXI. *To find the area of any curve, as near as possible, from some given equidistant ordinates thereof.*

Through the extremities of the ordinates describe a parabolic figure, and its area, which is found by the known methods, will be equal to the area of the proposed curve nearly. Q.E.I.

Scholium. For as much as it would be laborious always to have recourse to the parabola, I have computed the following table, which exhibits the area of a curve directly, from some given equidistant ordinates thereof.

A TABLE *of* AREAS.

$$3\left|\ \frac{A + 4B}{6} R\right.$$

$$5\left|\ \frac{7A + 32B + 12C}{90} R\right.$$

$$7\left|\ \frac{41A + 216B + 27C + 272D}{840} R\right.$$

$$9\left|\ \frac{989A + 5888B - 928C + 10496D - 4540E}{28350} R\right.$$

A TABLE *of* CORRECTIONS.

$$3\left|\ \frac{P - 4A + 6B}{180} R\right.$$

$$5\left|\ \frac{P - 6A + 15B - 20C}{470} R\right.$$

$$7\left|\ \frac{P - 8A + 28B - 56C + 70D}{930} R\right.$$

$$9\left|\ \frac{P - 10A + 45B - 120C + 210D - 252E}{1600} R\right.$$

In these tables A is the sum of the first and last ordinate, B of the second and last but one, C of the third and last but two, and so on, until you come to the ordinate in the midst of all, which is represented by the last of the letters A, B, C, &c. R is the base upon which the area lies, or the part of the abscissa intercepted between the first and last ordinate. P is the sum of two ordinates, whereof one stands before the first, the other after the last, at distances equal to that of the other ordinates. And the number of ordinates, which here is odd, is denoted at the sides of the tables. The expressions in the tables of areas are the areas contained between the base, curve, and the extream ordinates. And those in the table of

[114] Stirling [1749], pp. 98–99. What Stirling meant in his last remark is that the use of $(x)_n$ for finite differences is the natural analog of x^n for derivative methods. In the case of two middle ordinates Stirling had the Newton–Bessel formula.

corrections are of the same magnitude as the differences between the true areas and those produced by the table. Therefore if the first figure of the correction be found, and added when the correction is negative, or suducted when affirmative, we may safely conclude that the area, thus corrected, is true in that place of decimals in which the first figure of correction enter, but no further; wherefore by a table of corrections the area found is corrected, and at the same time the number of true figures is known.[115]

Thus Stirling has given in the first table the Newton–Cotes formulas. The curious expressions in the table of corrections are really differences. Thus, e.g., in the first of them $P - 4A + 6B$ is δ_0^4. To see this let our function $f(x)$ be tabulated at $x = -2, -1, 0, 1, 2$ and let $f(\alpha) = f_\alpha$. Then $A = f_{-1} + f_{+1}$, $B = f_0$, $R = 2$, $P = f_{-2} + f_{+2}$. Thus $P - 4A + 6B = f_2 + f_{-2} - 4f_{-1} - 4f_1 + 6f_0 = f_2 - 4f_1 + 6f_0 - 4f_{-1} + f_{-2} = \delta_0^4$.

To see how Stirling probably derived his correction table since he gives no discussion on this point we write our function as

$$f_x \sim f_0 + x\left(\mu\delta_0^1 + \frac{x}{2}\,\delta_0^2\right) + x(x^2 - 1)\left(\frac{1}{3!}\,\mu\delta_0^3 + \frac{x}{4!}\,\delta_0^4\right).$$

When we integrate this from -1 to $+1$, we find

$$\int_{-1}^{+1} f_x\,dx \sim 2f_0 + \frac{1}{3}\,\delta_0^2 - \frac{2}{180}\,\delta_0^4 = 2\,\frac{f_1 + 4f_0 + f_{-1}}{6} - \frac{2}{180}\,\delta_0^4$$
$$= \frac{(A + 4B)R}{6} - \frac{(P + 4A + 6B)R}{180},$$

since all odd order differences vanish. Similar remarks apply to the other entries in the table of corrections. Stirling said that "I had computed these tables further, but the expressions for eleven or more ordinates are useless by reason of the immense greatness of the numeral coefficients; but if nine ordinates do not give the area sufficiently accurate, divide the base into two or more parts, and by this the area is divided into as many parts; then if you seek each of them separately by nine ordinates, you will have the whole area as accurate as you will."[116]

He then gives a table showing areas expressed in terms of differences. "In this table A is the ordinate in the midst of them all, B is the second difference of three ordinates in the midst, C is the fourth difference of five ordinates in the middle, and so on to the last of the letters A, B, C, D, E, F, G, which is the last difference of all the ordinates."

---

[115] Stirling [1749], p. 134. A similar but more complete table of areas appears in a work of Cotes, *HAR*. This is a posthumous work which contains an important text on trigonometry from a fairly modern point of view (using series expansions) as well as several other papers by Cotes. One of these is on the differential method *De Methode Differentiali* NEWTONIANA and on p. 53 of that paper Cotes gave his formulas for the cases $n = 3, \ldots, 11$. The presentation is the same as that of Stirling, including notation. In the trigonometric part Cotes establishes the formula $\log(\cos\varphi + i\sin\varphi) = i\varphi$.

[116] Stirling [1749], pp. 135–136.

A TABLE *of Areas by the Differences of the Ordinates.*

1	A
3	$A + \frac{1}{6}B$
5	$A + \frac{2}{3}B + \frac{7}{90}C$
7	$A + \frac{3}{2}B + \frac{11}{20}C + \frac{41}{840}D$
9	$A + \frac{8}{3}B + \frac{86}{45}C + \frac{92}{189}D + \frac{3}{86}E$
11	$A + \frac{25}{6}B + \frac{175}{36}C + \frac{3445}{1512}D + \frac{4045}{9072}E + \frac{94}{3503}F$
13	$A + 6B + \frac{103}{10}C + \frac{158}{21}D + \frac{1833}{700}E + \frac{4813}{11550}F + \frac{66}{3050}G.$

Stirling says that "the last terms in the expressions for nine, eleven, and thirteen ordinates, is not true, but more simple than these, and sufficiently near the true ones."[117] His reason is that "... the middle ordinate *A*, and the differences, *B*, *C*, *D*, *E*, &c. constitute a converging series; and therefore it is not required that the last terms which enter the computation, be precisely accurate."

In passing it is interesting to read what de Moivre wrote about numerical integration. He said: "... I make use in this case of the Artifice of Mechanic Quadratures, first invented by Sir *Isaac Newton*, and since prosecuted by Mr. *Cotes*, Mr. *James Stirling*, myself, and perhaps others; it consists in determining the Area of a Curve nearly, from knowing a certain number of its Ordinates *A*, *B*, *C*, *D*, *E*, *F*, &c. placed at equal Intervals, the more Ordinates there are, the more exact will the Quadrature be; but here I confine myself to four, as being sufficient for my purpose:... ."[118] In these tables he gives the sum of the first n logarithms; thus, e.g., he gives the sum

$$\sum_{\alpha=1}^{n} \log \alpha$$

for $n = 10$ as 6.55976.30328.7678; for $n = 100$, 157.97001.30547.1585 and for $n = 900$, 2269.82947.61838.1577. (The second should be 157,97000.36547.155.)

[117] Stirling [1749], p. 136.

[118] De Moivre [1756], pp. 244ff. The first edition appeared in 1718 but is not nearly as complete as is the third. In his actuarial work, *A Treatise of Annuities on Lives* [1725], pp. 333–334, de Moivre quotes Stirling's Proposition XXVIII above in summing logarithms. He also gives tables in three of his works for the sums of the hyperbolic logarithms from $n = 10$ to 900 to fourteen decimal places. The tables appear in the treatise just cited, as well as in his *Miscellanea Analytica* [1730], pp. 103–104, and in *Miscellaneis Analyticis Supplementum*, which is a supplement to the previous work.

An interesting example in his book on probability theory is this: "If the Binomial 1 + 1 be raised to a very high Power denoted by n, the ratio which the middle Term has to the Sum of all the Terms, that is, to 2^n, may be expressed by the Fraction

$$\frac{2A \times \overline{n-1}^n}{n^n \times \sqrt{n-1}},$$

wherein A represents the Number of which the Hyperbolic Logarithm is $1/12 - 1/360 + 1/1260 - 1/1680$, &c. But because the Quantity $(n-1)^n/n^n$ or $(1 - 1/n)^n$ is very nearly given . . . the number of which the Hyperbolic Logarithm is -1; . . . and now suppose that B represents the Number of which the Hyperbolic Logarithm is $-1 + 1/12 - 1/360 + 1/1260 - 1/1680$, &c. that expression will be changed into $2/B\sqrt{n}$ . . . till my worthy and learned Friend, Mr. *James Stirling*, who had applied himself after me to that inquiry, found that the Quantity B did denote the Square-Root of the Circumference of a Circle whose Radius is Unity, so that if that Circumference be called c, the Ratio of the Middle Term to the Sum of all the Terms will be expressed by $2/\sqrt{nc}$."[119]

2.8. Leibniz

In concluding our account of Newton and contemporaries, it is appropriate to mention Leibniz, who was an outstanding figure, although his most important work was not in our field. It is quite true that he had an early interest in calculating machines and had impressed Newton with his high intellect, but in numerical matters his knowledge did not approach Newton's. There is an interesting study by J. E. Hofmann of Leibniz during the period 1672–1676, when Newton was most productive.[120]

Apparently, in the beginning Leibniz felt he could master the mathematics of his time with a minimal amount of effort, but was forced to face harsh reality. It was Huygens who first took him in hand and forced him into working with reasonable disciplines.

When Leibniz first came into contact with his British colleagues through his fellow countryman Oldenburg, he seems to have behaved immaturely. He asserted among other things "that he was the first who could interpolate and sum any numerical series — a boast considered by Pell to be mere empty

---

[119] de Moivre [1756], pp. 243–245. It is noteworthy that he suggests Stirling took up the study of the series $1 - 1/12 + 1/360 + \cdots$ after he did. (Abraham de Moivre (1667–1754) was a Huguenot refugee who fled to England in 1668). Stirling considered the same problem in his *Differential Method* [1749], pp. 109–114.

[120] Hofmann, *Leibniz*.

arrogance."[121] From this conversation with Pell it was clear that Leibniz was unaware of the work of Briggs, Gregory, Mengoli, Mercator or Mouton. Nevertheless, Pell felt that Leibniz must have plagiarized Mouton's work. Hofmann says: "Leibniz was deeply embarrassed. The very next day he looked at Mouton's book in the library of the Royal Society and found confirmation of Pell's assertion regarding Regnauld. Perhaps Oldenburg thereupon advised him to deposit an explanation of the affair with the papers of the Royal Society. The resulting document has manifestly been composed in great haste, but since Leibniz in his hurry could not trace the paper in question among his disordered notes, it gives us a fairly good picture of the mathematical knowledge he could at the time immediately command without referring to books or notes."[122]

Hofmann proceeds to tell us that at a later time Leibniz remarked on "how little he understood of mathematics at the time . . .; the only point of any importance in the letter was his summation there of the reciprocal figurate numbers."[123]

Leibniz and Oldenburg–Collins corresponded extensively after this point. Among other things Oldenburg (April, 1673) told Leibniz that the sums of the reciprocal figurate numbers

$$\frac{1}{3}+\frac{1}{6}+\frac{1}{10}+\cdots,\qquad \frac{1}{4}+\frac{1}{10}+\frac{1}{20}+\cdots,\qquad \frac{1}{5}+\frac{1}{15}+\frac{1}{35}+\cdots$$

had been formed by Mengoli in his *Novae quadraturae arithmeticae* (Bologna, 1650) but that he had not been able to form the sums $\sum 1/k$, $\sum 1/k^2$, $\sum 1/k^3$. He went on to say that Collins had evaluated these expressions. Leibniz was never able to find these sums although he did find that $\sum 1/k$ diverges only to be told by Collins that Mengoli had shown this too. Hofmann says: "The determination of $\sum 1/k$ occupied Leibniz all his life but the solution never came within his grasp."[124]

We do not have the time to pursue the topic of Leibniz further, but Hofmann's book gives a good account.

---

[121] Hofmann, *Leibniz*, p. 299. Hofmann tells us that Pell "was by nature a reserved character, always ready with a derogatory remark on other people's scientific achievements" (p. 26).

[122] Hofmann, *Leibniz*, p. 27.

[123] Hofmann, *Leibniz*, p. 29.

[124] Hofmann, *Leibniz*, p. 34. It should be remarked that Collin's assertions about the series "concerned only an approximation to the sum."

3. Euler and Lagrange

3.1. Introduction

The invention of the calculus and its early exploitation was certainly due to Newton and Leibniz, but the invention of classical analysis was very largely due to Leonhard Euler, of Basel, who worked mainly in St. Petersburg. He was the student of Johann or Jean Bernoulli the First, and maintained a justifiable lifelong admiration and affection for the family.

His enormous collected works are remarkable for their brilliance and scope. Even a cursory glance through a volume serves to illustrate how Euler differed from Newton. There are no geometrical figures present. Euler did analysis; he worked with functions and studied their properties in the modern manner. A small number of his proofs are fallacious, some are dubious, but the majority are extremely elegant.

We shall examine only a minute portion of Euler's total accomplishments in the following discussion. Most of the material we shall cover was originally published during Euler's days in Russia under the auspices of the Imperial Academy of Sciences of St. Petersburg. The interested reader can find a first-class bibliography of Euler's contributions to finite difference theory in Nörlund.[1]

Very early in Volume VIII of Euler's collected works we find him writing as follows:

We therefore let the radius of a circle, or the sinus totus, be 1; it is evident then that the circumference of this circle cannot be expressed as an exact rational number; by approximation however the semicircumference can be found to be

= 3,14159 26535 89793 23846 26433 83279 50288 41971 69399 37510
58209 74944 59230 78164 06286 20899 86280 34825 34211 70679
82148 08651 32823 06647 09384 46 +,

which for brevity we write

$$\pi;$$

[1] Nörlund [1924], pp. 480–481. In fact the whole bibliography from the eighteenth century onwards is very complete. It is on pp. 464–531 of Nörlund.

thus π = semicircumference of a circle of radius = 1, or the length of arc of 180°.[2]

Euler noted that since $a^0 = 1$, for ω very small, $a^\omega = 1 + \psi$, where ψ is also small. He reasoned, moreover, that ψ could be expressed as $k\omega$ and so $a^\omega = 1 + k\omega$. Thus if $a = 10$ and $k\omega = 0.000001$, $l(1 + 1/1000000) = l(1000001/1000000) = 0.00000043429 = \omega$, where the logarithms expressed as l are to the base 10. Then $1/k = 0.43429$ and $k = 2.30258$.

He next calculated $a^{i\omega}$ by the Binomial theorem and found

$$a^{i\omega} = (1 + k\omega)^i = 1 + \frac{i}{1}k\omega + \frac{i(i-1)}{1\cdot 2}k^2\omega^2 + \frac{i(i-1)(i-2)}{1\cdot 2\cdot 3}k^3\omega^3 + \text{etc.}$$

Set $i = z/\omega$, where z is any "finite number"; then

$$a^z = \left(1 + \frac{kz}{i}\right)^i = 1 + \frac{1}{1}kz + \frac{1(i-1)}{1\cdot 2i}k^2z^2 + \frac{1(i-1)(i-2)}{1\cdot 2i\cdot 3i}k^3z^3 + \frac{1(i-1)(i-2)(i-3)}{1\cdot 2i\cdot 3i\cdot 4i}k^4z^4 + \text{etc.},$$

where i is "infinitely large." Hence this series reduces to

$$a^z = 1 + \frac{kz}{1} + \frac{k^2z^2}{1\cdot 2} + \frac{k^3z^3}{1\cdot 2\cdot 3} + \frac{k^4z^4}{1\cdot 2\cdot 3\cdot 4} + \text{etc. in infinitum},$$

and for $z = 1$ Euler had the relation between a and k

$$a = 1 + \frac{k}{1} + \frac{k^2}{1\cdot 2} + \frac{k^3}{1\cdot 2\cdot 3} + \frac{k^4}{1\cdot 2\cdot 3\cdot 4} + \text{etc.}$$

For $a = 10$, he had $k = 2.30258$, as remarked above.

For $k = 1$, Euler was led to the value

$$a = 1 + \frac{1}{1} + \frac{1}{1\cdot 2} + \frac{1}{1\cdot 2\cdot 3} + \frac{1}{1\cdot 2\cdot 3\cdot 4} + \text{etc.},$$

which he found to be 2.71828182845904523536068. As early as 1728 he called this number by the letter e, and logarithms to this base he called natural or hyperbolic. He also noted the series expansions

$$e^z = 1 + \frac{z}{1} + \frac{z^2}{1\cdot 2} + \frac{z^3}{1\cdot 2\cdot 3} + \frac{z^4}{1\cdot 2\cdot 3\cdot 4} + \text{etc.},$$

which, he remarked, held for every z and

$$l(1 + x) = x - \frac{x^2}{2} + \frac{x^3}{3} - \frac{x^4}{4} + \frac{x^5}{5} - \frac{x^6}{6} + \text{etc.},$$

$$l\frac{1 + x}{1 - x} = \frac{2x}{1} + \frac{2x^3}{3} + \frac{2x^5}{5} + \frac{2x^7}{7} + \frac{2x^9}{9} + \text{etc.}^3$$

[2] Euler 1, VIII [1748], pp. 133–134. This value of π, according to the editors, Euler got from Th. F. deLagny [1719], p. 135. Recall that de Moivre, Maclaurin and Stirling used c for 2π. It was William Jones, whom we mentioned on p. 72 who first used the symbol π. Cf. Jones [1706], p. 243.

[3] Euler 1, VIII [1748], pp. 122–128.

We should not neglect to mention a dispute between John Bernoulli and Leibniz over the value of the logarithm of a negative number.[4] Bernoulli held that the logarithm of a negative number $-a$ is the logarithm of $+a$ (in Euler's symbolism, $l - a = l + a$). Bernoulli gave four "reasons." Typical of them is reason 1: the differential of $l - x$ ($= \log(-x)$) is $-dx/-x = dx/x$, which is the differential of $l + x$ ($= \log x$). Hence $l - x = l + x$. Here is reason 4: it is clear that $lp^n = nlp$ for every exponent n and every p. Set $p = -a$. Then $l(-a)^n = nl(-a)$; put $n = 2$ and note that $2l(-a) = l(-a)^2 = l(a)^2 = 2l(a)$. Thus $l(-a) = l(+a)$.

To the first reason Leibniz objected that the differentiation was only valid for x positive.

Next Euler restated the thesis of Leibniz, that logarithms of negative or imaginary numbers are imaginary. Leibniz gave three "reasons" why this is so. The first reason depended on the Taylor's series expansion of $l(1 + x)$ by setting $x = -2$. Then

$$l - 1 = -2 - \frac{1}{2}\cdot 4 - \frac{1}{3}\cdot 8 - \frac{1}{4}\cdot 16 - \frac{1}{5}\cdot 32 - \frac{1}{6}\cdot 64, \text{ etc.}$$

Leibniz remarked that since the series diverged, it clearly did not have the value 0. He concluded that $l - 1$ "will then be imaginary since it is clear, moreover, that it could not be real, i.e., positive or negative." His second reason is that the equation $e^y = -a$ is not solvable for y real. Euler points out the difficulties in the arguments of both men and then proceeds to his "dénouement," which depends on the following result.

"**Theorem.** *There are always an infinite number of logarithms associated with any given number; i.e., if y indicates the logarithm of a number x, I say that y can take on an infinity of different values.*"[5]

His proof is of some interest. He notes that for ω "infinitely small," $l(1 + \omega) = \omega$ and $l(1 + \omega)^n = n\omega$. Then for n "infinitely large," for $x = (1 + \omega)^n$, and for y = logarithm of x, $y = n\omega$, and hence

$$y = nx^{1/n} - n = lx.$$

Euler points out that as n increases this expression approaches lx and "for n infinite it gives the true value of the logarithm of x." But $x^{1/2}$ has two values, $x^{1/3}$ has three, etc., and so for n "infinite" there will be infinitely many values.

Euler goes on to show that if a is real and positive and A its real logarithm, then

$$la = A \pm 2n\pi\sqrt{-1} \qquad n = 0, 1, 2, \ldots .$$

---

[4] Euler 1, XVII]1749], pp. 195–232. Cf. p. 196*n* for references to the papers by Bernoulli and Leibniz.

[5] Euler 1, XVII [1749], p. 210.

Moreover,

$$l - a = A \pm (2p + 1)\pi\sqrt{-1} \qquad p = 0, 1, 2, \ldots \ .$$

He next shows how to find the logarithms of an imaginary number, and he closes his paper with Problem 4: "*Given a logarithm to find the number to which it corresponds.*"

To solve it he supposes the logarithm is $f + g\sqrt{-1}$ with f, g real. Then the desired number x will be $e^f(\cos g + \sqrt{-1}\cdot\sin g)$. He notes that

$$e^{f+g\sqrt{-1}} = e^f \cdot e^{g\sqrt{-1}} = e^f(\cos g + \sqrt{-1}\cdot\sin g).$$

In Chapter X of his *Introductio*, which deals with trinomial factors, Euler shows how to express power series as infinite products and makes use of the method in a large number of connections. Thus he establishes the beautiful expansions

$$\begin{aligned} \frac{e^x - e^{-x}}{2} &= x\left(1 + \frac{xx}{\pi\pi}\right)\left(1 + \frac{xx}{4\pi\pi}\right)\left(1 + \frac{xx}{9\pi\pi}\right)\left(1 + \frac{xx}{16\pi\pi}\right)\left(1 + \frac{xx}{25\pi\pi}\right) \text{ etc.} \\ &= x\left(1 + \frac{xx}{1\cdot 2\cdot 3} + \frac{x^4}{1\cdot 2\cdot 3\cdot 4\cdot 5} + \frac{x^6}{1\cdot 2\cdots 7} + \text{etc.}\right), \\ \frac{e^x + e^{-z}}{2} &= 1 + \frac{xx}{1\cdot 2} + \frac{x^4}{1\cdot 2\cdot 3\cdot 4} + \frac{x^6}{1\cdot 2\cdot 3\cdot 4\cdot 5\cdot 6} + \text{etc.} \\ &= \left(1 + \frac{4xx}{\pi\pi}\right)\left(1 + \frac{4xx}{9\pi\pi}\right)\left(1 + \frac{4xx}{25\pi\pi}\right)\left(1 + \frac{4xx}{49\pi\pi}\right) \text{ etc.}^6 \end{aligned} \tag{3.1}$$

Euler used de Moivre's theorem to deduce that (in his notation i is the "infinite number")

$$\cos.\, v = \frac{\left(1 + \frac{v\sqrt{-1}}{i}\right)^i + \left(1 - \frac{v\sqrt{-1}}{i}\right)^i}{2} = \frac{e^{+v\sqrt{-1}} + e^{-v\sqrt{-1}}}{2},$$

$$\sin.\, v = \frac{\left(1 + \frac{v\sqrt{-1}}{i}\right)^i - \left(1 - \frac{v\sqrt{-1}}{i}\right)^i}{2\sqrt{-1}} = \frac{e^{+v\sqrt{-1}} - e^{-v\sqrt{-1}}}{2\sqrt{-1}}.$$

These so-called Euler's formulas first appeared in 1743. But in letters of December 9, 1741, and May 8, 1742 he wrote his friend Christian Goldbach that he had found the results

$$\frac{2^{+\sqrt{-1}} + 2^{-\sqrt{-1}}}{2} = \text{Cos. Arc. } l2,$$

$$a^{p\sqrt{-1}} + a^{-p\sqrt{-1}} = 2 \text{ Cos. Arc. } pla.^7$$

---

[6] Euler 1, XVII [1749], pp. 167–169. Euler was preceded in the study of trinomial factors by Newton. It is perhaps worth remarking that in his discussion of trigonometric functions Euler, following de Moivre [1730], p. 1, notes that $(\cos z \pm \sqrt{-1}\cdot\sin z)^n = \cos nz \pm \sqrt{-1}\cdot\sin nz$.

[7] Cf. Euler 1, VIII [1748], pp. 140–148*n*.

To factor $e^x \pm e^{-x}$, Euler first considers how to factor the expressions $a^n \pm z^n$ into trinomial factors of the forms $a^2 - 2az\cos(2\alpha + 1)\pi/n + z^2$, with $\alpha = 0, 1, \ldots, [(n-1)/2]$ in the former case, and $a^2 - 2az\cos 2\alpha\pi/n + z^2$, with $\alpha = 0, 1, \ldots, [n/2]$ in the latter. (Actually not all of these are necessarily trinomial factors. E.g., in the latter case when n is even, the factors $x - 1$ and $x + 1$ occur only to the first degree; they correspond to the values $\alpha = 0$ and $\alpha = [n/2]$.)[8]

Look now at

$$e^x = \left(1 + \frac{x}{i}\right)^i,$$

where i is again, in his words, "the infinite number." Then set $a = 1 + x/i$, $n = i$, $z = 1$ in $a^n - z^n$. He thus finds that $e^x - 1$ has the factors

$$\left(1 + \frac{x}{i}\right)^2 - 2\left(1 + \frac{x}{i}\right)\cos\frac{2k}{i}\pi + 1,$$

where $2k$ takes on all even integer values. Now $2k = 0$ leads to the factor xx/ii and Euler concludes that $e^x - 1$ has x as a factor. Furthermore, he notes that

$$\cos\frac{2k}{i}\pi = 1 - \frac{2kk}{ii}\pi\pi,$$

provided that "arc $2k\pi/i$ is infinitely small." He therefore concludes that

$$\frac{xx}{ii} + \frac{4kk}{ii}\pi\pi + \frac{4kk\pi\pi}{i^3}x$$

is also a factor, and hence $e^x - 1$ is divisible by

$$1 + \frac{x}{i} + \frac{xx}{4kk\pi\pi}.$$

He applies this reasoning to

$$e^x - e^{-x} = \left(1 + \frac{x}{i}\right)^i - \left(1 - \frac{x}{i}\right)^i$$

and finds that it has factors of the form $a^2 - 2az\cos(2k\pi/n) + z^2$, where $n = i$, $a = (1 + x/i)$, $z = (1 - x/i)$. Thus these factors are $= 2 + 2xx/ii - 2(1 - xx/ii)\cos 2k\pi/i \sim 4xx/ii + 4kk\pi\pi/ii - 4kk\pi\pi xx/i^4$, since $\cos 2k\pi/i \sim 1 - 2kk\pi\pi/ii$; he finds that $e^x - e^{-x}$ is divisible by $1 + xx/kk\pi\pi - xx/ii$. He notes that the last term in this factor can be omitted since it is "infinitely

---

[8] To carry out this factoring note, with the help of de Moivre's theorem, that the equation $x^n = 1$ has as roots $x_\alpha = e^{2\pi\alpha i/n}$, $\alpha = 0, 1, \ldots, n - 1$; and hence for n even, $x^n - 1$ has the factors $(x - e^{2\pi\alpha i/n})(x - e^{-2\pi\alpha i/n}) = x^2 - 2x\cos(2\pi\alpha/n) + 1$ for $\alpha = 1, 2, \ldots, n/2 - 1$, and $x^2 - 1$ as the remaining factor; for n odd it has the factors $x^2 - 2x\cos(2\pi\alpha/n) + 1$ for $\alpha = 1, 2, \ldots, [n/2]$, and $x - 1$ as the remaining factor. A similar analysis shows how $x^n + 1$ may be factored. (Note that $[x]$ is largest integer in x.)

small." Thus he found the first of the relations (3.1). The second follows by very similar reasoning. When he replaces x by $z\sqrt{-1}$ he finds the corresponding infinite product expansions of $\sin z$ and $\cos z$.

3.2. Summation of Series

These expansions are then used by Euler in Chapter X of this volume (*Introductio*) to evaluate the series

$$\sum_{\alpha=1}^{\infty} \frac{1}{\alpha^n},$$

which he later (in 1, X [1755]) connects with the Euler–Maclaurin Summation formula.

Given an infinite series equal to an infinite product

$$1 + Az + Bz^2 + Cz^3 + \cdots = (1 + \alpha z)(1 + \beta z)(1 + \gamma z)\cdots,$$

Euler points out that

$$\begin{aligned} A &= \alpha + \beta + \gamma + \cdots, \\ B &= \alpha\beta + \alpha\gamma + \beta\gamma + \cdots, \\ C &= \alpha\beta\gamma + \cdots, \\ &\vdots \end{aligned}$$

He then defines some related symmetric functions

$$\begin{aligned} P &= \alpha + \beta + \gamma + \cdots, \\ Q &= \alpha^2 + \beta^2 + \gamma^2 + \cdots, \\ R &= \alpha^3 + \beta^3 + \gamma^3 + \cdots, \\ &\vdots \end{aligned}$$

and makes use of Newton's identities to show that

$$P = A, \qquad Q = AP - 2B, \qquad R = AQ - BP + 3C, \ldots .$$ [9]

Now Euler equates his expansions for $(e^x - e^{-x})/2$:

$$x\left(1 + \frac{x^2}{3!} + \frac{x^4}{5!} + \frac{x^6}{7!} + \cdots\right) = x\left(1 + \frac{x^2}{\pi^2}\right)\left(1 + \frac{x^2}{4\pi^2}\right)\left(1 + \frac{x^2}{9\pi^2}\right)\cdots$$

[9] These results are essentially Newtonian but were in the cases $n = 1, 2, 3, 4$ first derived by Albert Girard (1590–1633), of Lorraine (Girard [1629]). (Girard was also the first to write cube and fourth roots as $\sqrt[3]{\ }, \sqrt[4]{\ }$.) Newton's derivation was "an unproved generalization of his computations on the coefficients (expressed as symmetric functions of the roots) of an equation of the eighth degree. No rigorous demonstration of the rule was published till the contents of Maclaurin's letter to Stanhope on July 8, 1743 systematically establishing it was printed by 'the Publisher' (Patrick Murdoch?) in Part II, Chapter XII of Maclaurin's posthumous *A Treatise of Algebra* . . . , pp. 286–296." Cf. Newton, *Papers*, Vol. V, p. 361*n*.

and sets $x^2 = \pi^2 z$. This gives him

$$A = \frac{\pi^2}{6}, \qquad B = \frac{\pi^4}{120}, \qquad C = \frac{\pi^6}{5040}, \qquad D = \frac{\pi^8}{362880}, \ldots,$$

$$\alpha = 1, \qquad \beta = \frac{1}{4}, \qquad \gamma = \frac{1}{9}, \qquad \delta = \frac{1}{16}, \ldots,$$

and hence he finds by comparing corresponding coefficients that

$$1 + \frac{1}{2^2} + \frac{1}{3^2} + \frac{1}{4^2} + \frac{1}{5^2} + \text{etc.} = \frac{2^0}{1 \cdot 2 \cdot 3} \cdot \frac{1}{1} \pi^2,$$

$$1 + \frac{1}{2^4} + \frac{1}{3^4} + \frac{1}{4^4} + \frac{1}{5^4} + \text{etc.} = \frac{2^2}{1 \cdot 2 \cdot 3 \cdot 4 \cdot 5} \cdot \frac{1}{3} \pi^4,$$

$$1 + \frac{1}{2^6} + \frac{1}{3^6} + \frac{1}{4^6} + \frac{1}{5^6} + \text{etc.} = \frac{2^4}{1 \cdot 2 \cdot 3 \cdots 7} \cdot \frac{1}{3} \pi^6,$$

$$1 + \frac{1}{2^8} + \frac{1}{3^8} + \frac{1}{4^8} + \frac{1}{5^8} + \text{etc.} = \frac{2^6}{1 \cdot 2 \cdot 3 \cdots 9} \cdot \frac{3}{5} \pi^8.$$ [10]

He uses the expansions for $e^x + e^{-x}$ to find the values for sums of the form $\sum 1/(2\alpha + 1)^{2n}$, as well as many others (cf. p. 131 below). Then out of related trigonometric functions such as $\cos v/2 + \operatorname{cotan} \pi/4 \cdot \sin v/2$ Euler produces sums such as

$$\frac{\pi}{4} = 1 - \frac{1}{3} + \frac{1}{5} - \frac{1}{7} + \cdots, \qquad \frac{\pi^3}{32} = 1 - \frac{1}{3^3} + \frac{1}{5^3} - \frac{1}{7^3} + \cdots,$$

and a whole variety of related sums.

In Series 1, Volume X of his *Opera Omnia* — "Introductio calculi differentialis . . ." — Euler, along with other results, shows his version of the Summation formula and many of its applications.[11]

Close to the beginning of this volume Euler formulated the notions of finite differences and sums as inverse operations, like those of derivatives and integrals. He then establishes very much the same types of results that Stirling did, but we omit them here. One of his main purposes was to find the value for the sum $\sum x^n$, with n any positive integer. He states the result on pp. 55ff., but does not arrive at his proof until later, when he has established the Euler–Maclaurin Summation formula.[12] To understand his result we should know that he used the operator S to be the inverse of Δ. Thus,

---

[10] Euler 1, VIII [1748], pp. 177–181. In this place he goes up to $n = 26$. Later he took up the general case as we shall see below. The case $n = 2$ just given is, of course, one of the most celebrated expansions. In Section 3.1 below we see the case $n = 1$; cf. p. 131.

[11] Euler 1, X [1755]. Cf. also, Euler 1, XV [1738].

[12] Euler 1, X [1755], pp. 321ff. This is Chapter V.

e.g., if $\Delta x = x^2 + x + 1/6$, then $Sx = x^3/3 + x^2/2 + x/6$ plus a constant of summation.

3.3. Euler on the Euler–Maclaurin Formula

In terms of this indefinite sum Euler expressed his formula as

$$Sz = \int z\,dx + \frac{1}{2}z + \frac{\mathfrak{A}\,dz}{1\cdot 2\,dx} - \frac{\mathfrak{B}d^3z}{1\cdot 2\cdot 3\cdot 4\,dx^3} + \frac{\mathfrak{C}d^5z}{1\cdot 2\cdots 6\,dx^5} - \text{etc.} \quad (3.2)$$

where the Gothic letters stand for those numbers called by Euler *Bernoulli numbers*, in honor of James Bernoulli. In fact we have

$$\mathfrak{A} = |B_2'|, \qquad \mathfrak{B} = |B_4'|, \qquad \mathfrak{C} = |B_6'|, \ldots,$$

where B_1', B_2', B_4', B_6', ... are the nonzero Bernoulli numbers today (cf. pp. 127 below).

Let us see how Euler established his result. His proof is not unlike Maclaurin's but was apparently independent of it. He expands $y(\alpha - 1)$, with α integral, in a Taylor's expansion as

$$y(\alpha - 1) = v = y - \frac{dy}{dx} + \frac{d\,dy}{2\,dx^2} - \frac{d^3y}{6\,dx^3} + \text{etc.},$$

and sums both sides. Then

$$S\frac{dy}{dx} = y - A + S\frac{d\,dy}{2\,dx^2} - S\frac{d^4y}{6\,dx^3} - \text{etc.}$$

He now replaces y by $\int z\,dx$ and absorbs the constant A into this indefinite integral, finding (in his notation)

$$Sz = \int z\,dx + \frac{1}{2}S\frac{dz}{dx} - \frac{1}{6}S\frac{d\,dz}{dx^2} + \frac{1}{24}S\frac{d^3z}{dx^3} - \text{etc.} \quad (3.3)$$

To go from this partial result to that of (3.2) he repeatedly differentiates the relation (3.3) and substitutes the results *seriatim* back into (3.3). Thus

$$S\frac{dz}{dx} = z + \frac{1}{2}S\frac{d\,dz}{dx^2} - \frac{1}{6}S\frac{d^3z}{dx^3} + \frac{1}{24}S\frac{d^4z}{dx^4} - \text{etc.}; \quad (3.4)$$

and hence

$$Sz = \int z\,dx + \frac{1}{2!}\left(z + \frac{1}{2!}S\frac{d\,dz}{dx^2} - \frac{1}{3!}S\frac{d^3z}{dx^3} + \cdots\right) - \frac{1}{3!}S\frac{d\,dz}{dx^2} + \cdots$$

$$= \int z\,dx + \frac{1}{2}z + \frac{1}{12}S\frac{d\,dz}{dx^2} - \frac{1}{24}S\frac{d^3z}{dx^3} + \cdots \quad .$$

He now differentiates (3.4) to find Sd^2z/dx^2 and eliminate it. When this process is carried out repeatedly there results a series

$$Sz = \int z\,dx + \alpha z + \frac{\beta\,dz}{dx} + \frac{\gamma d\,dz}{dx^2} + \frac{\delta d^3 z}{dx^3} + \frac{\varepsilon d^4 z}{dx^4} + \text{etc.}$$

with

$$\alpha = \frac{1}{2}, \quad \beta = \frac{\alpha}{2} - \frac{1}{6} = \frac{1}{12}, \quad \gamma = \frac{\beta}{2} - \frac{\alpha}{6} + \frac{1}{24} = 0,$$

$$\delta = \frac{\gamma}{2} - \frac{\beta}{6} + \frac{\alpha}{24} - \frac{1}{120} = -\frac{1}{720}, \quad \varepsilon = \frac{\delta}{2} - \frac{\gamma}{6} + \frac{\beta}{24} - \frac{\alpha}{120} + \frac{1}{720} = 0, \text{ etc.}$$

It now remained for Euler to discuss the character of the coefficients. In this he went far beyond Maclaurin. To do this he introduced something like the modern notation of a generating function. (Later we shall see how Laplace made profound use of this concept.) He considered the function $V = u/(1 - e^{-u})$ and was able with the help of (3.2) to relate his coefficients $\alpha, \beta, \gamma, \ldots$ to the series expansion of V. The modern practice is to write

$$\frac{u}{e^u - 1} = \sum_{\alpha=0}^{\infty} a_\alpha u^\alpha. \tag{3.5}$$

Then

$$a_\alpha = \frac{1}{\alpha!} B'_\alpha,$$

where the B'_α are the Bernoulli numbers. (Cf. p. 131 below.) Euler's procedure with his function V was to write $V - u/2 = u(e^{u/2} + e^{-u/2})/2(e^{u/2} - e^{-u/2})$ and then to replace the numerator and denominator by their series expansions. He worked this out in his *Introductio* (1, VIII [1748], Chap. X).

By comparing coefficients he found the values for the Bernoulli numbers, and recognized their relationship to those previously found by Jacob Bernoulli. He said that when the symbols $\alpha, \beta, \gamma, \delta$, etc. found above are divided, respectively, by the odd numbers 3, 5, 7, 9, etc., the numbers customarily called Bernoullian, which were found by James Bernoulli, are seen to be

$$\frac{\alpha}{3} = \frac{1}{6} = \mathfrak{A}, \qquad \frac{\iota}{19} = \frac{43867}{798} = \mathfrak{J},$$

$$\frac{\beta}{5} = \frac{1}{30} = \mathfrak{B}, \qquad \frac{\kappa}{21} = \frac{174611}{330} = \mathfrak{K} = \frac{283 \cdot 617}{330},$$

$$\frac{\gamma}{7} = \frac{1}{42} = \mathfrak{C}, \qquad \frac{\lambda}{23} = \frac{854513}{138} = \mathfrak{L} = \frac{11 \cdot 131 \cdot 593}{2 \cdot 3 \cdot 23},$$

$$\frac{\delta}{9} = \frac{1}{30} = \mathfrak{D}, \qquad \frac{\mu}{25} = \frac{236364091}{2730} = \mathfrak{M},$$

$$\frac{\varepsilon}{11} = \frac{5}{66} = \mathfrak{E}, \qquad \frac{\nu}{27} = \frac{8553103}{6} = \mathfrak{N} = \frac{13 \cdot 657931}{6},$$

$$\frac{\zeta}{13} \quad \frac{691}{2730} = \mathfrak{F}, \qquad \frac{\xi}{29} \quad \frac{23749461029}{870} = \mathfrak{O},$$

$$\frac{\eta}{15} = \frac{7}{6} = \mathfrak{G}, \qquad \frac{\pi}{31} = \frac{8615841276005}{14322} = \mathfrak{P},$$

$$\frac{\theta}{17} = \frac{3617}{510} = \mathfrak{H}, \qquad \text{etc.}$$ [13]

This insight of Euler that the Bernoulli numbers and his coefficients were so intimately related is certainly a remarkable one and entirely characteristic of the man. The chapter entitled *Investigatio Summae Serierum ex Termino Generali* closes with the application of his summation formula to derive the value of Sx^n, thereby completing Jacob Bernoulli's theorem.

In passing we should note that George Boole gave a summation formula which allows us to expand a function f with the help of the so-called Euler polynomials and numbers. (See p. 136 below.)

Given a positive integer α we define $E_\alpha(x)$ as the polynomial of degree α such that

$$\nabla E_\alpha(x) = \frac{E_\alpha(x+1) + E_\alpha(x)}{2} = x^\alpha.$$

It is related to the Bernoulli polynomial $B_\alpha(x)$; in fact

$$E_{\alpha-1}(x) = \frac{2}{\alpha}\left[B_\alpha(x) - 2^\alpha B_\alpha\left(\frac{x}{2}\right)\right] = \frac{2^\alpha}{\alpha}\left[B_\alpha\left(\frac{x+1}{2}\right) - B_\alpha\left(\frac{x}{2}\right)\right],$$

$$\frac{dE_\alpha(x)}{dx} = \alpha E_{\alpha-1}(x).$$

We can see that $E_0(x) = 1$, $E_1(x) = x - 1/2$, $E_2(x) = x(x-1), \ldots$.

If $\nabla F(x) = [F(x+1) + F(x)]/2$, then Boole's formula is

$$f(x+h) = \sum_{\alpha=0}^{n-1} \frac{1}{\alpha!} E_\alpha(h) \nabla f^{(\alpha)}(x) + R_n.$$ [14]

---

[13] Euler 1, X [1755], p. 321.

[14] This result was first discovered by Euler. Cf., e.g., Nörlund [1924], pp. 23–26. The form of R_n has been discussed by a number of people, including Darboux and Hermite (cf. Nörlund [1924], p. 34.) The first 20 Eulerian numbers are tabulated by Nörlund [1924], p. 458. Boole's result appears in Boole [1880]. The so-called Boole polynomials are discussed at length by Jordan [1939], pp. 317ff. He also gives another form of a summation formula of Boole. It involves differences instead of derivatives and the

This is sometimes referred to as Boole's First formula. His second, without remainder term, is

$$f(x+h) = \sum_{\alpha=0}^{\infty} \xi_\alpha(h)\nabla\Delta^\alpha f(x),$$

where the ξ_α are defined in Footnote 14. Jordan deduces from this Boole's result

$$\Delta^{-1}[(-1)^x f(x)] = (-1)^{x+1} \sum_{\alpha=0}^{\infty} \frac{(-1)^\alpha}{2^{\alpha+1}} \Delta^\alpha f(x) + k.$$

3.4. Applications of the Summation Formula

In his next chapter, *De Summatione Progressionum per Series Infinitas*, pp. 332–367, Euler proceeds to a variety of interesting applications of his relation (3.2), starting out with the famous problem of summing the harmonic progression

$$s_x = \sum_{\alpha=1}^{x} \frac{1}{\alpha},$$

and finds that

$$\begin{aligned} s_x = lx &+ \frac{1}{2x} - \frac{\mathfrak{A}}{2x^2} + \frac{\mathfrak{B}}{4x^4} - \frac{\mathfrak{C}}{6x^6} + \frac{\mathfrak{D}}{8x^8} - \text{etc.} \\ &+ \frac{1}{2} + \frac{\mathfrak{A}}{2} - \frac{\mathfrak{B}}{4} + \frac{\mathfrak{C}}{6} - \frac{\mathfrak{D}}{8} + \text{etc.} \end{aligned} \tag{3.6}$$

He points out on p. 339 that the constant is

$$\frac{1}{2} + \frac{\mathfrak{A}}{2} - \frac{\mathfrak{B}}{4} + \frac{\mathfrak{C}}{6} - \frac{\mathfrak{D}}{8} + \frac{\mathfrak{E}}{10} - \text{etc.} = 0.57721\ 56649\ 01532\ 5.$$

(Kowalewski notes that the last digit should be 9.) This is the famous constant of Euler.[15]

Boole instead of the Euler polynomials. These Boole polynomials may be defined with the help of the relation

$$\nabla\xi_\alpha(x) = \frac{\xi_\alpha(x+1) + \xi_\alpha(x)}{2} = \binom{x}{\alpha}.$$

[15] Attached to Euler 1, XII [1768/69], pp. 415–542, are commentaries written by an Italian humanist and mathematician, Lorenzo Mascheroni (1750–1800), who first taught in a gymnasium in Ticino and then became professor of mathematics in Pavia (cf. Mascheroni [1790] and [1792]). On p. 431 of Euler 1, XII [1768/69], Mascheroni gives Euler's constant as 0.577215 664901 532860 61811 090082 39 (cf. also p. 442); later Gauss recalculated the constant and in 1878 J. C. Adams (Adams I [1878]) gave its value to 263 places. Mascheroni's value was wrong in the 20th, 21st, and 22nd places. It is interesting to see how Mascheroni expressed definite integrals. He wrote, e.g.,

$$\int \frac{dxlx}{\sqrt{(1-xx)}} \begin{pmatrix} ab & x=0 \\ ad & x=1 \end{pmatrix} = -\frac{1}{2}\pi l2.$$

He used this notation throughout whereas Euler wrote, e.g., the integral and then added the words *integratione ab* $x = 0$ *ad* $x = 1$.

Using this constant Euler could, e.g., find the values of s_x above for $x = 1{,}000$ and $1{,}000{,}000$ to 13 decimal places. He had

$$s_x = lx + \frac{1}{2x} - \frac{\mathfrak{A}}{2x^2} + \frac{\mathfrak{B}}{4x^4} - \frac{\mathfrak{C}}{6x^6} + \frac{\mathfrak{D}}{8x^8} - \cdots + C,$$

where C is his constant. In the former case, $x = 1{,}000$, he needed to keep only the term in $\mathfrak{A}/2x^2 = 0.00000008\ 333\ 33$, since all others are zero to this many places. He found $s_{1,000} = 7.4854708605503$. Then for the latter case, $x = 1{,}000{,}000$, he needed to keep only $1/2x = 5 \times 10^{-7}$, and found $s_{1,000,000} = 14.3927267228657$. He then gave a number of interesting expansions for log 2, log 3, ... such as

$$l2 = 1 - \frac{1}{2} + \frac{1}{3} - \frac{1}{4} + \frac{1}{5} - \frac{1}{6} + \frac{1}{7} - \frac{1}{8} + \text{etc.},$$

$$l3 = 1 + \frac{1}{2} - \frac{2}{3} + \frac{1}{4} + \frac{1}{5} - \frac{2}{6} + \frac{1}{7} + \frac{1}{8} - \frac{2}{9} + \text{etc.},$$

$$l4 = 1 + \frac{1}{2} + \frac{1}{3} - \frac{3}{4} + \frac{1}{5} + \frac{1}{6} + \frac{1}{7} - \frac{3}{8} + \text{etc.},$$

$$l5 = 1 + \frac{1}{2} + \frac{1}{3} + \frac{1}{4} - \frac{4}{5} + \frac{1}{6} + \frac{1}{7} + \frac{1}{8} + \frac{1}{9} - \frac{4}{10} + \text{etc.}$$

These follow easily from the sum (3.6) above by considering

$$\begin{aligned} s_{mx} - s_x = 1 + \frac{1}{2} + \cdots + \frac{1}{m} + \cdots + \frac{1}{2m} + \cdots + \frac{1}{3m} + \cdots + \frac{1}{mx} \\ - \frac{m}{m} \qquad - \frac{m}{2m} \qquad - \frac{m}{3m} \qquad - \frac{m}{mx}. \end{aligned}$$

This gives

$$\begin{aligned} lm + \frac{1}{2mx} - \frac{\mathfrak{A}}{2m^2x^2} + \frac{\mathfrak{B}}{4m^4x^4} - \text{etc.} \\ - \frac{1}{2x} + \frac{\mathfrak{A}}{2xx} - \frac{\mathfrak{B}}{4x^4} + \text{etc.} \end{aligned}$$

Now Euler sets $x = \infty$ and says the sum is therefore lm. Thus for $m = 2$ he has $l2$. The others follow by successive choices of m.

Next (on pp. 322ff.) he returned to the sums

$$\sigma_n = \sum \frac{1}{\alpha^n}$$

of nth powers of the reciprocals of the integers using this time his relation (3.2). (Recall this sum is $\zeta(n)$, the value of the Riemann ζ-function for positive integer arguments.) For n even Euler considered the expansion

$$\sum_{\alpha=1}^{\infty} \frac{1}{\alpha^2 - u^2} = \frac{1}{2u^2} - \frac{\pi}{2u} \cot \pi u = \sigma_2 + \sigma_4 u^2 + \sigma_6 u^4 + \cdots$$

and noted that this is very close in form to his generating function (3.5). Thus he found in effect that

$$\sigma_{2n} = \frac{1}{2}\frac{(-1)^{n+1}(2\pi)^{2n}}{(2n)!}B'_{2n}.$$

For n odd he found approximate values of the σ_n for $n = 1, 3, 5, 7, 9$ by extensive numerical calculations. Thus, e.g., he had

$$1 + \frac{1}{2} + \frac{1}{3} + \frac{1}{4} + \text{etc. in infin.} = \frac{\pi}{0.0000} = \infty,$$

$$1 + \frac{1}{2^3} + \frac{1}{3^3} + \frac{1}{4^3} + \text{etc. in infin.} = \frac{\pi^3}{25.79436}\ \text{prox.}^{16}$$

After a few other examples Euler takes up Stirling's approximation to the factorial. He finds the value of the constant C in that formula by using Wallis's expression for $\pi/2$ and concludes easily that $2C = \log 2\pi$. Hence he finds

$$l1 + l2 + l3 + l4 + \cdots + lx$$

$$= \frac{1}{2}l2\pi + \left(x + \frac{1}{2}\right)lx - x + \frac{\mathfrak{A}}{1\cdot 2x} - \frac{\mathfrak{B}}{3\cdot 4x^3} + \frac{\mathfrak{C}}{5\cdot 6x^5} - \frac{\mathfrak{D}}{7\cdot 8x^7} + \text{etc.}$$

He then evaluates $s = l1 + l2 + l3 + \cdots + l1000$, using Briggsian logarithms, and finds $s = 2567.6046442221328$, keeping only terms through x^{-3}. (I.e., he multiplies the result from his formula by $\log_{10} e$.)

He finishes the chapter with a discussion of

$$\sum_{\alpha=1}^{x} \sin \alpha a, \qquad \sum_{\alpha=1}^{x} \cos \alpha a.$$

But in the next chapter of Volume X, *Methodus Summandi Superior Ulterius Promota*, he continues his exploitation of result (3.2). To understand what he did, it is worth seeing in a little more detail how Euler handled alternating series and how he was led to the above results. As we saw before, he had formula (3.2) for forming Sz, where $z(x)$ was the general term in a series. Now he turned to the problem of evaluating $Sp^x y$. He first noted that

$$\sum_{\alpha=1}^{x} y(\alpha-1)p^{\alpha-1} = \frac{1}{p}\sum_{\alpha=1}^{x} y(\alpha-1)p^{\alpha} = \sum_{\alpha=1}^{x} y(\alpha)p^{\alpha} - y(x)p^{x} + y(0)$$

and that

$$y(\alpha-1) = y(\alpha) - \frac{dy(\alpha)}{d\alpha} + \frac{1}{2}\frac{d^2y(\alpha)}{d\alpha^2} - \frac{1}{6}\frac{d^3y(\alpha)}{d\alpha^3} + \cdots .$$

---

[16] Euler 1, X [1755], p. 350. Note that Euler used a bold-faced capital S lying on its side to denote infinity. Apparently it was Wallis who introduced this symbol, as we noted on p. 79.

Thus

$$\sum_{\alpha=1}^{x} y(\alpha - 1)p^{\alpha-1} = \frac{1}{p}\left[Syp^x - S\frac{dy}{dx}p^x + S\frac{d\,dy}{2\,dx^2}p^x - S\frac{d^3y}{6\,dx^3}p^x + \text{etc.}\right],$$

and so for $A = y(0)$

$$Syp^x - yp^x + A = \frac{1}{p}Syp^x - \frac{1}{p}S\frac{dy}{dx}p^x + \frac{1}{2p}S\frac{d\,dy}{dx^2}p^x - \frac{1}{6p}S\frac{d^3y}{dx^3}p^x + \frac{1}{24p}S\frac{d^4y}{dx^4}p^x - \text{etc.}$$

From this he infers directly that

$$Syp^x = \frac{1}{p-1}\left(yp^{x+1} - Ap - S\frac{dy}{dx}p^x + S\frac{d\,dy}{2\,dx^2}p^x - S\frac{d^3y}{6\,dx^3}p^x + \text{etc.}\right).$$

For the choice $y = x^n$ ($n > 0$), $A = y(0) = 0$, and Euler found a general formula for Sx^np^x which gave him

$$Sx^0p^x = p^x\cdot\frac{p}{p-1} - \frac{p}{p-1},$$

$$Sx^1p^x = p^x\left(\frac{px}{p-1} - \frac{p}{(p-1)^2}\right) + \frac{p}{(p-1)^2}$$

$$Sx^2p^x = p^x\left(\frac{px^2}{p-1} - \frac{2px}{(p-1)^2} + \frac{p(p+1)}{(p-1)^3}\right) - \frac{p(p+1)}{(p-1)^3}, \text{ etc.}$$

Euler returned to the general formula for Syp^x and replaced y *seriatim* by the successive derivatives of z. He found by repeated differentiation that

$$S\frac{p^x\,dz}{dx} = \frac{p^{x+1}}{p-1}\cdot\frac{dz}{dx} - \frac{1}{p-1}S\frac{p^xd\,dz}{dx^2} + \frac{1}{2(p-1)}S\frac{p^xd^3z}{dx^3} - \text{etc.},$$

$$S\frac{p^xd\,dz}{dx^2} = \frac{p^{x+1}}{p-1}\cdot\frac{d\,dz}{dx^2} - \frac{1}{p-1}S\frac{p^xd^3z}{dx^3} + \frac{1}{2(p-1)}S\frac{p^xd^4z}{dx^4} - \text{etc.},$$

$$S\frac{p^xd^3z}{dx^3} = \frac{p^{x+1}}{p-1}\cdot\frac{d^3z}{dx^3} - \frac{1}{p-1}S\frac{p^xd^4z}{dx^4} + \frac{1}{2(p-1)}S\frac{p^xd^5z}{dx^5} - \text{etc.}$$

He then repeatedly eliminated these from Sp^xz and found as before that

$$Sp^xz = \frac{p^{x+1}z}{p-1} - \frac{\alpha p^{x+1}}{p-1}\cdot\frac{dz}{dx} + \frac{\beta p^{x+1}}{p-1}\cdot\frac{d\,dz}{dx^2} - \frac{\gamma p^{x+1}}{p-1}\cdot\frac{d^3z}{dx^3} + \frac{\delta p^{x+1}}{p-1}\cdot\frac{d^4z}{dx^4} - \frac{\varepsilon p^{x+1}}{p-1}\cdot\frac{d^5z}{dx^5} + \text{etc.}$$

It remains to determine $\alpha, \beta, \gamma, \delta, \varepsilon$, etc. To do this Euler notes that

$$\frac{p^{x+1}z}{p-1} = Sp^x z + \frac{1}{p-1} S\frac{p^x\,dz}{dx} - \frac{1}{2(p-1)} S\frac{p^x d\,dz}{dx^2} + \frac{1}{6(p-1)} S\frac{p^x d^3z}{dx^3} - \text{etc.},$$

$$\frac{p^{x+1}\,dz}{(p-1)\,dx} = S\frac{p^x\,dz}{dx} + \frac{1}{p-1} S\frac{p^x d\,dz}{dx^2} - \frac{1}{2(p-1)} S\frac{p^x d^3z}{dx^3} + \text{etc.},$$

$$\frac{p^{x+1}d\,dz}{(p-1)\,dx^2} = S\frac{p^x d\,dz}{dx^2} + \frac{1}{p-1} S\frac{p^x d^3z}{dx^3} - \text{etc.},$$

$$\frac{p^{x+1}\,d^3z}{(p-1)\,dx^3} = S\frac{p^x d^3z}{dx^3} + \text{etc.}$$

Hence

$$\alpha = \frac{1}{p-1}, \quad \beta = \frac{1}{p-1}\left(\alpha + \frac{1}{2}\right), \quad \gamma = \frac{1}{p-1}\left(\beta + \frac{\alpha}{2} + \frac{1}{6}\right),$$

$$\delta = \frac{1}{p-1}\left(\gamma + \frac{\beta}{2} + \frac{\alpha}{6} + \frac{1}{24}\right), \quad \varepsilon = \frac{1}{p-1}\left(\delta + \frac{\gamma}{2} + \frac{\beta}{6} + \frac{\alpha}{24} + \frac{1}{120}\right), \text{ etc.}$$

With this in hand he went to the case where $p = -1$ and found a very neat generalization of the Euler–Maclaurin Summation formula (3.2). In this case the α, β, γ, etc., reduce to the Bernoulli numbers and he found the relation

$$\mp\left(\frac{1}{2}z + \frac{(2^2-1)\mathfrak{A}\,dz}{1\cdot 2\,dx} - \frac{(2^4-1)\mathfrak{B}d^3z}{1\cdot 2\cdot 3\cdot 4\,dx^3} + \frac{(2^6-1)\mathfrak{C}d^5z}{1\cdot 2\cdots 6\,dx^5} - \frac{(2^8-1)\mathfrak{D}d^7z}{1\cdot 2\cdots 8\,dx^7} + \text{etc.}\right) + \text{Const.}$$[17]

That is,

$$\sum_{\alpha=1}^{x} (-1)^{\alpha+1}z(\alpha) = (-1)^{x+1}\left[\frac{z(x)}{2} + \frac{(2^2-1)}{2!}\mathfrak{A}z' - \frac{(2^4-1)}{4!}\mathfrak{B}z''' + \cdots\right]. \tag{3.7}$$

This very pretty result seems to have disappeared from the literature. He gave another simple derivation of this result. The sum of terms $a + b + c + d + \cdots + z$ is given by Euler's relation (3.2). He then notes that the sum of the alternate terms of the series is given by replacing x by $x/2$ in that relation, thus by

$$\frac{1}{2}\int z\,dx + \frac{1}{2}z + \frac{2\mathfrak{A}\,dz}{1\cdot 2\,dx} - \frac{2^3\mathfrak{B}d^3z}{1\cdot 2\cdot 3\cdot 4\,dx^3} + \frac{2^5\mathfrak{C}d^5z}{1\cdot 2\cdots 6\,dx^5} - \text{etc.} \tag{3.8}$$

---

[17] Euler 1, X [1755], p. 382.

Then by subtraction of $2(b + d + f + \text{etc.})$ from $a + b + c + d + \cdots + z$ in the case where x is even, or of $a + b + c + d + \cdots + z$ from

$$2(a + c + \cdots + z)$$

when x is odd, Euler finds result (3.7).

Suppose now with Euler that $z = x^n$. Then the sum of the series is, in his notation,

$$1 - 2^n + 3^n - 4^n + \cdots \mp x^n =$$

$$\mp\frac{1}{2}\left\{\begin{aligned} & x^n + \frac{A}{2}nx^{n-1} - \frac{B}{4}\cdot\frac{n(n-1)(n-2)}{1\cdot 2\cdot 3}x^{n-3} \\ & \quad + \frac{C}{6}\cdot\frac{n(n-1)(n-2)(n-3)(n-4)}{1\cdot 2\cdot 3\cdot 4\cdot 5}x^{n-5} \\ & \quad - \frac{D}{8}\cdot\frac{n(n-1)\cdots(n-6)}{1\cdot 2\cdots 7}x^{n-7} + \text{etc.} + \text{Const.}\end{aligned}\right\}, \qquad (3.9)$$

where the upper sign is valid for x even, the lower for x odd and where

$$\begin{aligned} A &= 2\cdot 1\cdot 3\ \mathfrak{A} = 2(2^2 - 1)\mathfrak{A}, & E &= 2\cdot 31\cdot 33\ \mathfrak{E} = 2(2^{10} - 1)\mathfrak{E}, \\ B &= 2\cdot 3\cdot 5\ \mathfrak{B} = 2(2^4 - 1)\mathfrak{B}, & F &= 2\cdot 63\cdot 65\ \mathfrak{F} = 2(2^{12} - 1)\mathfrak{F}, \\ C &= 2\cdot 7\cdot 9\ \mathfrak{C} = 2(2^6 - 1)\mathfrak{C}, & G &= 2\cdot 127\cdot 129\ \mathfrak{G} = 2(2^{14} - 1)\mathfrak{G}, \\ D &= 2\cdot 15\cdot 17\ \mathfrak{D} = 2(2^8 - 1)\mathfrak{D}, & H &= 2\cdot 255\cdot 257\ \mathfrak{H} = 2(2^{16} - 1)\mathfrak{H}, \text{etc.}\end{aligned}$$ [18]

By means of this formula Euler is able to find very elegant expressions such as:

$$1 - 1 + 1 - 1 + \cdots \mp 1 = \mp\frac{1}{2}(1) + \frac{1}{2};$$

$$1 - 2 + 3 - 4 + \cdots \mp x = \mp\frac{1}{2}\left(x + \frac{1}{2}\right) + \frac{1}{4};$$

$$1 - 2^2 + 3^2 - 4^2 + \cdots \mp x^2 = \mp\frac{1}{2}(x^2 + x);$$

$$1 - 2^3 + 3^3 - 4^3 + \cdots \mp x^3 = \mp\frac{1}{2}\left(x^3 + \frac{3}{2}xx - \frac{1}{4}\right) - \frac{1}{8};$$

$$1 - 2^4 + 3^4 - 4^4 + \cdots \mp x^4 = \mp\frac{1}{2}(x^4 + 2x^3 - x); \text{ etc.}$$

Thus when the number of terms is even, the last series $= -\frac{1}{2}x^4 - x^3 + \frac{1}{2}x$ and when odd $= \frac{1}{2}x^4 + x^3 - \frac{1}{2}x$, etc. However, when he attempts to "let x go to infinity" he gets into real difficulties. We see here how perhaps for the first time a genius, using the beautiful apparatus at his disposal, could occasionally overreach himself. Euler argued that the terms above bearing $\mp$

---

[18] Euler 1, X [1755], pp. 381–382.

signs in front of them arose because x was either even or odd, which is of course correct. But he went on to say: "However, if x is the infinite number, which is neither even nor odd, these considerations are no longer valid and the terms with the $\mp$ signs no longer enter; the sums of the series to infinity are now given by the constant terms."[19] Hence he "found"

$$1 - 1 + 1 - 1 + \text{etc. in infinitum} = \frac{1}{2};$$

$$1 - 2 + 3 - 4 + \text{etc.} \quad \ldots \quad = \frac{A}{4} = +\frac{(2^2 - 1)\mathfrak{A}}{2};$$

$$1 - 2^2 + 3^2 - 4^2 + \text{etc.} \quad \ldots \quad = 0;$$

$$1 - 2^3 + 3^3 - 4^3 + \text{etc.} \quad \ldots \quad = -\frac{B}{8} = -\frac{(2^4 - 1)\mathfrak{B}}{4};$$

$$1 - 2^4 + 3^4 - 4^4 + \text{etc.} \quad \ldots \quad = 0; \text{ etc.}$$

Next Euler chose n in the relation (3.9) to be negative and found, e.g.,

$$\sum_{\alpha=1}^{x} (-1)^{\alpha+1}\frac{1}{\alpha} = \frac{(-1)^{x+1}}{2}\left[\frac{1}{x} - \frac{A}{2x^2} + \frac{B}{4x^4} - \frac{C}{6x^6} + \cdots\right] + \text{Const.}$$

To find the constant Euler set $x = 1$.

He developed a set of simple formulas for the sums of series such as $1 - 1/2^2 + 1/3^2 - 1/4^2 + \cdots$ by a simple device. Let

$$1 + \frac{1}{2^{2n}} + \frac{1}{3^{2n}} + \frac{1}{4^{2n}} + \frac{1}{5^{2n}} + \frac{1}{6^{2n}} + \text{etc.} = s;$$

then evidently

$$\frac{1}{2^{2n}} \qquad + \frac{1}{4^{2n}} \qquad + \frac{1}{6^{2n}} + \text{etc.} = \frac{s}{2^{2n}},$$

where s is given above on p. 131. From these two relations he had, e.g.,

$$1 + \frac{1}{3^{2n}} + \frac{1}{5^{2n}} + \frac{1}{7^{2n}} + \text{etc.} = \frac{(2^{2n} - 1)s}{2^{2n}}$$

and

$$1 - \frac{1}{2^{2n}} + \frac{1}{3^{2n}} - \frac{1}{4^{2n}} + \text{etc.} = \frac{(2^{2n} - 1)s - s}{2^{2n}} = \frac{(2^{2n-1} - 1)}{(2n)!}|B'_{2n}|\pi^{2n}.$$

[19] Euler 1, X [1755], p. 384. Thus, e.g., in the expression for $1 - 2^3 + \cdots$ the trinomial with the $\mp$ in front is to be discarded, leaving for the infinite sum $-1/8$. It was not until Gauss investigated the hypergeometric series that we find a true discussion of convergence. Later Cauchy, in his *Analyse Algébrique* (Paris, 1821) also gave a rigorous treatment of series convergence; it should be noted though that Leibniz did give a criterium for the convergence of alternating series. The interested reader may consult F. Cajori [1919], pp. 373–377.

Therefore,

$$1 - \frac{1}{2^2} + \frac{1}{3^2} - \frac{1}{4^2} + \frac{1}{5^2} - \text{etc.} = \frac{A - 2\mathfrak{A}}{1\cdot 2}\cdot\frac{\pi^2}{4} = \frac{(2-1)\mathfrak{A}}{1\cdot 2}\cdot\pi^2,$$

$$1 - \frac{1}{2^4} + \frac{1}{3^4} - \frac{1}{4^4} + \frac{1}{5^4} - \text{etc.} = \frac{B - 2\mathfrak{B}}{1\cdot 2\cdot 3\cdot 4}\cdot\frac{\pi^4}{4} = \frac{(2^3-1)\mathfrak{B}}{1\cdot 2\cdot 3\cdot 4\cdot}\cdot\pi^4,$$

$$1 - \frac{1}{2^6} + \frac{1}{3^6} - \frac{1}{4^6} + \frac{1}{5^6} - \text{etc.} = \frac{C - 2\mathfrak{C}}{1\cdot 2\cdots 6}\cdot\frac{\pi^6}{4} = \frac{(2^5-1)\mathfrak{C}}{1\cdot 2\cdots 6}\cdot\pi^6, \text{ etc.}$$

In Chapter VII, *Partis Posterioris*, "De usu calculi differentialis in formandis seriebus," Euler considers various ways of forming series and shows, e.g., that

$$\operatorname{cosec} x = \frac{1}{x} + \frac{2(2-1)\mathfrak{A}x}{1\cdot 2} + \frac{2(2^3-1)\mathfrak{B}x^3}{1\cdot 2\cdot 3\cdot 4} + \frac{2(2^5-1)\mathfrak{C}x^5}{1\cdot 2\cdots 6} + \text{etc.},$$

$$\sec x = \alpha + \frac{\beta}{1\cdot 2}xx + \frac{\gamma}{1\cdot 2\cdot 3\cdot 4}x^4 + \frac{\delta}{1\cdot 2\cdots 6}x^6 + \frac{\varepsilon}{1\cdot 2\cdots 8}x^8 + \text{etc.},$$

where

$$\alpha = 1, \quad \beta = 1, \quad \gamma = 5, \quad \delta = 61, \quad \varepsilon = 1385, \quad \xi = 50521,$$
$$\eta = 2702765, \quad \theta = 199360981,$$
$$\iota = 19391512145, \quad \kappa = 2404879661671.$$ [20]

These are the numerical values of the famous *Euler numbers* of even index in this sense: $\alpha = |E_0|$, $\beta = |E_2|$, $\gamma = |E_4|, \ldots$. The usual definition of these numbers is in terms of the coefficients of the powers of x in the Taylor's expansion of $\tan x$. If $\mathfrak{E}_n$ is the coefficient of $-(-1)^{[n/2]}x^n/n!$ in this series,

$$E_n = \sum_{i=0}^{n}\binom{n}{i}\mathfrak{E}_i \; (\mathfrak{E}_0 = 1).$$

It is not difficult to see that

$$E_n = 2^n E_n\left(\frac{1}{2}\right) \qquad n = 0, 1, \ldots,$$

where the function indicated on the right-hand side is the Euler function defined above (p. 128). As we saw in Chapter II, the Bernoulli polynomials are closely related to $\sum_{\alpha=0}^{x-1}\alpha^n$; in fact Raabe used that sum as his way of defining the polynomials. Likewise the Euler polynomials are closely related to a sum. Indeed, for integral x,

$$\sum_{\alpha=0}^{x-1}(-1)^\alpha\alpha^n = \frac{1}{2}\,[E_n(0) - (-1)^x E_n(x)].$$

[20] Euler 1, X [1755], p. 419. Kowalewski points out that κ should be 2404879675441.

This follows from the definition of the Euler polynomials. We have

$$\sum_{\alpha=0}^{x-1}(-1)^{\alpha}\alpha^{n} = \sum_{\alpha=0}^{x-1}(-1)^{\alpha}\nabla E_n(\alpha) = \frac{1}{2}\sum_{\alpha=0}^{x-1}(-1)^{\alpha}[E_n(\alpha+1)+E_n(\alpha)]$$

$$= \frac{1}{2}[E_n(0)-(-1)^{x}E_n(x)].$$

The original definitions of the Bernoulli and Euler functions were given by Raabe, as we mentioned earlier. He chose to define the former to be $\sum_{\alpha=1}^{x-1}\alpha^n$ and the latter to be $\sum_{\alpha=1}^{2x-1}(-1)^{\alpha}\alpha^n$.[21]

3.5. Euler on Interpolation

In Chapter IX of the *Partis Posterioris*, Volume X, Euler discusses a variety of ingenious schemes for finding the roots of equations numerically by variants on the ideas of Newton, but nothing here is very novel or worthy of comment. In fact it is not until Chapter XVII that we find some ideas which are basic to our interests. Here we see the genesis of functions such as the Gamma function. The problem posed by Euler was summed up in the chapter's title, *De Interpolatione Serierum*; thus e.g., he sought a function of a continuous variable x which equalled $n!$ when x was n. Nörlund tells us that Euler was urged to investigate this topic by Daniel Bernoulli and Goldbach.[22]

The relevant letter is dated October 13, 1729 and contains the relation (3.10) below. There are a number of others on the same topic. (There is an interesting discussion of the relation between Wallis's and Euler's inspiration for this sort of interpolation in N. Bourbaki [1960], pp. 226–227.) But his aim was general; he investigated a number of functions, as we shall see. Given $S_0 = 0$, $S_x = (\sum_{\alpha=1}^{x} 1/\alpha)$ where x is a nonnegative integer, Euler sought to find a function $\sum_x$ of a variable ω such that $\sum_x(\omega) = (\sum_{\alpha=1}^{x+\omega} 1/\alpha)$ if ω is a nonnegative integer. Euler's method is to define $\sum_x$ with the help of the formal relation

$$\textstyle\sum_x(\omega) = S_x + \displaystyle\sum_{\beta=x}^{\infty}\frac{1}{1+\beta} - \sum_{\beta=x}^{\infty}\frac{1}{1+\beta+\omega} = S_\infty - \sum_{\beta=x}^{\infty}\frac{1}{1+\beta+\omega}.$$

---

21 Raabe [1848]. Cf. also, [1851]. In this paper he relates the Bernoulli and Euler numbers to the tangent and secant coefficients. Cf. also, Glaisher [1914].

22 Nörlund [1924], p. 98. Euler's first letter on the Gamma function appears in P.–H. Fuss [1843]; the relevant letter is dated October 13, 1729 and contains the relation (3.10) below. There are a number of others on the same topic. (There is an interesting discussion of the relation between Wallis's and Euler's inspiration for this sort of interpolation in N. Bourbaki [1960], pp. 226–227.)

For $x = 0$, $S_0 = 0$ and Euler had $\sum_0 (\omega) = T$ where

$$T = +\frac{1}{1} + \frac{1}{2} + \frac{1}{3} + \frac{1}{4} + \text{etc.}$$

$$-\frac{1}{1+\omega} - \frac{1}{2+\omega} - \frac{1}{3+\omega} - \frac{1}{4+\omega} - \text{etc.},$$

and he expanded and rearranged terms to obtain

$$T = +\omega\left(1 + \frac{1}{2^2} + \frac{1}{3^2} + \frac{1}{4^2} + \frac{1}{5^2} + \text{etc.}\right)$$

$$-\omega^2\left(1 + \frac{1}{2^3} + \frac{1}{3^3} + \frac{1}{4^3} + \frac{1}{5^3} + \text{etc.}\right)$$

$$+\omega^3\left(1 + \frac{1}{2^4} + \frac{1}{3^4} + \frac{1}{4^4} + \frac{1}{5^4} + \text{etc.}\right)$$

$$-\omega^4\left(1 + \frac{1}{2^5} + \frac{1}{3^5} + \frac{1}{4^5} + \frac{1}{6^5} + \text{etc.}\right) \text{etc.}$$

Next he went back to his original series for T, set $\omega = 1/2$, and again rearranged terms so that he had

$$T = 1 - \frac{2}{3} + \frac{1}{2} - \frac{2}{5} + \frac{1}{3} - \frac{2}{7} + \frac{1}{4} - \frac{2}{9} + \text{etc.}$$

$$= 2\left(\frac{1}{2} - \frac{1}{3} + \frac{1}{4} - \frac{1}{5} + \frac{1}{6} - \text{etc.}\right) = 2 - 2l2.$$

He then evaluated T for $\omega = 1/2, 3/2, 5/2$, and $7/2$, and found

$$2 - 2l2, \quad 2 + \frac{2}{3} - 2l2, \quad 2 + \frac{2}{3} + \frac{2}{5} - 2l2, \quad 2 + \frac{2}{3} + \frac{2}{5} + \frac{2}{7} - 2l2 \text{ etc.}$$

Note that $\sum_0 (\omega) = S_x$ and $\sum_x (\omega) = S_{x+\omega}$ for x and ω positive integers.

In Section 400 (pp. 637ff.) Euler considers how to interpolate the sequence

$$\begin{array}{cccc} 1 & 2 & 3 & 4 \\ \frac{a}{b}, & \frac{a(a+c)}{b(b+c)}, & \frac{a(a+c)(a+2c)}{b(b+c)(b+2c)}, & \frac{a(a+c)(a+2c)(a+3c)}{b(b+c)(b+2c)(b+3c)} \text{ etc.} \end{array}$$

He finds

$$\sum = \frac{a(b+c\omega)}{b(a+c\omega)} \cdot \frac{(a+c)(b+c+c\omega)}{(b+c)(a+c+c\omega)} \cdot \frac{(a+2c)(b+2c+c\omega)}{(b+2c)(a+2c+c\omega)} \cdot \text{etc.}$$

If in the original sequence we set $a = 1$, $b = c = 2$ we find

$$\begin{array}{ccccc} 1 & 2 & 3 & 4 & 5 \\ \frac{1}{2}, & \frac{1\cdot 3}{2\cdot 4}, & \frac{1\cdot 3\cdot 5}{2\cdot 4\cdot 6}, & \frac{1\cdot 3\cdot 5\cdot 7}{2\cdot 4\cdot 6\cdot 8}, & \frac{1\cdot 3\cdot 5\cdot 7\cdot 9}{2\cdot 4\cdot 6\cdot 8\cdot 10} \text{ etc.,} \end{array}$$

and $\sum$ becomes

$$\sum = \frac{1(2+2\omega)}{2(1+2\omega)}\cdot\frac{3(4+2\omega)}{4(3+2\omega)}\cdot\frac{5(6+2\omega)}{6(5+2\omega)}\cdot\frac{7(8+2\omega)}{8(7+2\omega)}\text{ etc.}$$

For $\omega = 1/2$ this is

$$\sum = \frac{1\cdot 3}{2\cdot 2}\cdot\frac{3\cdot 5}{4\cdot 4}\cdot\frac{5\cdot 7}{6\cdot 6}\cdot\frac{7\cdot 9}{8\cdot 8}\cdot\frac{9\cdot 11}{10\cdot 10}\text{ etc.} = \frac{2}{\pi}.$$

Euler then shows that $\sum(1/2) = 2/\pi$, $\sum(3/2) = (2/3)(2/\pi)$, $\sum(5/2) = (2\cdot 4/3\cdot 5)(2/\pi)$, $\sum(7/2) = (2\cdot 4\cdot 6/3\cdot 5\cdot 7)(2/\pi)$, etc., which is in essence Wallis's result.

Another typical problem is Example I in Section 401 where the problem posed is to interpolate

$$\begin{matrix} 1 & 2 & 3 & 4 \\ 1, & 1\cdot 3, & 1\cdot 3\cdot 5, & 1\cdot 3\cdot 5\cdot 7\text{ etc.} \end{matrix}$$ [23]

To handle this Euler shows that for the sequence

$$\begin{matrix} 1 & 2 & 3 & 4 \\ a, & a(a+b), & a(a+b)(a+2b), & a(a+b)(a+2b)(a+3b)\text{ etc.,} \end{matrix}$$

a suitable interpolant is $\sum(\omega)$, where

$$\sum = a^{\omega}\cdot\frac{a^{1-\omega}(a+b)^{\omega}}{a+b\omega}\cdot\frac{(a+b)^{1-\omega}(a+2b)^{\omega}}{a+b+b\omega}\cdot\frac{(a+2b)^{1-\omega}(a+3b)^{\omega}}{a+2b+b\omega}\text{ etc.}$$

He then applied this to the sequence $1, 1\cdot 2, 1\cdot 2\cdot 3, \ldots$ and found

$$\sum = \frac{1^{1-\omega}\cdot 2^{\omega}}{1+\omega}\cdot\frac{2^{1-\omega}\cdot 3^{\omega}}{2+\omega}\cdot\frac{3^{1-\omega}\cdot 4^{\omega}}{3+\omega}\cdot\frac{4^{1-\omega}\cdot 5^{\omega}}{4+\omega}\text{ etc.} \tag{3.10}$$

This gives for $\omega = 1/2, 3/2, 5/2, 7/2$, etc., the values

$$\frac{\sqrt{\pi}}{2},\quad \frac{3}{2}\cdot\frac{\sqrt{\pi}}{2},\quad \frac{3\cdot 5}{2\cdot 2}\cdot\frac{\sqrt{\pi}}{2},\quad \frac{3\cdot 5\cdot 7}{2\cdot 2\cdot 2}\cdot\frac{\sqrt{\pi}}{2}\text{ etc.}$$

Recall that

$$\Gamma(x) = \int_0^{\infty} e^{-t}t^{x-1}\,dt \ (\operatorname{Re} x > 0)$$

and that $\Gamma(x) = (x-1)\Gamma(x-1)$. Thus $\Gamma(1/2) = \sqrt{\pi} = 2\sum(1/2)$, $\Gamma(3/2) = (1/2)\Gamma(1/2) = \sum(1/2)$, $\Gamma(5/2) = (3/2)\Gamma(3/2) = (3/2)\sum(1/2) = \sum(3/2)$, $\Gamma(7/2) = (5/2)\Gamma(5/2) = (5/2)\sum(3/2) = \sum(5/2)$. We see in general that

$$\sum\left(\frac{2n-1}{2}\right) = \Gamma\left(\frac{2n+1}{2}\right).$$

[23] Euler 1, X [1755], p. 640ff. Cf. Chap. XVII.

Let us go back to Euler's formula (3.10) and rewrite it in the form

$$\sum(\omega) = \lim_{n\to\infty} \frac{n!\cdot(n+1)^\omega}{(\omega+1)(\omega+2)\cdots(\omega+n)}.$$

From this we recognize the formula that is now called Gauss's Product for the Gamma function. We then see that $\sum(\omega) = \Gamma(\omega+1)$ for all ω.

Furthermore, in a letter to Goldbach in 1729, Euler wrote

$$\Gamma(z) = \frac{1}{z}\prod_{n=1}^{\infty}\left(1+\frac{1}{n}\right)^z\left(1+\frac{z}{n}\right)^{-1} \qquad (z \neq 0, -1, -2, \ldots).$$[24]

Although it is somewhat less relevant, we mention in passing that the Beta function

$$B(p,q) = \int_0^1 x^{p-1}(1-x)^{q-1}\,dx$$

was discussed by Euler, who wrote

$$\frac{1}{\mu+1}\cdot\frac{2}{\mu+2}\cdot\frac{3}{\mu+3}\cdots\frac{\nu-1}{\mu+\nu-1} = \mu u^\mu \int \frac{z^{\nu-1}\,dz}{(u+z)^{\mu+\nu}}$$

and meant the integral to be from $z = 0$ to $z = \infty$; note that the transformation $x = z/(u+z)$ gives

$$\int_0^\infty \frac{z^{\nu-1}\,dz}{(u+z)^{\mu+\nu}} = \frac{1}{u^\mu}\int_0^1 x^{\nu-1}(1-x)^{\mu-1}\,dx.$$[25]

Euler wrote that $\left(\frac{p}{q}\right) = B_n(p,q) = B_n(q,p)$, that $B_n(p, n-p) = \pi/n\cdot \csc(p\pi/n)$, and that

$$B_n(p,q) = \frac{p+q}{pq}\cdot\frac{n(p+q+n)}{(p+n)(q+n)}\cdot\frac{2n(p+q+2n)}{(p+2n)(q+2n)}\text{ etc.}$$

He studied these functions in considerable detail for $n = 1, 2, \ldots, 9$.[26]

---

[24] Whittaker, *W W*, p. 237. The letter is in Fuss [1843].

[25] Euler 1, XII [1768/69], pp. 269–270. This is a volume of his *Institutiones Calculi Integralis*. See also the footnote on p. 270 for further references. This function is also discussed in a number of papers contained in Vol. XVII starting on p. 233. He used the notation $\left(\frac{p}{q}\right)$ for the integral

$$\int \frac{x^{p-1}\,dx}{\sqrt[n]{(1-x^n)^{n-q}}} = \int x^{p-1}(1-x^n)^{-1+q/n}\,dx.$$

Cf. Euler 1, XVII [1762/65], pp. 268ff. (Here again the integral is to be understood as being evaluated at the limits $x = 0$ and $x = 1$, and n, p, q are positive integers.)

[26] Euler 1, XVII [1765/65], pp. 268–315. This is made possible by the previous paper, pp. 233–267. Cf. also, pp. 316ff. Note that

$$B_n(p,q) = \frac{1}{n}B\left(\frac{p}{n},\frac{q}{n}\right).$$

(Clearly they are a generalization of the Beta function.) For example, he shows that

$$\frac{1\cdot 2\cdot 3\cdots n}{(f+g)(f+2g)(f+3g)\cdots(f+ng)} = \frac{f}{g^n}\int x^{f-1}\,dx(1-x^g)^n,$$

where the integral is from $x = 0$ to $x = 1$. For $g = 1, f = p, n = q - 1$, we see that this is the familiar result

$$\frac{(p-1)!\,(q-1)!}{(p+q-1)!} = B(p, q).$$

Cauchy showed that $B(n, m+1) = (-1)^m\Delta^m(1/n)$. The proof is simple. We have

$$\frac{1}{n} = \int_0^\infty e^{-nx}\,dx, \qquad \Delta^m\frac{1}{n} = \int_0^\infty \Delta^m e^{-nx}\,dx.$$

But $\Delta^m e^{-nx} = e^{-nx}(e^{-x} - 1)^m$, and so if we set $t = e^{-x}$,

$$\Delta^m\frac{1}{n} = \int_0^1 t^{n-1}(t-1)^m\,dt = (-1)^m B(n, m+1).$$

In Euler 1, XI and 1, XII [1768/69] he showed how to integrate differential equations by infinite series and also began the study of the numerical integration of differential equations. In the former volume, Chapter VII is entitled *De Integratione Aequationum Differentialium per Approximationem* (pp. 424–434); in the latter, Chapter XII is *De Aequationum Differentio-Differentialium Integratione per Approximationes* (pp. 271–282). The ideas expressed there are only nascent, in a sense, but quite important, even though Euler did not carry them very far. He considered the differential equation

$$dy/dx = V(x, y), \qquad y(a) = b.$$

Basically, what he noted is that if $x = a + \omega$, then $y \sim b + V(a, b)\omega$. He then imagined this process proceeding as in his table

Ipsius	valores successivi							
x	$a,$	$a',$	$a'',$	$a''',$	$a^{\mathrm{IV}},$	$\ldots$	$'x,$	x
y	$b,$	$b',$	$b'',$	$b''',$	$b^{\mathrm{IV}},$	$\ldots$	$'y,$	y
V	$A,$	$A',$	$A'',$	$A''',$	$A^{\mathrm{IV}},$	$\ldots$	$'V,$	V

where $V(a, b) = A, b' = b + A(a' - a), A' = V(a', b'), \ldots$.

This led him to find more accurate formulas by the use of series expansions. He also gave solutions to differential equations in terms of power series expansions.[27]

[27] Euler 1, XI [1768/69], p. 429.

3.6. Lunar Theory

Euler worked in most branches of pure and applied mathematics (and, indeed, was himself the originator of many), including ballistics, naval architecture, lunar and planetary theories, rational mechanics, hydrodynamics, optics, etc. Hence, it is not possible in a book of this size to attempt to examine what he accomplished in all of these areas. Instead let us merely call attention here to the fact that his pioneering work in lunar theory resulted indirectly in the establishment of the British Nautical Almanac and Ephemeris in 1767. The great lunar astronomer E. W. Brown wrote thus:

> Euler's main contributions to the lunar theory are: — the application of moving rectangular axes; the method of the variation of arbitrary constants, as given in the appendix to his first theory; the use of indeterminate coefficients in the solution of the differential equations; a new method for the determination of the constants from observation; the formation and solution of equations of condition to determine the constants from observation when the number of unknowns is less than the number of equations; the final expression of the coordinates by means of angles of the form $\alpha + \beta t$. He also added to the subject in many other directions, and much of the progress which has since been made, may be said to be founded on his results.[28]

Euler's lunar theory appears in several volumes of his *Opera Omnia, Series Secunda.* In particular, Volume XXIII is entitled *Sol et Luna* I. It contains Euler's first lunar theory.[29]

The three great names of this period in lunar theory are Euler, Alexis Claude Clairaut (1713–1765), an infant prodigy who made important contributions to astronomy, and the remarkable Jean-le-Rond D'Alembert (1717–1783). Carlyle wrote of D'Alembert that he was "of great faculty, especially of great clearness and method; famous in Mathematics; no less so, to the wonder of some, in the intellectual provinces of Literature." All three of these scientists discovered, more or less simultaneously, solutions to the so-called three body problem which were applicable to the study of the moon's motion. Newton had also worked on the problem, before them, but all four calculated the motion of the moon's apogee to be only about half of what it actually was. This caused Clairaut and Euler each to doubt whether the inverse square law was correct. Eventually the difficulty was understood; Clairaut won the prize of the Academy of St. Petersbourg for his essay,

---

[28] Brown [1896], p. 241. For the most part, Brown's work, as improved in certain important respects by W. J. Eckert, is still the standard for calculating lunar ephemerides. (A certain emendation was also made by G. M. Clemence.) The interested reader may wish to consult Brouwer [1961] or *Ephemeris* [1952/59].

[29] Cf. Euler 2, XXIII [1753], pp. 64–336.

"Théorie de la lune." D'Alembert and Euler also published a theory of the moon's motion and tables.

The most important practical application of lunar theory and of the publication of accurate tables of the moon's position was to the navigation of ships. The problem of locating an object on the earth's surface is clearly reducible to giving its longitude and latitude at each instant of time. The latitude is trivially related to the height of the pole star. The longitude is much more complicated since it is essentially related to time. This point is really quite simple: the time at a point on the earth with longitude $\lambda°$ east of Greenwich is the time T at Greenwich plus 4λ minutes. Thus if one had a clock on board a ship which kept exact Greenwich time and observed the time it read at the instant the sun crossed the local meridian, one would know the longitude.

Another solution to the same problem would be to know the moon's position as a function of time, since this also is a very useful device for measuring longitude. An advantage of using the moon for this purpose is that it is visible almost nightly.

Actually the British Government in 1713 offered a substantial prize (£20,000) for a method of locating position to within a half a degree for solving the longitude problem. One solution was the invention of the chronometer by John Harrison, and he eventually received £13,000 for this great boon to navigation. A German astronomer Johann Tobias Mayer (1723–1762), who was professor of mathematics and political economy at Göttingen, adapted Euler's ideas and used accurate observations in the latter's first theory, which enabled him in 1755 to devise tables that James Bradley, third Astronomer Royal, declared to be within half a degree in accuracy. Both Mayer's widow in 1765 and Euler received substantial gifts from the British Government, the former for his practical work and the latter for his theoretical work.

The interested reader may pursue both Euler's first and second theories of the moon in Series 2, Volumes XXII and XXIII. There are very elegant and brief accounts of the two theories by F. Tisserand in Euler 2, XXIII, pp. xxix–lii.[30]

We arbitrarily conclude our discussion of this great mathematician's work by mentioning his investigations of Daniel Bernoulli's scheme for finding the extreme (largest or smallest) roots of an algebraic equation.[31] Euler's contributions appears in his 1, VIII [1748], Caput XVII, pp. 339–361, under the title *De Usu Serierum Recurrentium in Radicibus Aequationum Indagandis.* (We should also note his work on interpolation, which appears in Euler 1, XV [1783], pp. 435–497.)

---

[30] Cf. also, Tisserand [1889], Tome III, 1894, pp. 65–75, for the first theory, and pp. 76–88 for the second, and Brown [1896], pp. 239–241.

[31] Bernoulli, D. [1728].

Consider the problem of solving the equation

$$0 = f(z) = (1 - pz)(1 - qz)(1 - rz)(1 - sz)\ldots,$$

where the roots are all simple. We could write $1/f(z)$ in the form

$$\frac{A}{1-pz} + \frac{B}{1-qz} + \frac{C}{1-rz} + \frac{D}{1-sz} + \cdots,$$

as we well know. Moreover, since each term can easily be expanded in a power series, we may write

$$\frac{1}{f(z)} = \sum_{\alpha=0}^{\infty} (Ap^{\alpha} + Bq^{\alpha} + Cr^{\alpha} + Ds^{\alpha} + \cdots)z^{\alpha} = \sum_{\alpha=0}^{\infty} P_{\alpha} z^{\alpha}.$$

Now Euler observes that if, for example, $|p|$ is the largest of the numbers $|p|, |q|, \ldots$, then for α large $P_{\alpha} \sim Ap^{\alpha}$, $P_{\alpha+1} \sim Ap^{\alpha+1}$ and $P_{\alpha+1}/P_{\alpha} \sim p$.

As a first example of the utility of the method Euler considered the equation $f(x) = x^2 - 3x - 1 = 0$. Since he utilized various results of de Moivre, he preferred to write $x = 1/z$ and to expand

$$\frac{1-z}{1-3z-z^2} = 1 + 2z + 7z^2 + 23z^3 + 76z^4 + 251z^5 + 829z^6 + 2738z^7 + \cdots.$$[32]

Euler then remarks that $2738/829 \sim 3.3027744$ is a good approximation to $(3 + \sqrt{13})/2 \sim 3.3027756$, a root of the original equation. He considers a variety of similar problems but does not relate them to difference equations. In each case he seeks the largest root of an algebraic equation, except in his last example (p. 360), where he seeks the largest root of

$$\frac{1}{2} = z - \frac{1}{6}z^3 + \frac{1}{120}z^5 - \frac{1}{5040}z^7 + \cdots;$$

i.e., of $\sin z = 1/2$. He forms the expression

$$\frac{1}{1 - 2z + \frac{1}{3}z^3 - \frac{1}{60}z^5 + \cdots} = 1 + \left(2z - \frac{1}{3}z^3 + \frac{1}{60}z^5 + \cdots\right)$$
$$+ \left(2z - \frac{1}{3}z^3 + \cdots\right)^2 + (2z - \cdots)^3 + \cdots$$
$$= 1 + 2z + 4z^2 + \frac{23}{3}z^3 + \frac{44}{3}z^4$$
$$+ \frac{1681}{60}z^5 + \frac{2408}{45}z^6 + \cdots,$$

[32] de Moivre [1722], pp. 162–178 and [1730], pp. 4ff. It is fascinating to read on p. 178 of the first reference a remark by de Moivre that while his 1722 paper was in press (*Dum superiores paginae praelo subjiciebantur,* . . .) he accidentally discovered in the *Acta Leips.* 1702/03 that Leibniz had used similar methods for treating algebraic fractions.

and then concludes that the root is approximately

$$z = 1681 \cdot 45/2408 \cdot 60 \sim 0.52356.$$

(Actually 0.523567; moreover 30° ∼ 0.523599 radians; Euler gives 0.523598.)

3.7. Lagrange on Difference Equations

I am not able to find in Euler any suggestion that his method for finding a root of an algebraic equation is intimately related to linear difference equations. It appears that Lagrange was the first to have made the connection. In fact, in an early paper (1759) we see him solving linear difference equations with constant coefficients with the help of the so-called characteristic equation. Moreover, it is here that he introduces his "method of variation of parameters for solving a nonhomogeneous equation given solutions of the homogeneous one."[33]

Lagrange first considers the first-order differential equation (in his notation)

$$dy + yX\,dx = Z\,dx,$$

where X, Z are arbitrary functions of x, and replaces y by $y = uz$. He then finds his original equation replaced by

$$u\,dz + z\,du + uzX\,dx = Z\,dx.$$

But if u is regarded as a solution of the homogeneous equation $du + uX\,dx = 0$, then z must satisfy the relation $u\,dz = Z\,dx$. From these Lagrange concludes that

$$u = e^{-\int X\,dx}, \qquad z = \frac{1}{u}\int Z\,dx$$

and hence that

$$y = uz = e^{-\int X\,dx}\,dx \cdot \int Ze^{\int X\,dx}\,dx.$$

He then turns to difference equations and considers $dy + My = N$ where M, N are now arbitrary functions of x, but dy is now $dy = y(x + dx) - y(x)$. He again replaces y by uz but finds this time that

$$dy = u\,dz + z\,du + du\,dz.$$

He chooses u so that $du + Mu = 0$ and finds, in his notation, that $u = \omega(1 - M)$, where he expresses by $\omega(1 - M)$ the product of all the factors

[33] Lagrange I [1759], pp. 23–36.

$1 - M$.[34] Thus, e.g., $u(x_m) = \prod_{\alpha=1}^{m} [1 - M(x_\alpha)]$. He is then able to express z, with the help of the relation $u\,dz + du\,dz = N$, as

$$dz = \frac{N}{u + du},$$

or in his notation,

$$z = \int \frac{N}{u + du},$$

where his integral now stands for a finite sum.

He illustrates his procedure with the equation $y_1 = Ry + T$ or $y_{m+1} = Ry_m + T$, where $y_1 = y + dy$ and R, T are constant. Here $R = 1 - M$, $T = N$, and Lagrange finds

$$y = \omega R\left(A + \int \frac{T}{\omega R_1}\right)$$

where A is an arbitrary constant and $\omega R_1 = \prod_{\alpha=1}^{m+1} R(x_\alpha)$, $\omega R = \prod_{\alpha=1}^{m} R(x_\alpha)$. Thus

$$y_m = R^m\left(A + \int \frac{T}{R^{m+1}}\right) = R^m\left(A + T\sum_{\alpha=1}^{m} R^{-\alpha}\right) = R^m\left[A + T\frac{R^m - 1}{R^m(R-1)}\right]$$
$$= AR^m + T\frac{R^m - 1}{R - 1}.$$

He now proceeds to re-establish and generalize to finite differences a result of D'Alembert on differential equations. The theorem in reality relates the solutions of linear difference or differential equations with constant coefficients to the roots of the so-called characteristic or indicial equation.

D'Alembert proved that every linear differential equation with constant coefficients

$$y + A\frac{dy}{dx} + B\frac{d^2y}{dx^2} + C\frac{d^3y}{dx^3} + \cdots = X$$

is reducible to a system of equations, each of the form $z + H\,dz/dx = V$, where H is a constant and V, as well as X, is a function of x. (Actually Lagrange does not speak of a system but of an equation. His words are these: "... it is reduced to an equation of the form

$$z + H\frac{dz}{dx} = V,$$

where H is a constant and V a function of x; this equation is similar to the

---

[34] Lagrange I [1759], p. 25.

one we have shown how to integrate in the finite difference case.") Lagrange's proof is this: let

$$\frac{dy}{dx} = p, \qquad \frac{dp}{dx} = q, \qquad \frac{dq}{dx} = r, \ldots \quad .$$

Then, e.g., the original differential equation of the third order is transformed into

$$y + Ap + Bq + C\frac{dq}{dx} = X,$$

which may be written in the form

$$y + (A + a)p + (B + b)q - a\frac{dy}{dx} - b\frac{dp}{dx} + C\frac{dq}{dx} = X,$$

where a, b are as yet undetermined constants. Let us choose them so that, in Lagrange's notation,

$$dy + (A + a)\,dp + (B + b)\,dq = dy + \frac{b}{a}\,dp - \frac{C}{a}\,dq.$$

This means that

$$A + a = \frac{b}{a}, \qquad B + b = -\frac{C}{a},$$

and thus that

$$b = -\frac{C}{a} - B = Aa + a^2, \qquad a^3 + Aa^2 + Ba + C = 0; \tag{3.11}$$

but this is exactly the indicial equation.

Now Lagrange sets

$$z = y + (A + a)p + (B + b)q$$

for the properly chosen values of a, b, and he has

$$z - a\frac{dz}{dx} = X. \tag{3.12}$$

Now suppose a_1, a_2, a_3 are the roots of (3.11) above. Then there are in reality three equations (3.12) whose solutions Lagrange calls Z_1, Z_2, Z_3. Thus

$$y + (A + a_i)p + (B + b_i)q = Z_i \qquad (i = 1, 2, 3),$$

and Lagrange solves this system of linear equations in y, p, q for y in the form $y = FZ_1 + GZ_2 + HZ_3$. He then remarks that exactly the same procedure as that just outlined works equally well for linear difference or differential equations with constant coefficients of any order whatsoever.

It is of interest to note that he did not in this paper enquire into the case of "repeated" roots. It is also worth noting his use of subscripts on the a_1, a_2, Z_1, Z_2, Z_3 but not yet on the coefficients of the equation; also he does not use "&c." but ". . ." instead.[35] Thus by his period the modern notation in differential and integral calculus was almost completely in existence.

In section 8 of his paper Lagrange considers the difference equation $y_1 + Ay_2 + By_3 + Cy_4 + \cdots = X$, where, in his terms, $y_1 = y_2 + dy_2$, $y_2 = y_3 + dy_3$, $y_3 = y_4 + dy_4, \ldots$. He defines $y_2 = p_1$, $y_3 = q_1$, $y_4 = r_1$, $y_5 = s_1$ and so $y_6 = s_2$. This transforms

$$y_1 + Ay_2 + By_3 + Cy_4 + Dy_5 + Ey_6 = X \tag{3.13}$$

into

$$y_1 + Ap_1 + Bq_1 + Cr_1 + Ds_1 + Es_2 = X.$$

But $y_3 = p_2 = q_1$, $y_4 = q_2 = r_1$, $y_5 = r_2 = s_1$, and hence we may rewrite (3.13) as

$$y_1 + (A + a)p_1 + (B + b)q_1 + (C + c)r_1 + (D + d)s_1$$
$$- ay_2 - bp_2 - cq_2 - dr_2 + Es_2 = X.$$ [36]

He then chooses a, b, c, d to satisfy the algebraic equations

$$A + a = \frac{+b}{a}, \quad B + b = \frac{+c}{a}, \quad C + c = \frac{+d}{a}, \quad D + d = \frac{-E}{a}$$

$$a^5 + Aa^4 + Ba^3 + Ca^2 + Da + E = 0,$$

and sets

$$y_l + (A + a)p_l + (B + b)q_l + (C + c)r_l + (D + d)s_l = z_l \quad (l = 1, 2).$$

This gives him for each a the relation $z_1 - az_2 = X$; and, as before, he writes his solution y_m as a linear combination

$$y_m = FZ_1 + GZ_2 + HZ_3 + IZ_4 + KZ_5.$$

If X is a constant, Lagrange remarks that each Z_i is expressible as $Z = La^m + X(a^m - 1)/a^m(a - 1)$. Moreover, for $X = 0$ he notes that his solution is of the form

$$y_m = Fa_1^m + Ga_2^m + Ha_3^m + Ia_4^m + Ka_5^m + \cdots \quad .$$ [37]

---

[35] Lagrange I [1759], p. 28.
[36] Lagrange I [1759], p. 33. Note that he has reversed the usual order of the indices.
[37] Lagrange I [1759], p. 35.

3.8. Lagrange on Functional Equations

Let us leave this topic for now and see what else is of interest for us in Lagrange's work. Closely related is his analysis of the functional equation

$$\alpha\varphi[t + a(h + kt)] + \beta\varphi[t + b(h + kt)] + \gamma\varphi[t + c(h + kt)] + \cdots = T.^{38} \tag{3.14}$$

To do this we must first consider how Lagrange proposed to handle the equation

$$Ly + M\frac{dy}{dt} + N\frac{d^2y}{dt^2} + P\frac{d^3y}{dt^3} + \cdots = T, \tag{3.15}$$

where $L, M, N, \ldots, T$ are functions of t, the system being of order m. He multiplied both members by an as yet undetermined function z and integrated; this gave him

$$\int Lzy\,dt + \int Mz\frac{dy}{dt}\,dt + \int Nz\frac{d^2y}{dt^2}\,dt + \int Pz\frac{d^3y}{dt^3}\,dt + \cdots = \int Tz\,dt.$$

He then integrated each term "by parts" as many times as the order of its derivative. Thus

$$\int Lzy\,dt = \int Lzy\,dt, \qquad \int Mz\frac{dy}{dt}\,dt = Mzy - \int \frac{dMz}{dt}y\,dt,$$

$$\int Nz\frac{d^2y}{dt^2}\,dt = Nz\frac{dy}{dt} - \frac{dNz}{dt}y + \int\frac{d^2Nz}{dt^2}y\,dt.$$

This left him with a new equation of the form

$$\begin{aligned} y\left(Mz - \frac{dNz}{dt} + \frac{d^2Pz}{dt^2} + \cdots\right) \\ + \frac{dy}{dt}\left(Nz - \frac{dPz}{dt} + \cdots\right) + \frac{d^2y}{dt^2}(Pz - \cdots) + \cdots \\ + \int\left(Lz - \frac{dMz}{dt} + \frac{d^2Nz}{dt^2} - \frac{d^3Pz}{dt^3} + \cdots\right)y\,dt = \int Tz\,dt. \end{aligned} \tag{3.16}$$

He then chose z so that the integrand of the last term of the left-hand member was null, i.e.,

$$Lz - \frac{dMz}{dt} + \frac{d^2Nz}{dt^2} - \frac{d^3Pz}{dt^3}\cdots = 0. \tag{3.17}$$

---

[38] Lagrange I [1762], pp. 493–498. The study of this equation is central to Lagrange's work on "Fourier" series. It is in his famous method of "variation of parameters," which he may have evolved from a study of Euler's method for analyzing the perturbations in the orbits of the moon and planets caused by third bodies. It was used by Euler to study the famous "three-body problem" of astronomy. Cf. Euler 2, I [1736/42]. Lagrange hailed this as the first great application of analysis to dynamics.

This "trick" lowered the order of the equation given by (3.16) by one.[39] Lagrange remarked that in the case $T = 0$ a knowledge of m solutions y of (3.15) is precisely what is needed for a complete solution of the problem. He continued from this point in some detail, but it is not relevant to our interests. (In the course of these investigations he shows how to integrate the "équation de Riccati."[40])

Let us take up Lagrange's equation

$$Ay + B(h + kt)\frac{dy}{dt} + C(h + kt)^2\frac{d^2y}{dt^2} + D(h + kt)^3\frac{d^3y}{dt^3} + \cdots = T, \quad (3.18)$$

where A, B, C are constants and the equation is of order m. This is clearly a special case of the previous general problem. He set $z = (h + kt)^r$, and found that the relation (3.17) became in this case

$$P = A - Bk(r + 1) + Ck^2(r + 1)(r + 2) - Dk^3(r + 1)(r + 2)(r + 3)$$
$$+ \cdots \mp Vk^m(r + 1)(r + 2)\cdots(r + m) = 0; \quad (3.19)$$

moreover (3.16) reduced to an equation of the form

$$\alpha y + \beta(h + kt)\frac{dy}{dt} + \gamma(h + kt)^2\frac{d^2y}{dt^2} + \cdots$$
$$= (h + kt)^{-r-1}\int T(h + kt)^r\,dt = \vartheta, \quad (3.20)$$

where the $\alpha, \beta, \gamma, \ldots$ are functions of r whose form is given below in (3.24). Thus (3.20) is in reality a system of equations, one for each of $r_1, r_2, \ldots$, the roots of (3.19):

$$\alpha_i y + \beta_i(h + kt)\frac{dy}{dt} + \gamma_i(h + kt)^2\frac{d^2y}{dt^2} + \cdots = \vartheta_i \qquad (i = 1, 2, \ldots), \quad (3.21)$$

where there are exactly as many equations as there are values $y, y', y'', \ldots$. Thus Lagrange concluded that this system could be solved for y.

To do this he multiplied the ith equation in the system (3.21) by an as yet undetermined quantity M_i $(i = 1, 2, \ldots)$ and summed. (This is not in his original notation.) This gave the relation

$$\left(\sum \alpha_i M_i\right)y + \left(\sum \beta_i M_i\right)(h + kt)y' + \cdots = \sum \vartheta_i M_i.$$

He then chose the $M_1, M_2, \ldots, M_m$ so that

$$\sum \beta_i M_i = 0, \qquad \sum \gamma_i M_i = 0, \ldots \quad . \quad (3.22)$$

---

[39] Lagrange I [1762], pp. 473ff.

[40] The first major problem solved by Daniel Bernoulli was the solution of this equation. This had also been studied by Jacques Bernoulli. Riccati (1676–1754) himself was a member of a Venetian family that produced a number of well-known scientists. His works were published posthumously (Riccati [1758].)

From this he found his solution y to be

$$y = \frac{\sum \vartheta_i M_i}{\sum \alpha_i M_i}. \tag{3.23}$$

This reduced his problem to one of determining the M_i ($i = 1, 2, \ldots, m$) from the system (3.22). It should be remembered that the quantities $\alpha_i, \beta_i, \gamma_i, \ldots$ are functions of the roots of the equation in (3.19). In fact

$$\alpha = B - Ck(r + 2) + Dk^2(r + 2)(r + 3) - \cdots, \qquad \beta = C - Dk(r + 3) + \cdots,$$
$$\gamma = D - \cdots, \cdots, \tag{3.24}$$
$$\vartheta = (h + kt)^{-r-1} \int T(h + kt)^r \, dt.$$

Lagrange showed that the system (3.22) reduced to the very neat form

$$\sum_{i=1}^{m} r_j^i M_i = 0 \qquad (j = 0, 1, \ldots, m - 2), \tag{3.25}$$

and the denominator in (3.23) to the form

$$\pm Vk^{m-1}\left(\sum_{i=1}^{m} r_i^{m-1} M_i\right),$$

where V is the coefficient of $(h + kt)^m \, dy^m/dt^m$ in the equation given by (3.18). (The upper sign obtained for m odd, the lower for m even.) Start with the ultimate of the terms $\alpha, \beta, \gamma, \ldots$. It is clearly independent of r, and thus $\sum M_i = 0$. Then proceed to the penultimate one and find $\sum r_i M_i = 0$. Proceeding inductively, we can verify Lagrange's assertions without undue difficulty.

It is of interest to see how Lagrange handled the system (3.25) of $m - 1$ equations in m unknowns. He first considered the enlarged nonhomogeneous system

$$\sum_{i=1}^{m} r_i^k M_i = R_k \qquad (k = 0, 1, \ldots, m - 1),$$

in which the R_k are arbitrary constants which he soon fixed. He multiplied the kth equation by N_k and summed, obtaining thereby

$$\sum_{i=1}^{m} M_i \sum_{k=0}^{m-1} r_i^k N_k = \sum_{k=0}^{m-1} R_k N_k.$$

Now to find a typical value of M_i, say M_μ, he set the coefficients of all other M_i to null and obtained

$$M_\mu = \frac{\sum_{k=0}^{m-1} R_k N_k}{\sum_{k=0}^{m-1} r_\mu^k N_k} \tag{3.26}$$

together with the system

$$\sum_{k=0}^{m-1} r_i^k N_k = 0 \qquad (i \neq \mu;\ i = 0, 1, \ldots, m-1). \tag{3.27}$$

Recall that $r_1, r_2, \ldots, r_m$ are precisely the roots of (3.19), which he now wrote in the form

$$0 = A - Bk(r+1) + Ck^2(r+1)(r+2) - \cdots = \sum_{p=0}^{m} a_p r^p. \tag{3.28}$$

(Observe that $a_m = \mp Vk^m$.) He then noted that by "la théorie des équations" the polynomials

$$\sum_{k=0}^{m-1} r^k N_k = 0, \qquad \frac{1}{r - r_\mu} \sum_{p=0}^{m} a_p r^p = 0$$

have exactly the same $(m-1)$ roots: $r_1, r_2, \ldots, r_{\mu-1}, r_{\mu+1}, \ldots, r_m$, and therefore

$$\frac{1}{N_0} \sum_{k=0}^{m-1} N_k r^k = \frac{1}{a_0(1 - (r/r_\mu))} \sum_{p=0}^{m} a_p r^p. \tag{3.29}$$

From this he concluded easily that

$$N_q = \frac{N_0}{a_0 r_\mu^q} \sum_{k=0}^{q} a_k r_\mu^k \qquad (q = 1, 2, \ldots, m-1). \tag{3.30}$$

Now, by setting $R_k = 0$ $(k = 0, 1, \ldots, m-2)$, Lagrange found from (3.26) that M_μ was expressible as

$$M_\mu = \frac{N_{m-1} R_{m-1}}{\sum_{k=0}^{m-1} r_\mu^k N_k}. \tag{3.31}$$

If he now substituted the values (3.30) into this relation he would have M_μ as a function of R_{m-1} and r_μ. He did this in an elegant fashion: consider the limiting value of the relation (3.29) when r approaches r_μ. By l'Hôpital's rule (which Lagrange does not mention) he found

$$\frac{1}{N_0} \sum_{k=0}^{m-1} N_k r_\mu^k = -\frac{r_\mu}{a_0} \sum_{p=0}^{m-1} p a_p r_\mu^{p-1}.$$

Moreover, since

$$\sum_{k=0}^{m-1} a_k r_\mu^m = -a_m r_\mu^m,$$

he had by (3.30) $N_{m-1} = -N_0 a_m r_\mu^m / a_0 r^{m-1} = -N_0 a_m r_\mu / a_0$.

He now combined these facts into the relation for M_μ he desired: $M_\mu = uR_{m-1}/(a_1 + 2a_2 r_\mu + 3a_3 r_\mu^2 + \cdots)$; in this expression the $a_1, a_2, \ldots$ are defined by the relation (3.28); R_{m-1} is arbitrary and $u = \mp Vk^m$ ($-$ for m odd, $+$ for m even). This follows at once from (3.31) and the relations that succeed it. Now with the help of the function P defined in (3.19) he was able to write $M_\mu = (-1)^m Vk^m R_{m-1}/Q_\mu$, where $Q_i = dP/dr|^{r=r_i}$. Hence his solution (3.23) was expressible as $y = -k(\vartheta_1/Q_1 + \vartheta_2/Q_2 + \vartheta_3/Q_3 + \cdots)$. This last relation follows from the observation made just after the relations (3.25) that

$$\sum_{\alpha=1}^{m} \alpha_i M_i = (-1)^{m-1} Vk^{m-1}\left(\sum_{i=1}^{m} r_i^{m-1} M_i\right) = (-1)^{m-1} Vk^{m-1} R_{m-1}$$

and what has been said above. I.e.,

$$y = \frac{\sum_{i=1}^{m} \vartheta_i M_i}{\sum_{i=1}^{m} \alpha_i M_i} = -\frac{Vk^m R_{m-1} \sum_{i=1}^{m} \vartheta/Q_i}{Vk^{m-1} R_{m-1}} = -k \sum_{i=1}^{m} \frac{\vartheta_i}{Q_i}.$$

He considered also the case of repeated roots,[41] as well as the important case of constant coefficients, $h = 1$, $k = 0$. He argued that for this case the indicial equation is found as follows: "One supposes that k is infinitely small and r infinitely large so that kr will be equal to a finite quantity ρ; in this way one has

$$P = A - B\rho + C\rho^2 - D\rho^3 + \cdots = 0." \qquad (3.19')$$

Here $\vartheta = e^{-\rho t} \int Te^{\rho t}\, dt$. He then took up the cases of repeated roots as well as imaginary ones.

From our parochial point of view all this is preliminary to his consideration of the functional equation

$$\alpha\varphi[t + a(h + kt)] + \beta\varphi[t + b(h + kt)] + \gamma\varphi[t + c(h + kt)] + \cdots = T.$$

He reduced this to a differential equation of the form (3.18) by expanding $\varphi[t + a(h + kt)]$, $\varphi[t + b(h + kt)], \ldots$ into Taylor's series about t. He set $y = \varphi(t)$ and found that (3.14) reduced to

$$\begin{aligned}&(\alpha + \beta + \gamma + \cdots)y \\ &+ (\alpha a + \beta b + \gamma c + \cdots)(h + kt)\frac{dy}{dt} \\ &+ \frac{1}{2}(\alpha a^2 + \beta b^2 + \gamma c^2 + \cdots)(h + kt)^2 \frac{d^2y}{dt^2} \\ &+ \frac{1}{2\cdot 3}(\alpha a^3 + \beta b^3 + \gamma c^3 + \cdots)(h + kt)^3 \frac{d^2y}{dt^3} \\ &+ \cdots = T. \end{aligned} \qquad (3.32)$$

[41] Lagrange I [1762], pp. 488–490.

This is the same form as (3.18), provided we assume that $y = \varphi$ is a polynomial of degree m. (This Lagrange does not make explicit.) Now a comparison of (3.18) and (3.32) shows that

$$A = \alpha + \beta + \gamma + \cdots, \qquad C = \frac{1}{2}(\alpha a^2 + \beta b^2 + \gamma c^2 + \cdots),$$

$$B = \alpha a + \beta b + \gamma c + \cdots, \qquad D = \frac{1}{2\cdot 3}(\alpha a^2 + \beta b^3 + \gamma c^3 + \cdots), \cdots,$$

hence the indicial equation (3.19) reduces to

$$P = \alpha(1 + ka)^{-r-1} + \beta(1 + kb)^{-r-1} + \gamma(1 + kc)^{-r-1} + \cdots = 0.$$

Lagrange considers this in detail when only α, β are present. Thus $P = \alpha(1 + ka)^{-r-1} + \beta(1 + kb)^{-r-1} = 0$, and hence

$$r + 1 = \frac{\log(-(\alpha/\beta))}{\log((1 + ka)/(1 + kb))}.$$

He writes α/β in the form

$$-\frac{\alpha}{\beta} = \lambda(\cos\omega + \sin\omega\sqrt{-1}),$$

with λ real and positive. He says that one will find for ω an infinitude of different angles. Then

$$\log\left(-\frac{\alpha}{\beta}\right) = \log\lambda + \omega\sqrt{-1}$$

from which he concludes there are an infinite number of values for r. He goes on to assume α/β to be real and positive so that $\lambda = \alpha/\beta$, $\cos\omega = -1$. Then $\omega = (2\nu + 1)\pi$, and he writes

$$r + 1 = \frac{\log(\alpha/\beta) + (2\nu + 1)\pi\sqrt{-1}}{\log((1 + ka)/(1 + kb))},$$

for $\nu = 1, -1, 2, -2, \ldots$. He also treats the cases α/β negative and imaginary. (In none of the cases does he discuss the convergence of the solution function y in (3.32).)

3.9. Lagrange on Fourier Series

Although it is of oblique interest to us it is not possible to pass over in silence a section of Lagrange's "Solution de différents problèmes de calcul intégral."[42] In the following quotation he discusses the question of what

[42] Lagrange I [1762], pp. 514–516.

functions are representable by "Fourier Series": "The question which I need to examine here consists of determining if all the curves which are possible solutions of the vibrating string problem as set out in M. D'Alembert's theory are representable or not by the equation

$$y = \alpha \sin \frac{\pi x}{a} + \beta \sin \frac{2\pi x}{a} + \gamma \sin \frac{3\pi x}{a} + \cdots;$$

this is a question which this great geometer has actively discussed with MM. Bernoulli and Euler in the first mémoir of his *Opuscules mathématiques.*"[43]

Lagrange's discussion is quite elegant as far as it goes, but it does not touch upon the question of convergence. Let us look quickly at his work, which he said would resolve the outstanding "question in a direct and convincing manner." He starts with the shape of the vibrating string in the form $y = [\varphi(x + t) + \varphi(x - t)]/2$ and asks what φ must be like in order that $\varphi(t) + \varphi(-t) = 0$, $\varphi(a + t) + \varphi(a - t) = 0$. (Clearly the string is attached rigidly at the points $x = 0$ and $x = a$.) Thus he is left with the problem of solving a difference equation for φ an odd function

$$\varphi(t + a) - \varphi(t - a) = 0. \tag{3.33}$$

Now this is a very special case of the equation, (3.32), mentioned above where $\alpha = 1$, $\beta = -1$, γ and all following coefficients are zero, and $h = 1$, $k = 0$, $b = -a$. In this case the indicial equation, (3.19′), becomes

$$0 = \lim_{k \to 0} [(1 + ka)^{-1-\rho/k} - (1 - ka)^{-1-\rho/k}] = e^{-\rho a} - e^{+\rho a},$$

and Lagrange shows that this leads to the values

$$\rho_n = \frac{n\pi}{a}\sqrt{-1} \qquad (n = 0, 1, \ldots).$$

Now

$$\varphi_n(t) = e^{-\rho_n t} \int T e^{\rho_n dt}\, dt = c_n e^{-\rho_n t}$$

since $T = 0$ in the present case (3.33), and the indefinite integrals are then constants of integration.

Moreover $Q_n = dP/d\rho|^{\rho=\rho_n} = -ae^{-\rho_n a} - ae^{+\rho_n a}$. Thus, in Lagrange's notation,

$$\varphi(t) = Ae^{(\pi t/a)\sqrt{-1}} + A'e^{-(\pi t/a)\sqrt{-1}} + Be^{(2\pi t/a)\sqrt{-1}} + B'e^{-(2\pi t/a)\sqrt{-1}} + \cdots .$$

He rewrites this in the form

$$\varphi(t) = \alpha \sin \frac{\pi t}{a} + \alpha' \cos \frac{\pi t}{a} + \beta \sin \frac{2\pi t}{a} + \beta' \cos \frac{2\pi t}{a} + \cdots,$$

---

[43] Presumably it was Daniel Bernoulli to whom he referred. (Cajori [1919], p. 242. Cajori gives here a brief discussion of the topic.)

where the $\alpha, \alpha', \beta, \beta', \ldots$ are constants to be determined. Since $\varphi(t) = -\varphi(-t)$, there results an expression for φ of the form

$$\varphi(t) = \alpha \sin \frac{\pi t}{a} + \beta \sin \frac{2\pi t}{a} + \gamma \sin \frac{3\pi t}{a} + \cdots;$$

and consequently the initial shape of the string, he reasons, must be given in the form

$$y = \alpha \sin \frac{\pi x}{a} + \beta \sin \frac{2\pi x}{a} + \gamma \sin \frac{3\pi x}{a} + \cdots \quad .^{44}$$

In a later paper Lagrange made his method of variation of parameters much more explicit and easier to grasp.[45] In this paper (p. 159) Lagrange remarks: "The Marquis de Condorcet and M. de Laplace have previously remarked that this theorem on differential equations is also applicable to difference equations; and the latter has given a general and ingenious demonstration even though a little complicated (see Volume IV of the *Mémoires* de Turin and the *Mémoires* presented to the Academy of Sciences of Paris in 1773). It is this that has attracted me here to treat this matter by a novel and also as simple a method as one could desire."

The difference equation to be considered is

$$Ay_x + By_{x+1} + Cy_{x+2} + \cdots + Ny_{x+n} = 0 \tag{3.34}$$

with $A, B, C, \ldots, N$ constants. Lagrange sought to find as a solution a sequence $y_0, y_1, y_2, y_3, \ldots, y_x, y_{x+1}, y_{x+2}, \ldots$. He remarks that if $y = a\alpha^x$ is substituted into (3.34), there results the equation $A + B\alpha + C\alpha^2 + \cdots + N\alpha^n = 0$. He noted that a was arbitrary and that in general there were n different values of α. He then wrote his solution of (3.34) as $y_x = a\alpha^x + b\beta^x + c\gamma^x + \cdots$; he called this the *complete integral* of the equation in (3.34) *of the nth order*. Next he considered how to append initial conditions to his system (3.34). He supposed given $y_0, y_1, y_2, \ldots, y_{n-1}$, and asked what this would do to the constants $a, b, c, \ldots$. He found the values directly and went on to consider the case of repeated roots of the indicial equation.

Consider now the more complex problem

$$A_x y_x + B_x y_{x+1} + C_x y_{x+2} + \cdots + N_x y_{x+n} = X_x \tag{3.35}$$

where the coefficients now vary with x, and the equation is nonhomogeneous. Suppose with Lagrange that $y = a\alpha_x + b\beta_x + c\gamma_x + \cdots$ is a complete integral for the homogeneous equation given by (3.35); i.e., for $X_x = 0$. Further suppose that (3.35) has a solution of the form

$$y_x = a_x\alpha_x + b_x\beta_x + c_x\gamma_x + \cdots \tag{3.36}$$

---

[44] Lagrange I [1762], p. 516.

[45] Lagrange IV [1775], pp. 151–251.

where the $a_x, b_x, c_x, \ldots$ are functions of x to be determined. To simplify notations let us rewrite (3.35) in the form

$$\sum_{i=0}^{n} A_x^{(i)} y_{x+i} = X_x,$$

and (3.36) in the form

$$y_x = \sum_{i=1}^{n} a_x^{(i)} \alpha_x^{(i)}.$$

Then

$$y_{x+1} = \sum_{i=1}^{n} a_x^{(i)} \alpha_{x+1}^{(i)} + \sum_{i=1}^{n} \alpha_{x+1}^{(i)} \Delta a_x^{(i)}.$$

Lagrange arbitrarily set the second member of the right-hand side of this relation to 0,

$$\sum_{i-1}^{n} \alpha_{x+1}^{(i)} \Delta a_x^{(i)} = 0,$$

and could then write

$$y_{x+2} = \sum_{i=1}^{n} a_x^{(i)} \alpha_{x+2}^{(i)} + \sum_{i=1}^{n} \alpha_{x+2}^{(i)} \Delta a_x^{(i)}.$$

Once again he set the second member of the right-hand side to 0. Thus in general he arrived at the relations

$$y_{x+n} = \sum_{i=1}^{n} a_x^{(i)} \alpha_{x+n}^{(i)} + \sum_{i=1}^{n} \alpha_{x+n}^{(i)} \Delta a_x^{(i)},$$

together with the system of equations

$$\sum_{i=1}^{n} \alpha_{x+j}^{(i)} \Delta a_x^{(i)} = 0 \qquad (j = 1, 2 \ldots n-1),$$

$$\sum \alpha_{x+n}^{(i)} \Delta a_x^{(i)} = X_x / A_x^{(n)},$$

since $\alpha_x^{(i)}$ is a solution of the homogeneous equation $(i = 1, 2, \ldots, n)$. Lagrange now had n equations in unknowns $\Delta a_x^{(i)}$ $(i = 1, 2, \ldots, n)$. He did not tell us how he solved them, but merely remarked that they could be solved. He remarked further that he could eliminate differences by summing, and he wrote the $a_x^{(i)}$ in the form $a_x^{(i)} = \sum P_x^{(i)}$ so that his complete integral of (3.35) was (in his notation)

$$y_x = \alpha_x \sum P_x + \beta_x \sum Q_x + \gamma_x \sum R_x + \cdots \quad .$$

He then concluded that, in general, the equation given by (3.35) is always solvable if one can find n particular solutions of the homogeneous equation.

3.10. Lagrange on Partial Difference Equations

Lagrange then took up the problem of partial difference equations.[46] In Article II he considered two independent variables x and t; in Article III, three independent variables, and in IV, four independent variables. He imagined his equation to be (in his notation) of the form

$$\left.\begin{array}{l} Ay_{x,t} + By_{x+1,t} + Cy_{x+2,t} \quad + \cdots + Ny_{x+n,t} \\ \qquad + B'y_{x,t+1} + C'y_{x+1,t+1} + \cdots + N'y_{x+n-1,t+1} \\ \qquad\qquad + C''y_{x,t+2} \quad + \cdots + N''y_{x+n-2,t+2} \\ \qquad\qquad\qquad + \cdots \\ \qquad\qquad\qquad\qquad + N^{(n)}y_{x,t+n} \end{array}\right\} = 0,$$

where the coefficients are constants. He naturally sought $y_{x,t}$. He first considered the four-term case

$$Ay_{x,t} + By_{x+1,t} + B'y_{x,t+1} + C'y_{x+1,t+1} = 0, \tag{3.37}$$

and set $y_{x,t} = a\alpha^x\beta^t$, where a, α, β were to be determined. He found that $A + B\alpha + B'\beta + C'\alpha\beta = 0$, and so he could find either α or β in terms of the other. Thus he wrote

$$\beta = -\frac{A + B\alpha}{B' + C'\alpha}$$

and hence

$$y_{x,t} = a\alpha^x\left(-\frac{A + B\alpha}{B' + C'\alpha}\right)^t, \tag{3.38}$$

in which a and α are arbitrary. Lagrange then assumed that the expression in (3.38) was representable in the form

$$\left(-\frac{A + B\alpha}{B' + C'\alpha}\right)^t = T\alpha^{\mu t} + T'\alpha^{\mu t-1} + T''\alpha^{\mu t-2} + T'''\alpha^{\mu t-3} + \cdots,$$

where $T, T', T'', \ldots$ may be functions of t. This then yielded

$$y_{x,t} = Ta\alpha^{x+\mu t} + T'a\alpha^{x+\mu t-1} + T''a\alpha^{x+\mu t-2} + \cdots,$$

where a and α are arbitrary constants. Moreover, since there are an infinite number of different sets $(a, \alpha), (b, \beta), (c, \gamma), \ldots$ then $y_{x,t}$ is a linear combination of them. This led Lagrange to state that the general solution was of the form

$$y_{x,t} = Tf(x + \mu t) + T'f(x + \mu t - 1) + T''f(x + \mu t - 2) + \cdots .$$

Moreover, he saw that the form of f depends on the values of $y_{x,t}$ for $t = 0$.

[46] Lagrange IV [1775], pp. 165–215.

Thus for $t = 0$, $T = 1$, $T' = 0$, $T'' = 0, \ldots$ and $y_{x,0} = f(x)$. He concluded that

$$y_{x,t} = Ty_{x+\mu t,0} + T'y_{x+\mu t-1,0} + T''y_{x+\mu t-2,0} + \cdots \quad .$$

In the case where $y_{x,0} = 0$ for $x < 0$, Lagrange's solution assumed the form

$$y_{x,t} = Ty_{x+\mu t,0} + T'y_{x+\mu t-1,0} + T''y_{x+\mu t-2,0} + \cdots + T^{(x+\mu t)}y_{0,0}.$$[47]

But he recognized that, in general, the series was infinite.

To illustrate what is described in the previous paragraph he took the problem

$$y_{x+1,t+1} = y_{x,t+1} + y_{x,t},$$

$$y_{x,0} = 1 \qquad \text{for } x = 0, 1, 2, \ldots, \qquad y_{0,t} = 0 \qquad \text{for } t = 1, 2, 3, \ldots \quad .$$

This is the previous case with $A = 1 = B' = -C'$, $B = 0$. Hence

$$-\frac{A + B\alpha}{B' + C'\alpha} = \frac{1}{\alpha - 1},$$

and β^t is expressible as a power series in $1/\alpha$ as

$$\beta^t = \alpha^{-t} + t\alpha^{-t-1} + \frac{t(t+1)}{2}\alpha^{-t-2} + \frac{t(t+1)(t+2)}{2\cdot 3}\alpha^{-t-3} + \cdots \quad .$$

In this case Lagrange's $\mu = -1$, $T = 1$, $T' = t$, $T'' = t(t+1)/2, \ldots$ and

$$y_{x,t} = y_{x-t,0} + ty_{x-t-1,0} + \frac{t(t+1)}{2}y_{x-t-2,0} + \cdots \quad .$$

But we have the boundary conditions $y_{0,t} = 0$; by the relation above $0 = y_{0,t} = y_{-t,0}$ for $t = 1, 2, 3, \ldots$; and hence

$$y_{x,t} = y_{x-t,0} + ty_{x-t-1,0} + \frac{t(t+1)}{2}y_{x-t-2,0} + \cdots + \frac{t(t+1)\cdots(x-1)}{1\cdot 2\cdots(x-t)}y_{0,0}.$$

This relation, combined with the initial conditions $y_{x,0} = 1$ for $x = 0, 1, \ldots,$ implies that

$$y_{x,t} = 1 + t + \frac{t(t+1)}{2} + \frac{t(t+1)(t+2)}{2\cdot 3} + \cdots + \frac{t(t+1)\cdots(x-1)}{1\cdot 2\cdots(x-t)},$$

and thus $y_{x,t}$ is a binomial coefficient:

$$y_{x,t} = \frac{(t+1)(t+2)(t+3)\cdots x}{1\cdot 2\cdot 3\cdots(x-t)}.$$

Lagrange next discussed a variety of cases such as $C' = 0$ or $B' = 0$ and showed the forms for the solutions. Thus for $C' = 0$ in the equation given by (3.37),

$$-\frac{A + B\alpha}{B'} = p\alpha\left(1 + \frac{q}{\alpha}\right),$$

[47] Lagrange IV [1775], p. 168.

where $p = -B/B'$, $q = A/B$; and Lagrange had

$$\mu = 1, \qquad T = p^t, \qquad T' = tp^tq, \qquad T'' = \frac{t(t-1)}{2}p^tq^2, \ldots .$$

Thus his solution is of the form

$$y_{x,t} = p^t\left[y_{x+t,0} + tqy_{x+t-1,0} + \frac{t(t-1)}{2}q^2y_{x+t-2,0} + \cdots\right].$$

This series clearly terminates for t a positive integer. In case $B' = 0$ he has

$$-\frac{A + B\alpha}{C'\alpha} = p\left(1 + \frac{q}{\alpha}\right),$$

where $p = -B/C'$, $q = A/B$. Hence $\mu = 0$ and

$$y_{x,t} = p^t\left[y_{x,0} + tqy_{x-1,0} + \frac{t(t-1)}{2}q^2y_{x-2,0} + \cdots\right].$$

This series also is finite for t a positive integer, but when $t > x$ then values $y_{x,0}$ with x negative will appear. "Thus it does not suffice in this case to give only the values $y_{0,0}, y_{1,0}, \ldots$; it is apparently necessary also to give values $y_{-1,0}, y_{-2,0}, \ldots$."[48]

However, Lagrange showed how to deduce those values from the form of the solution. This is show he proceeded:

$$\begin{aligned} y_{0,1} &= p(y_{0,0} + qy_{-1,0}), \\ y_{0,2} &= p^2(y_{0,0} + 2qy_{-1,0} + q^2y_{-2,0}), \\ y_{0,3} &= p^3(y_{0,0} + 3qy_{-1,0} + 3q^2y_{-2,0} + q^3y_{-3,0}), \ldots . \end{aligned}$$

Evidently these equations for $y_{-1,0}, y_{-2,0}, \ldots$ may be easily solved and one finds

$$q^sy_{-s,0} = \frac{1}{p^s}y_{0,s} - \frac{s}{p^{s-1}}y_{0,s-1} + \frac{s(s-1)}{2p^{s-2}}y_{0,s-2} - \cdots .$$

In the case where neither B' nor C' is zero, "then it is, in general, impossible to find a finite expression for $y_{x,t}$ by the method of Section 7." In this case he has

$$\beta = -\frac{A + B\alpha}{B' + C'\alpha}.$$

Let $-\omega = B' + C'\alpha$; then $\alpha = -(\omega + B')/C'$, and

$$\beta = -\frac{B}{C'} + \left(A - \frac{BB'}{C'}\right)\frac{1}{\omega}.$$

Hence he obtains

$$\alpha = -\frac{\omega}{C'}\left(1 + \frac{B'}{\omega}\right), \qquad \beta = -\frac{B}{C'}\left[1 + \left(B' - \frac{AC'}{B}\right)\frac{1}{\omega}\right];$$

---

[48] Lagrange IV [1775], p. 172. I have paraphrased Lagrange here.

and he writes $\alpha^x\beta^t$ in the form

$$\alpha^x\beta^t = V\omega^x + V'\omega^{x-1} + V''\omega^{x-2} + V'''\omega^{x-3} + \cdots,$$

which has only a finite number of terms for x, t positive integers. Now since ω is "an undetermined constant", Lagrange concludes as before that

$$y_{x,t} = Vf(x) + V'f(x-1) + V''f(x-2) + V'''f(x-3) + \cdots,$$

where f is arbitrary. This form of the solution has the advantage of being finite, i.e., of containing only a finite number of terms.

There is a great deal more that is interesting in this paper but unfortunately space does not permit us to treat these topics in more depth. The applications of his difference techniques to probability theory are particularly elegant.

3.11. Lagrange on Finite Differences and Interpolation

Instead, let us look at Lagrange's work on the calculus of finite differences and interpolation. (Recall from Section 2.5 that it was Lagrange who rediscovered and appreciated Briggs's work on this subject.) While he was in Berlin he wrote on the topic.[49] This paper is of interest to us because it represents an attempt by Lagrange to make a careful analysis of Briggs's method of repeated extractions of square roots to find logarithms (cf. Chap. I above, pp. 13ff.). In this connection Lagrange, who was clearly very interested in the history of computation, credits Halley with being the first to express the reciprocal relations

$$y = \log(1+z) = z - \frac{1}{2}z^2 + \frac{1}{3}z^3 + \cdots,$$

$$z = e^y - 1 = \log(1+z) + \frac{1}{2!}(\log(1+z))^2 + \frac{1}{3!}(\log(1+z))^3 + \cdots$$

so that, given z, one could find $\log(1+z)$, and given the latter one could find the former.[50]

However, in a later paper, Lagrange takes up the problem of interpolation seriously.[51] In the paper in Volume V he considers a sequence of quantities $T_0, T_1, T_2, T_3, \ldots, T_n, T_{n+1}, T_{n+2}, \ldots$ and their differences, which he writes as $D_1, D_2, D_3, \ldots$ where $D_1 = T_1 - T_0$, $D_2 = T_2 - 2T_1 + T_0$, $D_3 = T_3 - 3T_2 + 3T_1 - T_0, \ldots$. He expresses these in general as

$$D_m = (T_1 - T_0)_m = T_m - mT_{m-1} + \frac{m(m-1)}{2}T_{m-2} + \frac{m(m-1)(m-2)}{2\cdot 3}T_{m-3} + \cdots \quad . \tag{3.39}$$

[49] Lagrange V [1783], pp. 517–532.

[50] Halley [1695]. Cf. also, Chap. II, pp. 59ff.

[51] Lagrange V [1792], pp. 663–684 and III [1772], pp. 441–476.

Incidentally, he remarks that this last relation "can be proved by induction." He defines $D_0 = T_0$ and notes that if m is negative, differences change to sums; and he then defines a sequence of sums $S_1, S_2, S_3, \ldots$ and, using the relation (3.39), defines $D_{-m} = S_m$, i.e.,

$$S_m = T_{-m} + mT_{-m-1} + \frac{m(m+1)}{2} T_{-m-2} + \cdots,$$

which he says is a known result.

Since $T_1 = D_0 + D_1$, he has

$$T_n = (D_0 + D_1)_n = D_0 + nD_1 + \frac{n(n-1)}{2} D_2 + \frac{n(n-1)(n-2)}{2\cdot 3} D_3 + \cdots, \tag{3.40}$$

which, he notes, is Newton's interpolation formula. (The derivation is given shortly.) He remarks that for n integral one can construct the formula by successive additions, but not for n fractional.

He now develops an elegant operational technique for manipulating differences. He first notes that $T_1 = D_0 + D_1 = D_0(1 + D_1)$, since

$$D_0 \times D_1 = D_{0+1} = D_1 \tag{3.41}$$

"by observing for the indices the laws of exponents."[52] Formalistically, he next writes

$$1 + D_1 = e^{\log(1+D_1)},$$

and remarks that this yields the relation $T_1 = D_0 \times e^{\log(1+D_1)}$; then he "raises both members to the power n," and finds that

$$T_n = D_0 \times e^{n\log(1+D_1)}.$$

This follows formally since $(D_0)_n = D_{0\times n} = D_0$.

Next he expands the exponential into a Taylor's series and has the relation

$$T_n = D_0 + n\log(1 + D_1) + \frac{n^2}{2}[\log(1 + D_1)]^2 + \frac{n^3}{2\cdot 3}[\log(1 + D_1)^3 + \cdots, \tag{3.42}$$

since in this formalism

$$D_0 \times [\log(1 + D_1)]^m = [\log(1 + D_1)]^m.$$

The reason for this may be seen with the help of the relation (3.41). In fact, we have $D_0 \times \log(1 + D_1) = D_0 \times (D_1 - \frac{1}{2}D_1^2 + \cdots) = D_1 - \frac{1}{2}D_2 + \cdots = \log(1 + D_1)$, with quite similar expressions for other powers of the

---

[52] Lagrange V [1792], p. 670. Evidently what he has done is to note that by (3.39), $D_0 = (T_1 - T_0)_0$, $D_1 = (T_1 - T_0)_1$ and hence that $(T_1 - T_0)_0 \times (T_1 - T_0)_1 = (T_1 - T_0)_1 = D_1$.

logarithm. If one expands the logarithms into Taylor's series, one has

$$\begin{aligned} T_n = D_0 &+ n\left(D_1 - \frac{1}{2}D_2 + \frac{1}{3}D_3 - \cdots\right) \\ &+ \frac{n^2}{2}\left(D_1 - \frac{1}{2}D_2 + \frac{1}{3}D_3 - \cdots\right)^2 \\ &+ \frac{n^3}{2\cdot 3}\left(D_1 - \frac{1}{2}D_2 + \frac{1}{3}D_3 - \cdots\right)^3 \\ &+ \cdots, \end{aligned}$$

which is essentially the relation (3.40).

Lagrange returns to the relation (3.42) and expresses it as

$$T_n = P_0 + nP_1 + \frac{n^2}{2}P_2 + \frac{n^3}{2\cdot 3}P_3 + \cdots \qquad (n = 0, 1, 2, 3, \ldots),$$

where

$$P_0 = D_0 = T_0, \qquad P_s = [\log(1 + D_1)]^s.$$

From these he finds that

$$P_1 = \log(1 + D_1), \qquad 1 + D_1 = e^{P_1} = 1 + P_1 + \frac{1}{2}P_2 + \frac{1}{2\cdot 3}P_3 + \cdots$$

and hence that

$$D_s = \left(P_1 + \frac{1}{2}P_2 + \frac{1}{2\cdot 3}P_3 + \frac{1}{2\cdot 3\cdot 4}P_4 + \cdots\right)^s.$$

This gives a way of calculating any order difference given the sequence $P_1, P_2, \ldots$. Consider now two sequences $T_0, T_1, T_2, \ldots, t_0, t_1, t_2, \ldots$ together with their corresponding difference sequences $D_0, D_1, D_2, \ldots, d_0, d_1, d_2, \ldots$. Moreover, suppose that $T_s = t_{sm}$ ($m = 2, 3, \ldots; s = 0, 1, \ldots$); thus, e.g., for $m = 3$ he has $T_0 = t_0$, $T_1 = t_3$, $T_2 = t_6$, $T_3 = t_9, \ldots$. His problem is now to relate the $D_0, D_1, D_2, \ldots$ and the $d_0, d_1, d_2, \ldots$. "By extracting the sth root," Lagrange has $T_1 = t_m$; or, in other words, $D_0 + D_1 = (d_0 + d_1)_m$. Moreover, since $T_0 = t_0$, we have $D_0 = d_0$, and this relation becomes $D_1 = (d_0 + d_1)_m - d_0$, or, in general,

$$D_s = \left[md_1 + \frac{m(m-1)}{2}d_2 + \frac{m(m-1)(m-2)}{2\cdot 3}d_3 + \cdots\right]_s \qquad (3.43)$$

Thus Lagrange was able to write

$$D_s = Ad_s + Bd_{s+1} + Cd_{s+2} + Ed_{s+3} + \cdots,$$

and he had

$$A = m^s, \quad B = \frac{s}{2}(m-1)A, \quad C = \frac{2s}{2}\frac{(m-1)(m-2)}{2\cdot 3}A + \frac{s-1}{2}\frac{m-1}{2}B,$$

$$E = \frac{3s}{3}\frac{(m-1)(m-2)(m-3)}{2\cdot 3\cdot 4}A + \frac{2s-1}{3}\frac{(m-1)(m-2)}{2\cdot 3}B$$

$$+ \frac{s-2}{3}\frac{m-1}{2}C, \ldots \quad .$$

Making use of another formalism,

$$d_0 + d_1 = (D_0 + D_1)_{1/m},$$

Lagrange writes

$$\begin{aligned} d_s &= [(D_0 + D_1)_{1/m} - D_0]_s \\ &= aD_s + bD_{s+1} + cD_{s+2} + eD_{s+3} + \cdots, \end{aligned} \tag{3.44}$$

with

$$a = \frac{1}{m^s}, \qquad b = s\frac{1-m}{2m}a,$$

$$c = \frac{2s}{2}\frac{(1-m)(1-2m)}{2\cdot 3m^2}a + \frac{s-1}{2}\frac{1-m}{2m}b, \ldots \quad .$$

He notes that these relations solve what he calls the "Problème de Mouton."[53]

If D_r is the last nonzero difference in the sequence $T_0, T_1, \ldots,$ then

$$d_r = aD_r \quad \text{and} \quad a = \frac{1}{m^r},$$

"which agrees with what Mouton found by induction." If $s = r - 1$,

$$d_{r-1} = aD_{r-1} + bD_r, \quad \text{and} \quad a = \frac{1}{m^{r-1}}, \qquad b = \frac{(r-1)(1-m)}{2m^r}.$$

Lagrange also gives the corresponding result for $s = r - 2$: $d_{r-2} = aD_{r-2} + bD_{r-1} + cD_r$.

Let us turn now with Lagrange to the application of his formula (3.43) to find the relations between derivatives and differences. Let us, like him, designate by Dy and Dx differences, and by dy and dx differentials. Then in (3.43) he takes $m = Dx/dx$ and views this as "infinitely large." This gives him the new formula

$$D^s y = \left(\frac{Dx\,dy}{dx} + \frac{Dx^2 d^2 y}{2\,dx^2} + \frac{Dx^3 d^3 y}{2\cdot 3\,dx^3} + \cdots\right)^s, \tag{3.45}$$

since terms such as $m(m-1)$ approach m^2 for m large. Thus Lagrange found

[53] Mouton [1670]. I.e., the subtabulation problem. Cf., also p. 201 below.

the result sometimes written symbolically [we will discuss this symbolism of Lagrange in (3.46) and later] as

$$D^s y = (e^{Dx \cdot dy/dx} - 1)^s.$$

In his relation (3.45) expressions such as $(d^m y)^n$ are understood to be $d^{mn}y$. With the help of this relation Lagrange now had the differences of all orders of a function in terms of its derivatives. If s is negative, he pointed out that this changed differences to sums and derivatives to integrals.

Conversely, by interchanging D and d, m and $1/m$, Lagrange found the corresponding formula expressing the derivatives of y in terms of its differences. Thus

$$\frac{d^s y}{dx^s} = \frac{1}{Dx^s}\left(Dy - \frac{1}{2}D^2 y + \frac{1}{3}D^3 y - \cdots\right)^s, \tag{3.46}$$

where the right hand member is to be expanded in powers of s. In symbolic terms this may be written as

$$\frac{d^s y}{dx^s} = \left[\frac{1}{Dx}\log(1 + Dy)\right]^s.$$

Again if s is negative, the relation (3.46) still remains valid when properly interpreted.

The simplest way to establish (3.45) precisely is to express y by a Taylor's series and to difference both members. Thus

$$\Delta^s y(x) = \sum_{n=s}^{\infty} \frac{y^{(n)}(a)}{n!}\Delta^s (x - a)^n,$$

and so

$$\Delta^s y(a) = \sum_{n=s}^{\infty} \frac{y^{(n)}(a)}{n!}\Delta^s x^n|^{x=0} = \sum_{n=s}^{\infty} \frac{y^{(n)}(a)}{n!} s!\, \mathfrak{S}_n^s \cdot (\Delta x)^n,$$

as we see from the definition (2.27) of Stirling's numbers of the second kind. This then is the precise form of (3.45).

To derive Lagrange's relation (3.46) note with the help of the definition (2.27′) of Stirling's numbers of the first kind and of a Newton's series expansion of y that

$$\frac{d^s y(a)}{dx^s} = \sum_{n=s}^{\infty} \frac{\Delta^n y(a)}{\Delta x^n}\frac{d^s}{dx^s}\binom{x}{n}_{\Delta x}\Bigg|^{x=0} = \sum_{n=s}^{\infty} \frac{s!}{n!}\frac{\Delta^n y(a)}{\Delta x^n} S_n^s.$$

In terms of these relations Lagrange's symbolic formulas become quite straightforward. But it is also of interest to examine (3.45) and (3.46) by symbolic means. Recall the operator E^h we discussed in connection with

Briggs's work (p. 27). It is defined so that $E^h f(x) = f(x + h)$. Then we shall show that

$$(e^{hx} - 1)^s = x^s + \frac{x^{s+1}}{(s+1)!} \underset{h}{\Delta}^s x^{s+1}\big|^{x=0} + \frac{x^{s+2}}{(s+2)!} \underset{h}{\Delta}^s x^{s+2}\big|^{x=0} + \cdots \quad (3.47)$$

[54]

The proof is simple. Let us consider a term in the binomial expansion of the left-hand member, $e^{h\alpha x}$. We have

$$e^{(\xi + h\alpha)x} = 1 + x(\xi + h\alpha) + \frac{x^2}{2!}(\xi + h\alpha)^2 + \cdots$$

$$= 1 + xE^{h\alpha}\xi + \frac{x^2}{2!}E^{h\alpha}\xi^2 + \cdots,$$

and hence

$$e^{h\alpha x} = 1 + xE^{h\alpha}\xi\big|^{\xi=0} + \frac{x^2}{2!}E^{h\alpha}\xi^2\big|^{\xi=0} + \cdots .$$

Thus

$$h^{-s}(e^{hx} - 1)^s = h^{-s}\sum_{\alpha=0}^{s}(-1)^{\alpha+s}\binom{s}{\alpha}e^{h\alpha x} = \sum_{n=0}^{\infty}\frac{x^n}{n!}h^{-s}(E^h - I)^s\xi^n\big|^{\xi=0}.$$

But, by definition, formally,

$$(E^h - I)^s = h^s \underset{h}{\Delta}^s$$

and hence

$$h^{-s}(e^{hx} - 1)^s = \sum_{n=0}^{\infty}\frac{x^n}{n!}\underset{h}{\Delta}^s\xi^n\big|^{\xi=0} = \sum_{n=s}^{\infty}\frac{x^n}{n!}\underset{h}{\Delta}^s\xi^n\big|^{\xi=0}. \quad (3.48)$$

From this the relation (3.45) and the one immediately after it follow by substituting the operator

$$Dx \cdot \frac{d}{dx}$$

for $h \cdot x$.

Lagrange's paper in Volume III of his Works (Lagrange III [1772]), contains the genesis of his operator methods. He mentions in the Preface (p. 441*n*) Leibniz's observations on the analogy between derivatives of all orders of a product of two or more variables and the powers of the same orders of binomials or polynomials made up of the sum of these same variables. (For the Leibniz paper see Leibniz XXVII, pp. 377–382.) Lagrange also says that Leibniz remarked on the same analogy obtaining between negative powers

---

[54] This is a special case of what is known as Herschel's theorem. (Milne–Thomson [1933], p. 32. Cf. also, Boole [1880].)

and integrals in Leibniz XVII. He also remarks on a contribution to the field by John Bernoulli in a letter to Leibniz (Bernoulli, Jo. [1695], pp. 195–205).

Lagrange considers a function $u(x, y)$ and expands $u(x + \xi, y + \psi)$ in a Taylor's expansion

$$\begin{aligned} &u + p\xi + q\psi \\ &\quad + p'\xi^2 + q'\xi\psi + r'\psi^2 \\ &\quad + p''\xi^3 + q''\xi^2\psi + r''\xi\psi^2 + s''\psi^3, \ldots, \end{aligned}$$

where the $p, p', \ldots, q, q', \ldots, r', r'', \ldots, s'', \ldots$ are partial derivatives of u. He then says that the differential calculus consists of finding directly from u the partial derivatives $p, p', p'', \ldots, q, q', q'', \ldots, r', r'', \ldots$ while the integral calculus consists of finding u with the help of these latter functions. He then points out that the coefficient of $x^\mu y^\nu z^\omega t^\rho, \ldots$ in the expansion of

$$(x + y + z + t + \cdots)^{\mu+\nu+\omega+\rho+\cdots}$$

is given by

$$\frac{1\cdot 2\cdot 3\cdot 4\cdot 5 \cdots (\mu + \nu + \omega + \rho + \cdots)}{1\cdot 2\cdot 3\cdots \mu \times 1\cdot 2\cdot 3\cdots \nu \times 1\cdot 2\cdot 3\cdots \omega \times 1\cdot 2\cdot 3\cdots \rho \times \cdots}.$$

He relates this to the derivative $d^{\mu+\nu+\omega+\rho+\cdots}u/dx^\mu\, dy^\nu\, dz^\omega\, dt^\rho \cdots$. This enables him to evaluate

$$\begin{aligned} &\frac{x + y + z + t + \cdots}{1} + \frac{(x + y + z + t + \cdots)^2}{1\cdot 2} \\ &\quad + \frac{(x + y + z + t + \cdots)^3}{1\cdot 2\cdot 3} + \cdots \\ &= e^{x+y+z+t+\cdots} - 1 \\ &= e^x \times e^y \times e^z \times e^t \times \cdots - 1 \\ &= \left(1 + \frac{x}{1} + \frac{x^2}{1\cdot 2} + \frac{x^3}{1\cdot 2\cdot 3} + \cdots\right) \times \left(1 + \frac{y}{1} + \frac{y^2}{1\cdot 2} + \frac{y^3}{1\cdot 2\cdot 3} + \cdots\right) \\ &\quad \times \left(1 + \frac{z}{1} + \frac{z^2}{1\cdot 2} + \frac{z^3}{1\cdot 2\cdot 3} + \cdots\right) \\ &\quad \times \left(1 + \frac{t}{1} + \frac{t^2}{1\cdot 2} + \frac{t^3}{1\cdot 2\cdot 3} + \cdots\right) \cdots - 1. \end{aligned}$$

He then considers

$$\Delta u = \exp\left(\frac{du}{dx}\xi + \frac{du}{dy}\psi + \frac{du}{dz}\zeta + \cdots\right) - 1,$$

which he interprets as a series in powers of du, where he replaces expressions such as du^λ by $d^\lambda u$. If $x, y, z, \ldots$ are increased to $x + \xi, y + \eta, z + \zeta, \ldots$

then Δu is exactly the expression above and

$$\Delta^\lambda u = \left(\exp\left(\frac{du}{dx}\xi + \frac{du}{dy}\psi + \frac{du}{dz}\zeta + \cdots\right) - 1\right)^\lambda.$$

Next he says that λ can be negative as well as positive, in which case

$$d^{-1} = \int, \quad d^{-2} = \int^2, \ldots, \quad \Delta^{-1} = \sum, \quad \Delta^{-2} = \sum^2, \ldots,$$

and the preceding relation becomes

$$\sum^\lambda u = \frac{1}{(\exp(du/dx)\xi + (du/dy)\psi + (du/dz)\zeta + \cdots) - 1)^\lambda}.$$

He then discusses the case of one variable in more detail by setting $du/dy = 0$, $du/dz = 0, \ldots$. Consider now

$$(e^\omega - 1)^\lambda = \omega^\lambda(1 + A\omega + B\omega^2 + C\omega^3 + D\omega^4 + \cdots),$$

and take the logarithmic derivative of both sides. There results

$$\lambda\left(\frac{e^\omega}{e^\omega - 1} - \frac{1}{\omega}\right) = \frac{A + 2B\omega + 3C\omega^2 + 4D\omega^4 + \cdots}{1 + A\omega + B\omega^2 + C\omega^3 + D\omega^4 + \cdots};$$

but

$$\frac{e^\omega}{e^\omega - 1} = \frac{1}{1 - e^{-\omega}} = \frac{1}{\omega - (\omega^2/2) + (\omega^3/2\cdot 3) - (\omega^4/2\cdot 3\cdot 4) + \cdots}.$$

When this is substituted and coefficients compared, Lagrange finds

$$A = \frac{\lambda}{2},$$

$$2B = \frac{(\lambda + 1)A}{2} - \frac{\lambda}{2\cdot 3},$$

$$3C = \frac{(\lambda + 2)B}{2} - \frac{(\lambda + 1)A}{2\cdot 3} + \frac{\lambda}{2\cdot 3\cdot 4},$$

$$4D = \frac{(\lambda + 3)C}{2} - \frac{(\lambda + 2)B}{2\cdot 3} + \frac{(\lambda + 1)A}{2\cdot 3\cdot 4} - \frac{\lambda}{2\cdot 3\cdot 4\cdot 5}, \ldots,$$

and hence

$$\Delta^\lambda u = \frac{d^\lambda u}{dx^\lambda}\xi^\lambda + A\frac{d^{\lambda+1}u}{dx^{\lambda+1}}\xi^{\lambda+1} + B\frac{d^{\lambda+2}u}{dx^{\lambda+2}}\xi^{\lambda+2} + C\frac{d^{\lambda+3}u}{dx^{\lambda+3}}\xi^{\lambda+3} + \cdots \quad .$$

If λ is negative, replace λ by $-\lambda$ and we have

$$\sum^\lambda u = \frac{\int^\lambda u\,dx^\lambda}{\xi^\lambda} - \alpha\frac{\int^{\lambda-1} u\,dx^{\lambda-1}}{\xi^{\lambda-1}} + \beta\frac{\int^{\lambda-2} u\,dx^{\lambda-2}}{\xi^{\lambda-2}} - \gamma\frac{\int^{\lambda-3} u\,dx^{\lambda-3}}{\xi^{\lambda-3}} + \cdots,$$

where

$$\alpha = \frac{\lambda}{2}, \quad 2\beta = \frac{(\lambda - 1)\alpha}{2} + \frac{\lambda}{2\cdot 3}, \quad 3\gamma = \frac{(\lambda - 2)\beta}{2} + \frac{(\lambda - 1)\alpha}{2\cdot 3} + \frac{\lambda}{2\cdot 3\cdot 4}, \ldots .$$

Now if $\lambda = 1$, this result gives us the relation

$$\sum u = \frac{\int u\,dx}{\xi} - \alpha u + \beta \frac{du}{dx}\xi - \gamma \frac{d^2u}{dx^2}\xi^2 + \delta \frac{d^3u}{dx^3}\xi^3 - \cdots,$$

since

$$\int^{-1} u\,dx^{-1} = \frac{du}{dx}, \qquad \int^{-2} u\,dx^{-2} = \frac{d^2u}{dx^2}, \ldots .$$

(This is the Euler–Maclaurin formula.) Lagrange remarks that this is a known result: If $u = \varphi(x)$,

$$\varphi(x - \xi) + \varphi(x - 2\xi) + \varphi(x - 3\xi) + \cdots$$
$$= \frac{`\varphi(x)}{\xi} - \alpha\varphi(x) + \beta\varphi'(x)\cdot\xi - \gamma\varphi''(x)\cdot\xi^2 + \cdots,$$

where he expresses the integral of φ by $`\varphi(x)$. He goes on to generalize this, but says correctly it has already been treated by Maclaurin in his *Traité des Fluxions* and by Euler in his *Institutions du Calcul différential.* To generalize their summation formula Lagrange then writes

$$\frac{du}{dx}\xi + \frac{du}{dy}\psi + \frac{du}{dz}\zeta + \cdots = \log(1 + \Delta u)$$
$$= \Delta u - \frac{\Delta^2 u}{2} + \frac{\Delta^3 u}{3} - \frac{\Delta^4 u}{4} + \cdots,$$

and therefore finds when $du/dy = 0$, $du/dz = 0, \ldots$ that

$$\frac{d^\lambda u}{dx^\lambda}\xi^\lambda = \Delta^\lambda u + M\Delta^{\lambda+1}u + N\Delta^{\lambda+2}u + P\Delta^{\lambda+3}u + \cdots,$$

and

$$\frac{\int^\lambda u\,dx^\lambda}{\xi^\lambda} = \sum^\lambda u + \mu \sum^{\lambda-1} u + \nu \sum^{\lambda-2} u + \varpi \sum^{\lambda-3} u + \cdots,$$

where the parameters are given as

$$M = -\frac{\lambda}{2},$$

$$\mu = \frac{\lambda}{2},$$

$$2N = -\frac{(\lambda + 1)M}{2} + \frac{\lambda}{2\cdot 3},$$

$$2\nu = \frac{(\lambda - 1)\mu}{2} - \frac{\lambda}{2\cdot 3},$$

$$3P = -\frac{(\lambda + 2)N}{2} + \frac{(\lambda + 1)M}{2\cdot 3} - \frac{\lambda}{3\cdot 4},$$

$$3\varpi = \frac{(\lambda - 2)\nu}{2} - \frac{(\lambda - 1)\mu}{2\cdot 3} + \frac{\lambda}{3\cdot 4},$$

$$4Q = -\frac{(\lambda + 3)P}{2} + \frac{(\lambda + 2)N}{2\cdot 3} - \frac{(\lambda + 1)M}{3\cdot 4} + \frac{\lambda}{4\cdot 5},$$

$$4\chi = \frac{(\lambda - 3)\varpi}{2} - \frac{(\lambda - 2)\nu}{2\cdot 3} + \frac{(\lambda - 1)\mu}{3\cdot 4} - \frac{\lambda}{4\cdot 5}, \text{ etc.}$$

Hence for $\lambda = 1$ Lagrange unknowingly refound Gregory's formula

$$\frac{\int u\,dx}{\xi} = \sum u + \mu u + \nu\Delta u + \varpi\Delta^2 u + \chi\Delta^3 u + \cdots .$$

Lagrange points out that Cotes, Stirling and others had found a formula for computing areas from sums and differences of equidistant coordinates, but that the formula above is different and in his opinion preferable, since it is couched in terms of successive differences which ordinarily diminish. By way of illustration Lagrange sets $u = 1/x$ and finds

$$\log x = \sum \frac{1}{x} + \frac{\mu}{x} + \nu\Delta\frac{1}{x} + \varpi\Delta^2\frac{1}{x} + \chi\Delta^3\frac{1}{x} + \cdots .$$

But

$$\sum \frac{1}{x} = \frac{1}{x-1} + \frac{1}{x-2} + \frac{1}{x-3} + \cdots,$$

$$\Delta\frac{1}{x} = \frac{1}{x+1} - \frac{1}{x} = -\frac{1}{x(x+1)}, \ldots,$$

$$\Delta^\lambda \frac{1}{x} = \pm\frac{1\cdot 2\cdot 3\cdots\lambda}{x(x+1)(x+2)\cdots(x+\lambda)},$$

and thus

$$\log x = \frac{1}{x-1} + \frac{1}{x-2} + \frac{1}{x-3} + \cdots$$
$$+ \frac{\mu}{x} - \frac{\nu}{x(x+1)} + \frac{2\varpi}{x(x+1)(x+2)}$$
$$- \frac{2\cdot 3\cdot\chi}{x(x+1)(x+2)(x+3)} + \cdots .$$

Out of this he deduces interesting results for $\log x/(x - 1)$ and $\log(1 + 1/x)$. These are discussed in Chapter IV (see p. 202).

He then carries out the same analysis for the case of several variables and takes up a few other topics.

Although we could discuss the work of Lagrange in much greater detail, we will close this discussion with a reference to his well-known interpolation formula, which he attributes to Newton. Lagrange discovered this interpolation formula in 1794/5.[55] The proof is straightforward and does not warrant repeating here.

3.12. Lagrange on Hidden Periodicities

Lagrange clearly understood the importance of "Fourier series" to astronomy. He remarks, e.g., that the so-called Equation of center, which he writes as

$$-2\varepsilon \sin \varphi + \frac{5\varepsilon^2}{4} \sin 2\varphi + \frac{\varepsilon^3}{4} \sin \varphi - \frac{13}{12} \varepsilon^3 \sin 3\varphi + \cdots,$$

is of such form. (Here ε is the eccentricity and φ is the so-called mean anomaly. Note that this is the negative of the present "equation".) Moreover, he shows how the epicyclical technique of the ancients will lead to this Equation of center by the use of a sequence of epicycles upon epicycles. This was perhaps the first modern mathematical demonstration of the essential reasonableness of Ptolemy's procedure. This was probably due to the fact that it was Lagrange who found the "Fourier series" expansion for the Equation of center.[56]

He concerned himself seriously with the use of trigonometrical series to analyze numerical data and was, as far as I can discover, the first to write on the subject. He wrote two very important papers on this topic; one in 1772 and the other in 1778. In a sense the second was a refinement and improvement on the first one. We will therefore mention only a few details from the former.[57]

In both his first and second paper on trigonometric analysis of data Lagrange considers the fundamental problem of how to find "hidden" periodicities in a table of data. Thus as an illustration of his methods he takes Tobias Mayer's *Tabulae Solares, etc.* (p. III) and extracts Mayer's so-called Equation of time. This table is quite irregular in character:

$$\begin{array}{llllllllll} 456, & -168, & 274, & -933, & 220, & 631, & -232, & 349, & -823, & -72, \\ 772, & -237, & 358, & -657, & -360, & 860, & -181, & 305, & -457, & -616, \ldots . \end{array}$$

These entries correspond to a constant interval of 70° of true solar longitude and are recorded in seconds of arc. (Recall that the Equation of time is the

[55] Lagrange VII [1795], pp. 284–287. (In the German translation by Niedermüller the theorem is on pp. 112–116.)

[56] Lagrange III [1771], p. 113.

[57] Lagrange VI [1772], pp. 509–627.

right ascension of the mean sun less the right ascension of the true sun.) Lagrange then pretends that he is ignorant of the analytical form of the equation and deduces it from the tabular data. He finds by his numerical work that in fact it is approximately expressible as

$$451'' \sin (60°;28 + m\cdot 69°;49) + 591'' \sin (140°;7 + m\cdot 140°;1).$$[58]

By theory he calculates that it is of the form

$$462'' \sin (61° + m\cdot 70°) + 593'' \sin (140° + m\cdot 140°),$$

plus some higher order terms of very small amplitude. In explanation we note that the first two terms of the Equation of time are $-462 \sin(\varphi - \alpha) - 593 \sin 2\varphi$ where α, the longitude of apogee, is 99° and φ is the true longitude of the sun. Now the entry 772 in Mayer's table corresponds to 340°. Thus $\varphi = 340° + m\cdot 70°$ and $\varphi - \alpha = 241° + m\cdot 70°$; hence $\sin(\varphi - \alpha) = -\sin(61° + m\cdot 70°)$. Moreover $2\varphi = 680° + m\cdot 140°$ and

$$\sin 2\varphi = -\sin(140° + m\cdot 140°).$$

Lagrange develops, by means of continued fractions, a very detailed and somewhat involved method for seeking hidden periodicities in recurrent series in his first paper. His method in the second paper is rather more direct.[59] He said: "I have decided some day to apply the method of this memoir to examine the law of errors in Halley's Tables for the oppositions of Saturn and of Jupiter."[60]

He considers his data to be representable in the form

$$A_x = a \sin(\alpha + x\varphi) + b \sin(\beta + x\theta) + c \sin(\gamma + x\psi) + \cdots \tag{3.49}$$

where he writes

$$\cdots, {}_3A, {}_2A, {}_1A, A, A_1, A_2, A_3, \ldots \tag{3.50}$$

as the values corresponding to $x = \cdots, -3, -2, -1, 0, 1, 2, 3, \ldots$. His problem is now this: given a sequence (3.50) to find amplitudes $a, b, c, \ldots$ and angles $\alpha, \beta, \gamma, \ldots, \varphi, \psi, \ldots$ so that (3.49) furnishes a "good fit" to the data (3.50). To start he assumes for simplicity that (3.49) contains only one term, and so

$$\begin{gathered} {}_3A = a \sin(\alpha - 3\varphi), \quad {}_2A = a \sin(\alpha - 2\varphi), \quad {}_1A = a \sin(\alpha - \varphi), \\ A = a \sin \alpha, \quad A_1 = a \sin(\alpha + \varphi), \\ A_2 = a \sin(\alpha + 2\varphi), \quad A_3 = a \sin(\alpha + 3\varphi), \ldots \end{gathered}$$

---

[58] Lagrange VI [1772], pp. 604–606. Cf. also, Mayer [1770]. The use of these tables is explained on pp. 97–99 of that work.

[59] Lagrange VII [1778], pp. 541–553.

[60] Lagrange VII [1778], p. 553. He also said on p. 547 that his method was free of the theory of continued fractions since that subject is not always familiar to astronomers.

These he differences and finds, in his notation,

$$\ldots, {}_3B = 2a \sin\frac{\varphi}{2}\cos\left[\alpha - \left(2 + \frac{1}{2}\right)\varphi\right],$$

$${}_2B = 2a \sin\frac{\varphi}{2}\cos\left[\alpha - \left(1 + \frac{1}{2}\right)\varphi\right],$$

$${}_1B = 2a \sin\frac{\varphi}{2}\cos\left(\alpha - \frac{\varphi}{2}\right),$$

$$B_1 = 2a \sin\frac{\varphi}{2}\cos\left(\alpha + \frac{\varphi}{2}\right),$$

$$B_2 = 2a \sin\frac{\varphi}{2}\cos\left[\alpha + \left(1 + \frac{1}{2}\right)\varphi\right],$$

$$B_3 = 2a \sin\frac{\varphi}{2}\cos\left[\alpha + \left(2 + \frac{1}{2}\right)\varphi\right], \ldots;$$

$${}_2C = -4a \sin^2\frac{\varphi}{2}\sin(\alpha - 2\varphi),$$

$${}_1C = -4a \sin^2\frac{\varphi}{2}\sin(\alpha - \varphi),$$

$$C = -4a \sin^2\frac{\varphi}{2}\sin\alpha,$$

$$C_1 = -4a \sin^2\frac{\varphi}{2}\sin(\alpha + \varphi),$$

$$C_2 = -4a \sin^2\frac{\varphi}{2}(\alpha + 2\varphi), \ldots,$$

where $B, C, D, \ldots$ are Lagrange's notation for the successive differences. (Note that $B = ({}_1B + B_1)/2$, $D = ({}_1D + D_1)/2, \ldots$; moreover, ${}_3B = {}_2A - {}_3A, \ldots$.) By a little trigonometric manipulation he concludes that the even order differences $A, C, \ldots$ are expressible as

$$A = a\sin\alpha, \qquad C = -a\sin\alpha\left(2\sin\frac{\varphi}{2}\right)^2, \qquad E = a\sin\alpha\left(2\sin\frac{\varphi}{2}\right)^4, \ldots$$

and the odd ones $B, D, \ldots$ as

$$B = a\cos\alpha\sin\varphi, \qquad D = -a\cos\alpha\sin\varphi\left(2\sin\frac{\varphi}{2}\right)^2,$$

$$F = a\cos\alpha\sin\varphi\left(2\sin\frac{\varphi}{2}\right)^4,$$

i.e., each is a geometrical progression with ratio $-(2\sin\varphi/2)^2$.

If now A_x is more generally expressible as a sum of terms in the form (3.49), then the sequences $A, C, E, \ldots$ will contain several geometric progressions, as will the $B, D, F, \ldots$. Lagrange remarked that, in general, when a number of different geometrical progressions with ratios

$$-\left(2\sin\frac{\varphi}{2}\right)^2, \qquad -\left(2\sin\frac{\theta}{2}\right)^2, \qquad -\left(2\sin\frac{\psi}{2}\right)^2, \ldots$$

are present, then it is easy to separate out the largest, since the others tend to zero relative to that one. Thus if we have a large number of terms in the series $A, C, E, \ldots, B, D, F, \ldots$ and look at the high order differences, we should see they are in nearly constant ratio to each other.

By this means the parameters A, B above can be found for the most important periodicity. This term can be subtracted out. If the last terms are not in a nearly constant ratio, then "one can conclude that the form of the general term A is not composed of sines of uniformly increasing arcs."[61]

Of course the point to Lagrange's argument depends upon the angle φ being considerably closer to 180° than are the others: $\theta, \psi, \ldots$. In particular, in the case of Mayer's Equation of time, we note that $\varphi = 140°$, $\theta = 70°$. Hence $(2\sin\varphi/2)^2 = 3.532$, $(2\sin\theta/2)^2 = 1.316$, and their ratio is 0.3726. To achieve accuracy to three decimal places it is necessary to raise this latter number to the power 8. ($0.3726^8 = 3.7 \times 10^{-4}$.) In other words, we need to difference Mayer's table to find the differences Lagrange would call Q and R. This is tedious but not difficult.

Suppose that the differences for a given A_x have been carried sufficiently far to reveal the value of $\sin\varphi/2 = \sqrt{-q/2}$. Then the other parameters a and α are obtainable from the relations

$$a = \sqrt{A^2 + \frac{B^2}{\sin^2\varphi}}, \qquad \tan\alpha = \frac{A\sin\varphi}{B},$$

where, as we recall, $B = a\cos\alpha\sin\varphi$.

Lagrange now discussed a more general procedure based on the quantities $A, C, E, \ldots, B, D, F, \ldots$ forming a recurrent sequence of arbitrary order. (We discuss such sequences below on pp. 177ff.) This means, in Lagrange's notation, that there are unknowns (0), (1), (2), (3), ... such that

$$\begin{aligned}
A(0) + C(1) + E(2) + \cdots &= 0,\\
B(0) + D(1) + F(2) + \cdots &= 0,\\
C(0) + E(1) + G(2) + \cdots &= 0,\\
D(0) + F(1) + H(2) + \cdots &= 0,\\
E(0) + G(1) + L(2) + \cdots &= 0,\\
F(0) + H(1) + K(2) + \cdots &= 0, \ldots \quad .
\end{aligned}$$

Lagrange suggests an elimination method to solve this homogeneous system

[61] Lagrange VII [1778], p. 546.

but recognizes it cannot always succeed.[62] He takes $(0) = 1$ and assumes first that T, T' are two successive terms of the series $A, C, E, \ldots$ or $B, D, F, \ldots$. He then asks if he has the relation $T + T'(1) = 0$, which gives him the linear equation $z + (1) = 0$. If this is not satisfied by the series, he tries a three term law $T + T'(1) + T''(2) = 0$, finding the quadratic $z^2 + (1)z + (2) = 0$. He proceeds in this way: if he needs an $(n + 1)$-term law he finds the equation

$$z^n + (1)z^{n-1} + (2)z^{n-2} + \cdots + (n) = 0.$$

(There is such a law if and only if the sequence is recurrent.) He points out that the roots of this equation will be the ratios of the different geometric progressions which make up the series $A, C, \ldots$ and $B, D, \ldots$, and consequently are equal to

$$-\left(2 \sin \frac{\varphi}{2}\right)^2, \quad -\left(2 \sin \frac{\theta}{2}\right)^2, \quad -\left(2 \sin \frac{\psi}{2}\right)^2, \ldots .$$ [63]

Thus the defining equation for the recurrent sequence gives not only the angles $\varphi, \theta, \ldots$ but also their number n. It is then very easy to find the other parameters $a, b, \ldots$ as well as the angles $\alpha, \beta, \ldots$ with the help of the relations

$$\begin{aligned}
A &= a \sin \alpha + b \sin B + c \sin \gamma + \cdots, \\
\frac{A_1 + {}_1A}{2} &= a \sin \alpha \cos \varphi + b \sin \beta \cos \theta + c \sin \gamma \cos \chi + \cdots, \\
\frac{A_2 + {}_2A}{2} &= a \sin \alpha \cos 2\varphi + b \sin \beta \cos 2\theta + c \sin \gamma \cos 2\psi + \cdots, \\
&\vdots \\
\frac{A_1 - {}_1A}{2} &= a \cos \alpha \sin \varphi + b \cos \beta \sin \theta + c \cos \gamma \sin \psi + \cdots, \\
\frac{A_2 - {}_2A}{2} &= a \cos \alpha \sin 2\varphi + b \cos \beta \sin 2\theta + c \cos \gamma \sin 2\psi + \cdots, \\
\frac{A_3 - {}_3A}{2} &= a \cos \alpha \sin 3\varphi + b \cos \beta \sin 3\theta + c \cos \gamma \sin 3\psi + \cdots, \\
&\vdots
\end{aligned} \tag{3.51}$$

The first group of equations may be used to find the quantities $a \sin \alpha$, $b \sin \beta$, $c \sin \gamma, \ldots$ and the second $a \cos \alpha$, $b \cos \beta$, $c \cos \gamma, \ldots$. From these we can easily determine a, b, c as well as $\alpha, \beta, \gamma, \ldots$. Thus all quantities are determined. Lagrange also discussed the case of repeated roots, as well as

---

[62] Lagrange VII [1778], p. 548. Here Lagrange has given for the homogeneous case the now well-known method of "Gaussian" elimination for solving linear systems. We say more about this later on p. 216.

[63] Lagrange VII [1778], pp. 549–550.

the case when a value $q = -(2 \sin \varphi/2)^2$ gives a φ which is imaginary. In this connection he mentions that the expressions

$$\frac{(e^{x\omega} + e^{-x\omega})}{2}, \qquad \frac{(e^{x\omega} - e^{-x\omega})}{2}$$

are called *hyperbolic cosine* and *sine* and that they had been tabulated by Lambert (Johann H., 1727–1777) in an Appendix to his *Tables logarithmiques et trigonométriques*. (It was Lambert who introduced the notations *cosh, sinh,* but not the concept; this was due to Vincenzo Riccati (1707–1775), a son of J. Riccati.)

Up to this point, we have more or less avoided discussing recurring or recurrent series or sequences. Since they played such a large role in the work of Stirling, de Moivre, Euler, etc., and since they are inherently important, perhaps we should say more on the topic. We use Lagrange's earlier paper for this purpose.[64] Suppose that $T, T', T'', T''', \ldots, T^{(m)}, T^{(m+1)}, \ldots$ represent the successive terms of a sequence with

$$T^{(m)} = A \sin (a + m\alpha) + B \sin (b + m\beta) + C \sin (c + m\gamma) + \cdots .$$

Then consider the infinite series

$$s = T + T'x + T''x^2 + T'''x^3 + \cdots + T^{(m)}x^m + T^{(m+1)}x^{m+1} + \cdots . \tag{3.52}$$

(Lagrange raises no convergence questions.) Lagrange expresses the coefficients of this series in the form

$$T^{(m)} = Kp^m + Lq^m + Mr^m + Ns^m + Pt^m + Qu^m + \cdots,$$

where $p = e^{\alpha\sqrt{-1}}$, $q = e^{-\alpha\sqrt{-1}}$, and

$$K = \frac{Ae^{\alpha\sqrt{-1}}}{2\sqrt{-1}}, \qquad L = -\frac{Ae^{\alpha\sqrt{-1}}}{2\sqrt{-1}}, \ldots;$$

then, e.g., $A \sin (a + m\alpha) = Kp^m + Lq^m$, $B \sin (b + m\beta) = Mr^m + Ns^m$.

Next Lagrange assumed the series (3.52) to be of the form

$$T + T'x + T''x^2 + T'''x^3 + T^{\text{IV}}x^4 + \cdots = \frac{K}{1 - px} + \frac{L}{1 - qx} + \frac{M}{1 - rx} + \frac{N}{1 - sx} + \cdots$$

where there are n terms in the right-hand member. It is then expressible, in his terms, as

$$\frac{[0] + [1]x + [2]x^2 + [3]x^3 + \cdots + [n-1]x^{n-1}}{(0) + (1)x + (2)x^2 + (3)x^3 + \cdots + (n)x^n}, \tag{3.53}$$

[64] Lagrange VI [1772], pp. 511ff.

where the [ ] and ( ) represent functions of the $K, L, \ldots, p, q, \ldots$. In fact,

$$(0) = 1, \qquad (1) = -p - q - r - s - \cdots,$$
$$(2) = pq + pr + ps + \cdots + qr + qs + \cdots,$$
$$(3) = -pqr - pqs - qrs - \cdots \quad .$$

Lagrange calls the expression (3.53) the *generating fraction* of the series (3.52). He defines what we call a *recurring* or *recurrent* series or sequence as one in which any given term, ignoring some initial ones, is a linear combination of a certain number of its predecessors.[65] He now goes on to show that the $T, T', T'', \ldots$ form such a sequence. [A sequence is recurrent if and only if its series (3.52) has a generating fraction (3.53).] In particular we note that

$$T = [0], \qquad T' = -(1)T + [1],$$
$$T'' = -(1)T' - (2)T + [2], \ldots,$$
$$T^{(n-1)} = (-1)T^{(n-2)} - (2)T^{(n-3)} - \cdots - (n-1)T + [n+1],$$
$$T^{(n)} = -(1)T^{(n-1)} - (2)T^{(n-2)} - \cdots - (n)T,$$

and, in general, $T^{(m)} = -(1)T^{(m-1)} - (2)T^{(m-2)} - \cdots - (n)T^{(m-n)}$, where the coefficients $-(1), -(2), -(3), \ldots, -(n)$ form what de Moivre called "*l'échelle de la série récurrente*", the scale of the recurrent series.

A little analysis shows that $(1 - px)(1 - qx) = 1 - 2x \cos \alpha + x^2, \ldots,$ and hence that the denominator of the fraction (3.53) will be the product of factors of the above form (here Lagrange assumes n is even). Moreover $(n) = 1, (n - 1) = (1), \ldots$.

Now in his Proposition II Lagrange proposes a problem: given any sequence of terms to find if it is recurrent and to determine the form of the general term.[66] He supposes the sequence to be $T, T', \ldots$ and again forms the series (3.52). He then asks whether s is representable as a rational function. If, for example, the series is recurrent of the first order, then, by definition.

$$s = \frac{a'}{a + bx} \quad \text{and} \quad \frac{1}{s} = \frac{a + bx}{a'} = p + qx.$$

Thus if we form s^{-1}, we know it is a linear function. Lagrange then goes in detail through examples of various orders, and concludes in his Proposition II that in general s must be representable by means of a continued fraction in this way:

$$s = \cfrac{1}{p + qx + \cfrac{x^2}{p' + q'x + \cfrac{x^2}{p'' + q''x + \cfrac{x^2}{p''' + q'''x + \ddots + \cfrac{x^2}{p^{(n-1)} + q^{(n-1)}x}}}}}.$$

[65] Lagrange VI [1772], p. 513.

[66] Lagrange VI [1772], p. 525.

Moreover, if s is of this form, it can be reduced to the form (3.53) and the recurrence relations found.

He then considers the sequence 1, 2, 3, 3, 7, 5, 15, 9, 31, 17, 63, 33, 127, 65, ... as an example. He asks if this sequence is recurrent and if it is what is the law of recurrence. Set s equal to the series (3.52) as before. I.e., $s = 1 + 2x + 3x^2 + 3x^3 + \cdots$. Then form $1/s$. This gives $1 - 2x$ and a remainder of $x^2 + 3x^3 - x^4 + \cdots = x^2(1 + 3x - x^2 + \cdots) = x^2 \cdot s'$. Next divide s by s' and find a quotient of $1 - x$ and a remainder of $x^2 \cdot s'' = x^2(7 - 7x + \cdots)$. Continuing one finds a quotient of $1/7 + 4x/7$ and a remainder of 0. It is then evident that the generating fraction is

$$\frac{1 + 3x + 3x^2}{1 + x - 2x^2 - 2x^3};$$

and de Moivre's "scale" is $-1, 2, 2$. Thus if t, t', t'', t''' are any three consecutive terms in the sequence, then $t''' = -t'' + 2t' + 2t$.

Lagrange summarizes on p. 537 another of his algorithms in this rule: "Divide unity by the given series and continue the division until the quotient contains two terms that are consecutive powers of x, such as $px^\lambda + qx^{\lambda+1}$; after this divide the series s by the remainder from the division, and under the same conditions; after this divide the second remainder by the first, and continue in this fashion to divide the new remainder by the preceding one so that each quotient always contains two terms of the preceding form; if the series s is recurrent, some division will be exact; in that event call

$$px^\lambda + qx^{\lambda+2}, \qquad p'x^\mu + q'x^{\mu+1},$$
$$p''x^\nu + q''x^{\nu+1}, \ldots, p^{(n-1)}x^\sigma + q^{(n-1)}x^{\sigma+1}$$

the quotients found in the n divisions, then

$$s = \cfrac{1}{px^\lambda + qx^{\lambda+1} + \cfrac{1}{p'x^\mu + q'x^{\mu+1} + \cfrac{1}{p''x^\nu + q''x^{\nu+1} + \ddots + \cfrac{1}{p^{(n-1)}x^\sigma + q^{(n-1)}x_\rho{}^{+1}}}}."$$

He applies this rule to the sequence 1, 4, 10, 19, 31, 46, 64, 85, 109, 136, 166, 199, ... and finds $s = 1 + 4x + 10x^2 + \cdots$; then $1/s$ has a quotient of $1 - 4x$ and remainder $s' = 6x^2 + 21x^3 + 45x^4 + 78x^5 + 120x^6 + 171x^7 + 231x^8 + 300x^9 + \cdots$; s/s' yields a quotient of $1/6x^2 + 1/12x$ and a remainder of $s'' = 3x^2/4 + 9x^3/4 + \cdots$. After doing the work, he finds the sequence has as its generating fraction

$$\frac{1 + x + x^2}{1 - 3x + 3x^2 - x^3},$$

Lagrange was clearly very pleased with his results and remarked on their novelty. He proceeded beyond this point, however, and developed many other new and elegant properties of recurrent sequences. For example, in Proposition V Lagrange proposed another problem defined in terms of so-called *reciprocal* polynomials. He says a polynomial is *reciprocal* when its coefficients (0), (1), ..., (n) are such that (0) = (n), (1) = (n − 1), (2) = (n − 2), ... He now supposes that the generating fraction for a recurrent series has both its numerator and denominator reciprocal polynomials of even degrees — the order of the denominator is at least two units more than that of the numerator. Then the problem if Proposition V is to find an algorithm easier than that of Proposition II which will yield the generating fraction directly.[67] In this case the change of variables

$$y = \frac{x}{1 + x^2}$$

results in a continued fraction expansion of the form

$$\cfrac{1}{p + qy + \cfrac{y^2}{p' + q'y + \cfrac{y^2}{p'' + q''y + \cfrac{y^2}{p''' + q'''y + \ddots + \cfrac{y^2}{p^{(\mu-1)} + q^{(\mu-1)}y}}}}} \tag{3.54}$$

This transforms back into the expansion — apart from a factor $1 + x^2$ —

$$\cfrac{1}{p(1+x^2)+qx+\cfrac{x^2}{p'(1+x^2)+q'x+\cfrac{x^2}{p''(1+x^2)+q''x+\ddots+\cfrac{x^2}{p^{(\mu-1)}(1+x^2)+q^{(\mu-1)}x}}}} \tag{3.54}$$

His method of attack is quite direct and straightforward. If the degree of the denominator exceeds that of the numerator by 2μ, then his new procedure terminates after μ divisions whereas his previous procedure (Proposition II) took 2μ. It is of interest to note that as early as Lagrange people were concerned with finding economical algorithms. It is also noteworthy that it was he who seems to have been the first to observe that the nonlinear operations are the relevant ones in estimating the amount of computational work involved in carrying out an algorithm.

Having set up these algorithms, Lagrange now considered various schemes for applying them to the trigonometric case we discussed earlier. His first method relies on some of his earlier results. He gives three methods in all (pp. 584ff.), and applies them to Mayer's table cited above.

We consider here only his first one which he particularly liked.

"Let T be one of the terms in the middle of the proposed series: T_1, T_2, T_3, ... those following and ${}_1T$, ${}_2T$, ${}_3T$, ... those preceding; and from these

[67] Lagrange VI [1772], pp. 565ff.

form the series of sums

$$(T + {}_1T) + (T_1 + {}_2T)x + (T_2 + {}_3T)x^2 + (T_3 + {}_4T)x^3 + \cdots \quad .$$

To make the calculations easier first divide out of all terms the first coefficient $T + {}_1T$, which I shall call p, so that the resulting series which I call s will have its constant term unity.

Having done this, divide $1 - x$ by s and in place of 1 in the quotient write $1 + x^2$; continue the division until the quotient contains a term of the form qx; after this the remainder will have a first term of the form $p'x^2$.

Divide this first term out of the remainder leaving a polynomial called s' with constant term unity; divide s by s' and in place of 1 again write $1 + x^2$ in the quotient; then continue the division as usual finding a term $q''x$; then the remainder will have its first term in the form $p'''x^2, \ldots$.

Operate upon the remainder as before and continue this process until a remainder is reached that is either exactly zero or very small; in the former case one has an exact solution and in the latter an approximate one.

Now let n be the number of quotients; then there are two sequences

$$p, p', p'', p''', \ldots, p^{(n-1)}, \qquad q, q', q'', q''', \ldots, q^{(n-1)};$$

from these form

$$\begin{aligned}
\psi &= 1,\\
\psi' &= 1 + q^{(n-1)}y,\\
\psi'' &= (1 + q^{(n-2)}y)\psi' + p^{(n-1)}y^2\psi,\\
&= (1 + (q^{(n-1)} + q^{(n-2)})y + (q^{(n-1)}q^{(n-2)} + p^{(n-1)})y^2,\\
\psi''' &= (1 + q^{(n-3)}y)\psi'' + p^{(n-2)}y^2\psi',\\
&= 1 + (q^{(n-1)} + q^{(n-2)} + q^{(n-3)})y\\
&\quad + (q^{(n-1)}q^{(n-2)} + q^{(n-1)}q^{(n-3)} + q^{(n-2)}q^{(n-3)})y^2\\
&\quad + (q^{(n-1)}q^{(n-2)}q^{(n-3)} + q^{(n-1)}p^{(n-2)} + q^{(n-3)}p^{(n-1)})y^3,\\
&\ \vdots\\
\psi^{(n)} &= (1 + qy)\psi^{(n-1)} + p'y^2\psi^{(n-2)}.
\end{aligned}$$

Then consider the fraction $p\psi^{(n-1)}/\psi^{(n)}$, where the denominator $\psi^{(n)}$ will always be a polynomial in y of degree n with constant term unity; it can then be resolved into n simple factors of the form $1 - \varpi y, 1 - \rho y, \ldots$. These factors can be found by known methods and the fraction can then be decomposed by partial fractions into the form

$$\frac{F}{1 - \varpi y} + \frac{G}{1 - \rho y} + \cdots \quad .$$

When this is done form the series of differences

$$(T - T') + (T' - {}''T)x + (T'' - {}'''T)x^2 + (T''' - {}^{\mathrm{IV}}T)x^3 + \cdots,$$

and treat it in the same fashion as the preceding series with the sole difference that instead of $1 - x$ as the first dividend it is now necessary to use $1 + x$.

In following the same procedure one finds also a fraction $p(\psi)^{(n-1)}/(\psi)^{(n)}$, whose denominator $(\psi)^{(n)}$ will be exactly, or nearly so, the same as the one found for the first series; this provides a check on the goodness of the calculation. Thus it can be decomposed into n partial fractions

$$\frac{(F)}{1-\bar{\omega}y}+\frac{(G)}{1-\rho y}+\cdots \quad .$$

Having found in this way the values for the quantities

$$\bar{\omega}, \rho, \ldots, F, G, \ldots, (F), (G), \ldots,$$

we then have

$$\cos\alpha=\frac{\bar{\omega}}{2}, \qquad \tan a=\frac{F+(F)}{(F)\cot\frac{\alpha}{2}-F\tan\frac{\alpha}{2}},$$

$$A=\frac{1}{2}\sqrt{F^2\,\text{séc}^2\frac{\alpha}{2}+(F)^2\,\text{coséc}^2\frac{\alpha}{2}},$$

$$\cos\beta=\frac{\rho}{2}, \qquad \tan b=\frac{G+(G)}{(G)\cot\frac{\beta}{2}-G\tan\frac{\beta}{2}},$$

$$B=\frac{1}{2}\sqrt{G^2\,\text{séc}^2\frac{\beta}{2}+(G)^2\,\text{coséc}^2\frac{\beta}{2}},$$

$$\vdots$$

and we have for the general term T_m the expression

$$T^{(m)}=A\sin(a+m\alpha)+B\sin(b+m\beta)+\cdots \quad .\text{"}^{68}$$

Lagrange chooses for T the term $+772$ corresponding to 340° of longitude in Mayer's table and forms the sums and differences $T \pm {}_1T$, $T_1 \pm {}_2T$, $T_2 \pm {}_3T, \ldots$, where ${}_1T=-72$, $T_1=-227, \ldots$. (Recall that Mayer's sequence was 456, −168, 274, −933, 220, 631, −232, 349, −823, −72, 772, −237, 358, −657, −360, 860, −181, 305, −457, −616, ... This gives him the sequences 700, −1060, 707, −889, 271, 1080, −1114, 579, −625, −160, . . . and 844, 586, 9, −425, −991, 640, 752, 31, −289, −1072, ... His series are then

$$700-1060x+707x^2-889x^3+271x^4+\cdots=700s,$$
$$844+586x+9x^2-425x^3-991x^4+\cdots=844(s),$$

and he expresses them in the form (3.54). For example, he has for his first series approximately

$$700\,\frac{1+x^2}{1-x}\,s=\frac{700}{1+0.51428y-\dfrac{1.23123y^2}{1+0.32522y}}$$

$$=\frac{F}{1-\bar{\omega}y}+\frac{G}{1-\rho y}.$$

---

[68] Lagrange II [1767], pp. 539–578 and II [1768], pp. 581–652.

Clearly $p = 700$, $p' = -1.23123$, $q = 0.51428$, $q' = 0.32522$. Moreover for these series the quantities ϖ, ρ, F, G are given by the relations

$$\varpi = -\frac{q+q'}{2} + \frac{\sqrt{(q-q')^2 - 4p'}}{2}, \qquad \rho = -\frac{q+q'}{2} - \frac{\sqrt{(q-q')^2 - 4p'}}{2}.$$

and

$$F = -\frac{p(q'+\varpi)}{\rho - \varpi} = \frac{p}{2}\left[1 - \frac{q-q'}{\sqrt{(q-q')^2 - 4\rho'}}\right],$$

$$G = -\frac{p(q'+\rho)}{\varpi - \rho} = \frac{p}{2}\left[1 + \frac{q-q'}{\sqrt{(q-q')^2 - 4p'}}\right].$$

Hence the first sequence yields these results:

$$\varpi = 0.69387, \qquad \rho = -1.53337, \qquad F = 320.30, \qquad G = 379.70,$$

and the second

$$\varpi = 0.69113, \qquad \rho = -1.53248, \qquad (F) = 465.64, \qquad (G) = 379.36$$

Lagrange argues that ϖ and ρ are not very accurate and therefore chooses the averages of both, $\varpi = 0.69250$, $\rho = -1.53292$. From these he finds $\alpha = 69°;45$, $\beta = 140°;2$, $a = 60°;30$, $b = 140°;6$; and hence $A = 451$, $B = 591$ in the expression $A \sin(a + m\alpha) + B \sin(b + m\beta)$.

3.13. Lagrange on Trigonometric Interpolation

It is difficult to know where to stop in analyzing Lagrange's contributions to our field. However, let us close with his work on trigonometrical interpolation.[69] In this paper Lagrange was not actually concerned directly with the topic. Instead he was trying to gain an understanding of "Fourier" series. He was discussing the initial shape of a vibrating string. He said: "Now it is clear that for any initial curve it is always possible to approximate to that curve by one of the form

$$y = \alpha \sin \pi x + \beta \sin 2\pi x + \cdots,$$

in such a way that the difference between the two curves will be as small as one wishes, even though this difference will only be exactly zero in the case when the initial curve is also of this form; in all other cases the initial curve will be a kind of asymptote to which the generating curve will approach as they go to infinity, without their ever completely coinciding."[70] Lagrange considers the function

$$y = \frac{2Y_1}{n+1}\sin x\pi + \frac{2Y_2}{n+1}\sin 2x\pi + \frac{2Y_3}{n+1}\sin 3x\pi + \cdots + \frac{2Y_n}{n+1}\sin nx\pi,$$

[69] Lagrange I [1762], pp. 552ff.

[70] Lagrange I [1762], p. 552.

where

$$Y_m = Y' \sin \frac{m\pi}{n+1} + Y'' \sin \frac{2m\pi}{n+1} + Y''' \sin \frac{3m\pi}{n+1} + \cdots + Y^{(n)} \sin \frac{nm\pi}{n+1}.$$

Now he shows that for $x = s/(n+1)$, $y = Y^{(s)}$. His proof is not quite easy but it is essentially this:

$$\begin{aligned} y\left(\frac{s}{n+1}\right) &= \frac{2}{n+1} \sum_{\alpha=1}^{n} Y_\alpha \sin\left(\frac{\alpha s\pi}{n+1}\right) \\ &= \frac{2}{n+1} \sum_{\beta=1}^{n} Y^{(\beta)} \left[\sum_{\alpha=1}^{n} \sin\left(\frac{\alpha s\pi}{n+1}\right) \sin\left(\frac{\alpha\beta\pi}{n+1}\right)\right] \\ &= \frac{1}{n+1} \sum_{\beta=1}^{n} Y^{(\beta)} \times \sum_{\alpha=1}^{n} \left[\cos \frac{(s-\beta)\alpha\pi}{n+1} - \cos \frac{(s+\beta)\alpha\pi}{n+1}\right]; \end{aligned}$$

but we know that

$$\sum_{\alpha=1}^{n} \cos \frac{k\pi}{n+1}\alpha = \begin{cases} n & (k=0) \\ \dfrac{\cos(k\pi/2)\sin(k\pi n/2(n+1))}{\sin(k\pi/2(n+1))} = -\cos^2 \dfrac{k\pi}{2} & (k = 1, 2, \ldots). \end{cases}$$

Hence it follows that

$$\begin{aligned} y\left(\frac{s}{n+1}\right) = \frac{1}{n+1} \Bigg[(n+1)\, Y^{(s)} + \sum_{\beta \neq s} \Bigg(-\cos^2 \frac{(s-\beta)\pi}{2} \\ + \cos^2 \frac{(s+\beta)\pi}{2}\Bigg) Y^{(\beta)}\Bigg]^{(s)} = Y^{(s)} \end{aligned}$$

since the summand above is $-\sin s\pi \sin \beta\pi = 0$. Thus Lagrange found a trigonometric polynomial passing through the points $(s/(n+1), Y^{(s)})$ on a given curve. Clairaut also considered the same problem.[71] Neither Clairaut nor Lagrange indicates any awareness of the other's work in this connection.

Although it is peripheral to our subject, we mention a dispute between Lagrange and D'Alembert about the convergence of Fourier series, since it illustrates how little was understood as yet about infinite processes.[72] Lagrange considered the series

$$\cos x + \cos 2x + \cos 3x + \cdots \tag{3.55}$$

as a sum of exponentials by replacing each cosine by its exponential counterpart. He saw that this yielded

$$\cos x + \cos 2x + \cdots + \cos mx = \frac{\cos mx - \cos(m+1)x}{2(1-\cos x)} - \frac{1}{2}.$$

[71] Clairaut [1759]; see especially, pp. 544–564.

[72] These matters appear in three long papers. Cf. Lagrange I [1759′], I [1760], and I [1761]. They occupy pp. 39–332 of Vol. I.

Then he said: "But in the case where m is an infinite number the number 1 vanishes relative to m, and hence the term $\cos(m + 1)x$ becomes equal to $\cos mx$, and

$$\frac{\cos mx - \cos(m + 1)x}{2(1 - \cos x)} = 0;$$

but when $x = 0$, the denominator also vanishes. . . . The value of the series in this case is

$$m + \frac{1}{2} - \frac{1}{2} = m,$$

precisely as we saw above."[73] Thus Lagrange concluded that the value of the series was $-1/2$ for all $x \neq 0$. Similarly he attempted to sum the series

$$\sin \varphi \sin \theta + \sin 2\varphi \sin 2\theta + \cdots \quad .$$

To show that Lagrange's methods were not valid d'Alembert put $x = 45°$ in the infinite series (3.55) and found

$$\frac{1}{\sqrt{2}}, \quad 0, \quad -\frac{1}{\sqrt{2}}, \quad -1, \quad -\frac{1}{\sqrt{2}}, \quad 0, \quad +\frac{1}{\sqrt{2}}, \quad +1.$$

"But, he said, the sum of this finite series is $1/\sqrt{2}$, or 0, or -1, or $-1-1/\sqrt{2}$, according to whether one takes more or less terms. Of course the sum of the entire series is also $1/\sqrt{2}$, or 0, or $-1-1/\sqrt{2}$, depending upon the number m of terms one takes, whether this number is finite or infinite, and the sum will be zero only when $m \times 45°$ is equal to an infinite number times 360° or to 135° plus such a quantity."[74] In rebuttal Lagrange wrote: "I reply that with the same reasoning it could also be argued that

$$\frac{1}{1 + x}$$

is not the general expression for the sum of the infinite series $1 - x + x^2 - x^3 + \cdots$, because by letting $x = 1$ one has $1 - 1 + 1 - 1 + \cdots$, which is either 0 or 1 depending on whether one takes an even or odd number of terms, whereas the value of

$$\frac{1}{1 + x}$$

is $\frac{1}{2}$. But I do not believe that any geometer would grant this conclusion."[75]

---

[73] Lagrange I [1759′], pp. 111–112. He found this value, $m + 1/2 - 1/2$, using l'Hôpital's rule.

[74] Lagrange I [1761], p. 323.

[75] Lagrange I [1761], p. 323.

4. Laplace, Legendre, and Gauss

4.1. Introduction

Laplace's efforts in our field appear both in his work on celestial mechanics and on probability theory. Moreover these contributions, which overlap each other and those of Gauss very much, are central to the interests of Laplace. One of his chief tools was that of the generating function. He was perhaps the first to exploit fully the generating function of a sequence $y_0, y_1, y_2, \ldots$. He wrote the function as

$$u = y_0 + y_1 t + y_2 t^2 + y_3 t^3 + \cdots + y_x t^x + y_{x+1} t^{x+1} + \cdots + y^\infty t_\infty$$

without any consideration of convergence. Let us accept his formalistic approach and see how the generating function tool, which he used so powerfully, could produce various interpolation formulas.[1]

4.2. Laplace on Interpolation

He says that if u is the generating function of y_x, then u/t^r is that of y_{x+r} (i.e., of the sequence $y_r, y_{r+1}, \ldots$) and the coefficient of t^x in $u \cdot (1/t - 1)$ is $y_{x+1} - y_x = \Delta y_x$. Moreover, the generating function of $\Delta^i y_x$ is $u \cdot (1/t - 1)^i$. Consider this last relation,

$$\Delta^i y_x = u\left(\frac{1}{t} - 1\right)^i = u\left[\frac{1}{t^i} - \frac{i}{1!\,t^{i-1}} + \frac{i(i-1)}{2!\,t^{i-2}} - \cdots\right]$$

$$= y_{x+i} - i y_{x+i-1} + \frac{i(i-1)}{2!} y_{x+i-2} - \cdots .$$

---

[1] Cf. Laplace VII [1820], pp. 7ff. The series u is said to be the generating function for y_x in the sense that the coefficient of t^x in u is y_x.

Let us now express u/t^i as

$$\frac{u}{t^i} = u\left(1 + \frac{1}{t} - 1\right)^i = u\left\{\begin{array}{l} 1 + i\left(\frac{1}{t} - 1\right) + \frac{i(i-1)}{1\cdot 2}\left(\frac{1}{t} - 1\right)^2 \\ + \frac{i(i-1)(i-2)}{1\cdot 2\cdot 3}\left(\frac{1}{t} - 1\right)^3 + \cdots \end{array}\right\}.$$ [2]

Then by what we saw earlier this implies at once that

$$y_{x+i} = y_x + i\Delta y_x + \frac{i(i-1)}{2!}\Delta^2 y_x + \frac{i(i-1)(i-2)}{3!}\Delta^3 y_x + \cdots$$

which is of course a standard interpolation formula. Laplace next set out to derive some other interpolation formulas. To do this he makes use of an implicit-function expansion he established in his *Mécanique Céleste*, Première partie, Livre deuxième, Section 21.[3] He wrote $1/t = 1 + \alpha/t^r$, and asked for an expansion of $1/t^i$ in powers of $\alpha = t^{r-1} - t^r$. He found

$$\frac{u}{t^i} = u\left\{\begin{array}{l} 1 + i\alpha + \frac{i(i+2r-1)}{1\cdot 2}\alpha^2 + \frac{i(i+3r-1)(i+3r-2)}{1\cdot 2\cdot 3}\alpha^3 \\ + \frac{i(i+4r-1)(i+4r-2)(i+4r-3)}{1\cdot 2\cdot 3\cdot 4}\alpha^4 + \cdots \end{array}\right\}.$$

Let us look into this expansion, which is called the Theorem of Laplace in some calculus texts. It is this: given $t = f[x + \alpha\varphi(t)]$, it is required to expand any function $F(t)$ in powers of α. The expansion is an immediate consequence of the so-called Theorem of Lagrange, which is concerned with expanding F in powers of α when the function $f(s)$ above is s.[4]

Suppose that t is given as a function of α and x by the implicit relation

$$t = x + \alpha\varphi(t) \tag{4.1}$$

and that we wish to expand a function $w(t)$ in powers of α. Now let $t = T(\alpha, x)$ be the solution of the equation given by (4.1) above in a neighborhood of $(\alpha, x) = (0, 0)$, and let $W(\alpha, x) = w[T(\alpha, x)]$. Then

$$\frac{\partial T}{\partial \alpha} = \phi(T) + \alpha\phi'(T)\frac{\partial T}{\partial \alpha}, \qquad \frac{\partial T}{\partial x} = 1 + \alpha\phi'(T)\frac{\partial T}{\partial x}.$$

---

[2] Laplace VII [1820], pp. 11ff.

[3] Laplace I [1805], pp. 188ff. Cf. also, another proof in IX [1777], pp. 330ff. Lagrange's theorem appears in Lagrange III [1771], pp. 207ff. It is a special case of Laplace's results. Laplace says: "If $\varphi(a + \alpha z)$ is $a + \alpha z$, we have the elegant theorem that Lagrange found by induction in the *Berlin Mémoires* for 1769, and if we let $z = 1$, we have Taylor's theorem. . . ."

[4] Williamson [1889], pp. 151–154.

We infer from these, by eliminating φ' between them, that

$$\frac{\partial T}{\partial \alpha} = \phi(T)\frac{\partial T}{\partial x}. \tag{4.2}$$

Moreover, we have

$$\frac{\partial W}{\partial \alpha} = w'(T)\frac{\partial T}{\partial \alpha}, \qquad \frac{\partial W}{\partial x} = w'(T)\frac{\partial T}{\partial x}.$$

If we eliminate w' between these and use (4.1), we find at once that

$$\frac{\partial W}{\partial \alpha} = \phi(T)\frac{\partial W}{\partial x} = Z\frac{\partial W}{\partial x},$$

where $Z = \varphi(T)$. Now

$$\frac{\partial^2 W}{\partial \alpha^2} = \frac{\partial}{\partial \alpha}\left(Z\frac{\partial W}{\partial x}\right) = \frac{\partial}{\partial x}\left(Z\frac{\partial W}{\partial \alpha}\right)$$

and by induction

$$\frac{\partial^n W}{\partial \alpha^n} = \frac{\partial^{n-1}}{\partial x^{n-1}}\left(Z^n\frac{\partial W}{\partial \alpha}\right).$$

It follows from this that

$$\begin{aligned} W(\alpha, x) &= W(0, x) + \frac{\alpha}{1!}\frac{\partial W}{\partial \alpha}\bigg|_{\alpha=0} + \frac{\alpha^2}{2!}\frac{\partial^2 W}{\partial \alpha^2}\bigg|_{\alpha=0} + \cdots \\ &= w(x) + \frac{\alpha}{1!}\phi(x)w'(x) + \frac{\alpha^2}{2!}\frac{d}{dx}[\phi^2(x)w'(x)] + \cdots \quad . \end{aligned}$$

This is Lagrange's theorem. If now $t = f[x + \alpha\varphi(t)]$, let us set $\tau = x + \alpha\varphi(t)$; then $t = f(\tau)$ and we have $\tau = x + \alpha\varphi[f(\tau)]$. This is essentially our previous relation (4.1) above. Hence we have

$$\begin{aligned} w[f(\tau)] = W(\alpha, x) &= w[f(x)] + \frac{\alpha}{1!}\phi[f(x)]w'[f(x)] \\ &+ \frac{\alpha^2}{2!}\frac{d}{dx}\{\phi[f(x)]^2 w'[f(x)]\} + \cdots, \end{aligned}$$

which is Laplace's result.

In our case we have $\varphi(\tau) = \tau^r$, $\tau = 1/t$, $w(\tau) = \tau^i$, and at the end we wish to evaluate our expansion for $x = 1$. Clearly $\varphi(x) = x^r$, $w(x) = x^i$. Hence

$$\begin{aligned} \tau^i &= 1 + \frac{i\alpha}{1!} + \frac{i\alpha^2}{2!}\frac{dx^{i+2r-1}}{dx}\bigg|_{x=1} + \frac{i\alpha^3}{3!}\frac{d^2x^{i+3r-1}}{dx^2}\bigg|_{x=1} + \cdots \\ &= 1 + i\alpha + \frac{i(i+2r-1)}{2!}\alpha^2 + \frac{i(i+3r-1)(i+3r-2)}{3!}\alpha^3 + \cdots \quad . \end{aligned}$$

where $\alpha = t^{-r} - 1$. This expansion for u/t^i yields an interesting interpolation formula:

$$y_{x+i} = y_x + i\Delta y_{x-r} + \frac{i(i + 2r - 1)}{2!}\Delta^2 y_{x-2r} + \frac{i(i + 3r - 1)(i + 3r - 2)}{3!}\Delta^3 y_{x-3r} + \cdots .$$[5]

Let us next look at another example of Laplace's dexterity with his generating functions. To this end we shall examine his derivation of the Laplace–Everett and the Newton–Stirling formulas. He sets

$$z = t\left(\frac{1}{t} - 1\right)^2, \tag{4.3}$$

and asks for the value of $1/t^i$ as a power series in z. He remarks that $1/t^i$ is the coefficient of θ^i in the expansion of the fraction $1/(1 - \vartheta/t)$; hence he multiplies the numerator and denominator of this fraction by $1 - \vartheta t$. This transforms the fraction into the form

$$\frac{1 - \vartheta t}{1 - \vartheta(1/t + t) + \vartheta^2}. \tag{4.4}$$

But by (4.3) above this becomes

$$\frac{1 - \vartheta t}{(1 - \vartheta)^2 - z\vartheta}.$$

Laplace now expanded the denominator into the series

$$f(\vartheta) = \frac{1}{(1 - \vartheta)^2 - z\vartheta} = \frac{1}{(1 - \vartheta)^2[1 - (z\vartheta/(1 - \vartheta)^2)]}$$

$$= \frac{1}{(1 - \vartheta)^2} + \frac{z\vartheta}{(1 - \vartheta)^4} + \frac{z^2\vartheta^2}{(1 - \vartheta)^6} + \cdots . \tag{4.5}$$

He then looked at each term to see what the coefficient of ϑ^i is, using the fact that the coefficient of ϑ^r in the expansion of $(1 - \vartheta)^{-s}$ is

$$\frac{s(s + 1)(s + 2)\cdots(s + r - 1)}{1\cdot 2\cdot 3\cdots r}.$$

Thus if Z is the coefficient of ϑ^i in the expansion (4.5) above, then

$$Z = i + 1 + \frac{i(i + 1)(i + 2)}{1\cdot 2\cdot 3}z + \frac{(i - 1)i(i + 1)(i + 2)(i + 3)}{1\cdot 2\cdot 3\cdot 4\cdot 5}z^2 + \frac{(i - 2)(i - 1)i(i + 1)(i + 2)(i + 3)(i + 4)}{1\cdot 2\cdot 3\cdot 4\cdot 5\cdot 6\cdot 7}z^3 + \cdots,$$

[5] Laplace VII [1820], p. 13.

which Laplace writes as

$$Z = (i+1)\left\{1 + \frac{[(i+1)^2 - 1]z}{1\cdot 2\cdot 3} + \frac{[(i+1)^2 - 1][(i+1)^2 - 4]z^2}{1\cdot 2\cdot 3\cdot 4\cdot 5} + \cdots\right\}.$$

Next he calls Z' the coefficients of ϑ^i in the development of $\vartheta f(\vartheta)$ and notes that this merely involves replacing i by $i - 1$ in Z. Thus the coefficient of ϑ^i in the fraction (4.4) above is $Z - tZ'$, and so it is the desired expression for $1/t^i$. Accordingly $u/t^i = u(Z - tZ')$. But the coefficient of t^x in u/t^i is y_{x+i}, and the corresponding coefficient in uZ arising from a term $kuz^r = kut^r(1/t - 1)^{2r}$ is $k\Delta^{2r}y_{x-r}$. Similarly, the coefficient arising from utZ' is $k\Delta^{2r}y_{x-r-1}$. Hence Laplace arrived at the result (in his notation)

$$y_{x+i} = (i+1)\left\{\begin{aligned} &y_x + \frac{(i+1)^2 - 1}{1\cdot 2\cdot 3}\Delta^2 y_{x-1} \\ &\quad + \frac{[(i+1)^2 - 1][(i+1)^2 - 4]}{1\cdot 2\cdot 3\cdot 4\cdot 5}\Delta^4 y_{x-2} + \cdots \end{aligned}\right\}$$

$$- i\left[y_{x-1} + \frac{i^2 - 1}{1\cdot 2\cdot 3}\Delta^2 y_{x-2} + \frac{(i^2-1)(i^2-4)}{1\cdot 2\cdot 3\cdot 4\cdot 5}\Delta^4 y_{x-3} + \cdots\right].$$

This formula of Laplace's was rediscovered in 1900 by Everett [1900], p. 648, and goes by the latter's name. The point is discussed by Lidstone.[6]

To transform this into a more symmetrical form Laplace let Z'' be the value of Z' when i is changed into $i - 1$. Thus $1/t^i = Z - tZ'$ becomes $1/t^{i-1} = Z' - tZ''$, $1/t^i = Z'/t - Z'' = Z - tZ'$; averaging the last two relations we have

$$\frac{1}{t^i} = \frac{1}{2}Z - \frac{1}{2}Z'' + \frac{1}{2}(1+t)\left(\frac{1}{t} - 1\right)Z'.$$

Moreover,

$$\frac{1}{2}Z - \frac{1}{2}Z'' = 1 + \frac{i^2}{1\cdot 2}z + \frac{i^2(i^2-1)}{1\cdot 2\cdot 3\cdot 4}z^2 + \frac{i^2(i^2-1)(i^2-4)}{1\cdot 2\cdot 3\cdot 4\cdot 5\cdot 6}z^3 + \cdots,$$

and so

$$\frac{u}{t^i} = u\left[1 + \frac{i^2}{1\cdot 2}t\left(\frac{1}{t} - 1\right)^2 + \frac{i^2(i^2-1)}{1\cdot 2\cdot 3\cdot 4}t^2\left(\frac{1}{t} - 1\right)^4 + \cdots\right]$$

$$+ \frac{i}{2}u(1+t)\left\{\begin{aligned} &\frac{1}{t} - 1 + \frac{i^2 - 1}{1\cdot 2\cdot 3}t\left(\frac{1}{t} - 1\right)^3 \\ &\quad + \frac{(i^2-1)(i^2-4)}{1\cdot 2\cdot 3\cdot 4\cdot 5}t^2\left(\frac{1}{t} - 1\right)^5 + \cdots \end{aligned}\right\}.$$

---

[6] Lidstone [1922], pp. 21–26, Whittaker, *WR*, pp. 40–41, and Laplace VII [1820], p. 14. Whittaker and Robinson call the result the Laplace–Everett formula.

This implies directly that

$$y_{x+i} = y_x + \frac{i^2}{1\cdot 2}\Delta^2 y_{x-1} + \frac{i^2(i^2-1)}{1\cdot 2\cdot 3\cdot 4}\Delta^4 y_{x-2} + \frac{i^2(i^2-1)(i^2-4)}{1\cdot 2\cdot 3\cdot 4\cdot 5\cdot 6}\Delta^6 y_{x-3} + \cdots + \frac{i}{2}(\Delta y_x + \Delta y_{x-1}) + \frac{i}{2}\frac{i^2-1}{1\cdot 2\cdot 3}(\Delta^3 y_{x-1} + \Delta^3 y_{x-2}) + \frac{1}{2}\frac{(i^2-1)(i^2-4)}{1\cdot 2\cdot 3\cdot 4\cdot 5}(\Delta^5 y_{x-2} + \Delta^5 y_{x-3}) + \cdots .$$

This is the so-called Newton–Stirling formula. (Recall that $\Delta^i y_x = u(1/t - 1)^i$.)[7]

Laplace points out that this formula enables us to interpolate among $2x + 1$ equidistant quantities; y_x is the mean of $y_0, y_1, y_2, \ldots, y_{2x}$, and i is its distance from this mean. The corresponding result for the even case is obtained by changing i into $i + 1$ in the last expression above for u/t^i; i.e., the expression $u/t^{i+1} - u/t^i = u(1/t - 1)/t^i$ is divided by $(1/t - 1)$ and there results

$$\frac{u}{t^i} = \frac{u}{2}(1+t)\left\{1 + \frac{(i+\frac{1}{2})^2 - \frac{1}{4}}{1\cdot 2}t\left(\frac{1}{t}-1\right)^2 + \frac{[(i+\frac{1}{2})^2 - \frac{1}{4}][(i+\frac{1}{2})^2 - \frac{9}{4}]}{1\cdot 2\cdot 3\cdot 4}t^2\left(\frac{1}{t}-1\right)^2 + \cdots\right\}$$
$$+ \left(i + \frac{1}{2}\right)ut\left(\frac{1}{t}-1\right)\left\{1 + \frac{(i+\frac{1}{2})^2 - \frac{1}{4}}{1\cdot 2\cdot 3}t\left(\frac{1}{t}-1\right)^2 + \frac{[(i+\frac{1}{2})^2 - \frac{1}{4}][(i+\frac{1}{2})^2 - \frac{9}{4}]}{1\cdot 2\cdot 3\cdot 4\cdot 5}t^2\left(\frac{1}{t}-1\right)^4 + \cdots\right\}.$$

This then yields the interpolation relation

$$y_{x+i} = \frac{1}{2}(y_x + y_{x-1}) + \frac{(i+\frac{1}{2})^2 - \frac{1}{4}}{1\cdot 2}\frac{1}{2}(\Delta^2 y_{x-1} + \Delta^2 y_{x-2}) + \frac{[(i+\frac{1}{2})^2 - \frac{1}{4}][(i+\frac{1}{2})^2 - \frac{9}{4}]}{1\cdot 2\cdot 3\cdot 4}\frac{1}{2}(\Delta^4 y_{x-2} + \Delta^4 y_{x-3}) + \cdots$$
$$+ \left(i + \frac{1}{2}\right)\left\{\Delta y_{x-1} + \frac{(i+\frac{1}{2})^2 - \frac{1}{4}}{1\cdot 2\cdot 3}\Delta^3 y_{x-2} + \frac{[(i+\frac{1}{2})^2 - \frac{1}{4}][(i+\frac{1}{2})^2 - \frac{9}{4}]}{1\cdot 2\cdot 3\cdot 4\cdot 5}\Delta^5 y_{x-3} + \cdots\right\}.$$ [8]

---

[7] Laplace VII [1820], p. 16.

[8] Laplace VII [1820], pp. 13–17. This result is essentially the Newton–Bessel formula. Cf. Lidstone [1920], p. 24*n*.

4.3. Laplace on Finite Differences

Instead of following the development of this topic by Laplace, let us briefly study his operator methods.[9] Clearly we have

$$u\left(\frac{1}{t^i} - 1\right)^n = u\left[\left(1 + \frac{1}{t} - 1\right)^i - 1\right]^n, \tag{4.6}$$

and we know that the coefficient of t^x in the left-hand member of this relation is the nth difference of y_x, where x varies now by i unitsteps. Let us set $'\Delta y_x = y_{x+i} - y_x$. This is consistent, since the coefficient of t^x in $u(1/t^i - 1)$ is $y_{x+i} - y_x$. Moreover if we develop the right-hand member of (4.6) in powers of $(1/t - 1)$, we see that the coefficients of t^x in each of the terms are $\Delta y_x, \Delta^2 y_x, \ldots$. Further, when we combine these, we find the total coefficient of t^x to be $[(1 + \Delta y_x)^i - 1]^n$, and hence

$$'\Delta^n y_x = [(1 + \Delta y_x)^i - 1]^n. \tag{4.7}$$

Similarly, Laplace defines finite sums $\sum^n$ and $'\sum^n$ by the formal equality

$$'\textstyle\sum^n y_x = [(1 + \Delta y_x)^i - 1]^{-n}. \tag{4.8}$$

To see why this is so let u be the generating function of the sequence y_x and z of $\sum^r y_x$, where

$$\sum y_x = \sum_{\alpha=0}^{x-1} y_\alpha, \qquad \textstyle\sum^r y_x = \sum \sum^{r-1} y_x.$$

Now Laplace asserts that the coefficient of t^x in the expansion of $u(1/t - 1)^{-1}$ is, apart from sign, $\sum y_x$. This is quite easy to see by direct calculation. Moreover, if x varies by increments i instead of 1, then the coefficient of t^x in the expansion of $u(1/t^i - 1)^{-1}$ is, by definition, $'\sum y_x$. The higher order sums are derived analogously.

Laplace examines the relations (4.7) and 4.8) above with a view to relating differences and derivatives, finite sums and integrals. To do this, note that these relations hold even if in Δy_x the x changes by an increment ω instead of by 1, and so in $'\Delta y_x$ by $i\omega$ instead of by i. He now chooses $\omega = \Delta x$ and $i = \alpha/\omega$. Then (4.7) becomes

$$'\Delta^n y_x = \left[\alpha \frac{\Delta y_x}{\Delta x} + \frac{\alpha(\alpha - \Delta x)}{2!} \frac{\Delta^2 y_x}{\Delta x^2} + \cdots\right]^n,$$

and in the limit as Δx approaches 0 this expression may be written symbolically as

$$'\Delta^n y_x = (e^{\alpha(dy/dx)} - 1)^n, \tag{4.9}$$

where it is understood that α remains fixed. The symbolic relation (4.9)

[9] Laplace VII [1820], pp. 38ff. Cf. also p. 201 below

enables Laplace to relate differences and derivatives. Thus if we choose $n = 1$ we have $'\Delta y_x = e_x^{\alpha y'} - 1$ or

$$\frac{\alpha\, dy_x}{dx} = \log(1 + '\Delta y_x),$$

where $y'_x = dy/dx$ and $'\Delta y_x = y_{x+\alpha} - y_x$. (Laplace sometimes used c where we today use e for the base. Later he systematically adopted e.) In the same way the symbolic relation (4.8) may be written as

$$'\textstyle\sum^n y_x = (e_x^{iy'} - 1)^{-n}; \qquad (4.10)$$

and for $n = 1$ we have $'\sum y_x = (e_x^{iy'} - 1)^{-1}$. Let us return to (4.7) and (4.8). This time let us choose i to be Δx and see what happens. We have in (4.7)

$$\frac{\Delta^n y_x}{\Delta x^n} = \left[\frac{e^{\Delta x \cdot \log(1 + \Delta y_x)} - 1}{\Delta x}\right]^n$$

and so in the limit

$$\frac{d^n y_x}{dx^n} = [\log(1 + \Delta y_x)]^n;$$

similarly (4.8) becomes, in Laplace's notation,

$$\int^n y_x\, dx^n = \frac{1}{[\log(1 + \Delta y_x)]^n}.$$ [10]

Laplace proceeded in succeeding chapters to discuss generating functions and interpolation formulas in more than one variable.

4.4. Laplace Summation Formula

In his study of the variation of the elements of a comet's orbit he needed a formula to express an integral approximately as a sum. To obtain it he rediscovered Gregory's formula. Apparently he knew neither this nor the Maclaurin–Euler formula; at least he makes no reference to either of them. His derivation is another application of his powerful method of generating functions.[11] As before, he expresses y_x as a series

$$y_x = y_0 + x\Delta y_0 + \frac{x(x-1)}{2!}\Delta^2 y_0 + \cdots,$$

[10] Laplace VII [1820], pp. 40ff.

[11] Laplace IV [1805], pp. 205–208.

which he integrates from $x = 0$ to $x = 1$. He has

$$\int_0^1 y_x\,dx = y_0 + \frac{1}{2}\Delta y_0 - \frac{1}{12}\Delta^2 y_0 + \frac{1}{24}\Delta^3 y_0 - \frac{19}{720}\Delta^4 y_0 + \frac{3}{160}\Delta^5 y_0 - \frac{863}{60480}\Delta^6 y_0 + \cdots .$$

Similarly he finds the areas between 1 and 2, etc. Summing these he infers that

$$\int_0^n y_x\,dx = \sum_{\alpha=0}^{n-1} y_\alpha + \frac{1}{2}\sum_{\alpha=0}^{n-1}\Delta y_\alpha - \frac{1}{12}\sum_{\alpha=0}^{n-1}\Delta^2 y_\alpha + \cdots . \tag{4.11}$$

Now he notes that, by definition,

$$\sum_{\alpha=0}^{n-1}\Delta y_\alpha = y_n - y_0, \qquad \sum_{\alpha=0}^{n-1}\Delta^2 y_\alpha = \Delta y_n - \Delta y_0, \ldots,$$

and so he has

$$\int_0^n y_x\,dx = \frac{1}{2}y_0 + y_1 + y_2 + \cdots + y_{n-1} + \frac{1}{2}y_n - \frac{1}{12}[\Delta y_n - \Delta y_0] + \frac{1}{24}[\Delta^2 y_n - \Delta^2 y_0] - \cdots .$$

Then Laplace remarks that the differences $\Delta y_n, \Delta^2 y_n$ depend upon values $y_{n+1}, y_{n+2}, \ldots$ outside the range of the integration. He now proceeded to remove this difficulty. The coefficient of t^n in the expansion of $u(1/t - 1)^r$ is, as we saw earlier, $\Delta^r y_n$; but this expansion may be carried out in the following way: Firstly, notice that

$$u\left(\frac{1}{t} - 1\right)^r = u(1 - t)^r[1 - (1 - t)]^{-r} = u(1 - t)^r\left[1 + r(1 - t) + \frac{r(r + 1)}{2!}(1 - t)^2 + \cdots\right].$$

Secondly, note that the coefficient of t^n in the expansion of $u(1 - t)^q$ is $\Delta^q y_{n-q}$, and hence

$$\Delta^r y_n = \Delta^r y_{n-r} + r\Delta^{r+1} y_{n-r-1} + \frac{r(r + 1)}{2!}\Delta^{r+2} y_{n-r-2} + \cdots .$$

Now this relation enables Laplace to replace his formula (4.11) by the equivalent:

$$\int_0^n y_x\,dx = \frac{1}{2}y_0 + y_1 + y_2 + \cdots + y_{n-1} + \frac{1}{2}y_n - \frac{1}{12}[\Delta y_{n-1} - \Delta y_0] - \frac{1}{24}[\Delta^2 y_{n-2} + \Delta^2 y_0]$$

$$-\frac{19}{720}[\Delta^3 y_{n-3} - \Delta^3 y_0] - \frac{3}{160}[\Delta^4 y_{n-4} + \Delta^4 y_0] + \cdots .$$

Laplace does not indicate a formula for the general term, nor does he consider any convergence questions. The general term in this series is actually

$$(-1)^r b_r[\Delta^r y_{n-r} - (-1)^{r-1}\Delta^r y_0],$$

where the b_n are coefficients of the Bernoulli polynomial of the second kind.[12] They are

$$b_n = \frac{1}{n!}\sum_{m=1}^{u} \frac{S_n^m}{m+1} = \int_0^1 \binom{x}{n} dx,$$

in which the S_n^m are Stirling's numbers of the first kind as we see directly. The polynomials themselves may be defined by the relations

$$\frac{d\psi_n}{dx} = \binom{x}{n-1}, \qquad \Delta\psi_n = \psi_{n-1}, \qquad \psi_0 = 1. \tag{4.12}$$

This completely defines the ψ_n $(n = 0, 1, \ldots)$. If we express ψ_n in the form

$$\psi_n(x) = b_0\binom{x}{n} + b_1\binom{x}{n-1} + \cdots + b_n, \tag{4.13}$$

then these coefficients $b_0, b_1, \ldots, b_n$ are the Bernoulli numbers of the second kind given above.[13] To see the form of the b_n we recall that

$$(x)_n = \sum_{\alpha=1}^{n} \frac{x^\alpha}{\alpha!} \frac{d^\alpha(x)_n}{dx^\alpha}\bigg|_{x=0} = \sum_{\alpha=1}^{n} x^\alpha S_n^\alpha,$$

by the definition of the Stirling numbers. This relation expresses $(x)_n$ as a polynomial in x. If we integrate term by term, we find directly that

$$\psi_{n+1}(x) = \frac{1}{n!}\sum_{\alpha=1}^{n} S_n^\alpha \frac{x^{\alpha+1}}{\alpha+1} + b_{n+1},$$

which becomes, for $x = 1$,

$$\psi_{n+1}(1) - b_{n+1} = \frac{1}{n!}\sum_{\alpha=1}^{n} \frac{S_n^\alpha}{\alpha+1}.$$

To evaluate the left-hand member of this equation we use the second of the relations (4.12) and the expression (4.13) for ψ_n. Clearly $\psi_n(0) = b_n$, $\psi_{n+1}(1) - b_{n+1} = \psi_n(0) = b_n$, and hence

$$b_n = \frac{1}{n!}\sum_{\alpha=1}^{n} \frac{S_n^\alpha}{\alpha+1}.$$

Next, by the first of the relations (4.12), we have

---

[12] Jordan [1939], p. 147. The result now goes by the name of the *Laplace summation formula*. See pp. 77ff above

[13] Jordan [1939], pp. 265ff.

$$b_n = \psi_{n+1}(1) - \psi_{n+1}(0) = \int_0^1 \binom{x}{n}\, dx.$$

Finally we should note that the signs of the $\binom{x}{n}$ on (0, 1) are easy to determine. Clearly the sign of $\binom{x}{n}$ is $(-1)^{n+1}$, so that $b_{2\alpha} < 0$, $b_{2\alpha+1} > 0$. This tells us that all coefficients in the Laplace summation formula are negative.

A major achievement of Laplace was his work in probability theory. It is outside the scope of this book to discuss his accomplishments in this field in general, but it is essential to mention some of them. Credit must be given jointly for one of the most impressive results to de Moivre and Laplace. They were responsible for the recognition and use of the so-called Gaussian or Normal distribution function.[14] Based on this, Laplace gave an excellent discussion of the Method of least squares. Since this topic intimately involves not only Laplace, but also Legendre and Gauss, we will leave the discussion of "least squares" until later, while we treat some other aspects of Laplace's work in our field.

4.5. Laplace on Functional Equations

Laplace made major contributions to the theory of difference equations, particularly to linear partial ones. Much of this material appeared in connection with his interest in probability theory.[15]

There is such a wealth of material in these papers that it is difficult to be selective. Let us start, however, with Laplace's solution in 1772 of a problem posed by Monge. This problem was also solved by Condorcet in 1771 by a different method. The problem is this: given functions $\varphi(x)$, $\psi(x)$, H_x, X_x, to find a function f such that

$$f[\varphi(x)] = H_x f[\psi(x)] + X_x.$$

Laplace's solution is of interest to study. He arbitrarily set $u_z = \psi(x)$, $u_{z+1} = \varphi(x)$, and solved the first for x, finding $x = \Gamma(u_z)$; hence the second becomes $u_{z+1} = \Pi(u_z)$. But this is a difference equation of the first order which is solvable. Its solution $u_z = P(z)$ gives $x = T(z) = \Gamma[P(z)]$. If this value of x as a function of z is substituted into H_x, X_x, we find functions L_z and Z_z; and, if f exists, we must have $f[\varphi(x)] = f(u_{z+1})$ and $f[\psi(x)] = f(u_z)$. Laplace now chooses $f(u_z) = y_z$, and then finds the linear difference equation

$$y_{z+1} = L_z y_z + Z_z, \tag{4.14}$$

whose solution he has previously discussed. He notes here that the "constants" of integration may, as remarked by Euler, be supposed to be any

---

[14] Feller, *Prob.*, Vol. I, pp. 168ff. There is some discussion in that book of Laplace's contributions.

[15] There are also several major papers on the subject. Laplace VII [1772] and VIII [1776]. Cf. also, VII [1820].

functions of sin $2\pi z$, cos $2\pi z$, where "π expresses the ratio of the circumference to the diameter." [Thus the symbol π is just coming into fashion.][16] Laplace's solution of (4.14) above is very simple.[17] He sets $\nabla L_z = L_1 \cdot L_2 \cdots L_z$, and defines a function w_z so that

$$y_z = w_z \nabla L_{z-1}.$$

Then evidently $w_{z+1}\nabla L_z = L_z w_z \nabla L_{z-1} + Z_z = w_z \nabla L_z + Z_z$. Thus $\Delta w_z \cdot \nabla L_z = Z_z$, and so

$$w_z = A + \sum_{\alpha=0}^{z-1} \frac{Z_\alpha}{\nabla L_\alpha} = A + \sum \frac{Z_z}{\nabla L_z}$$

so that

$$f(u_z) = y_z = \nabla L_{z-1}\left(A + \sum \frac{Z_z}{\nabla L_z}\right).$$

Laplace completes his discussion with the remark: "Now if in the expression for y_x we substitute for z its value in x, we have $f[\psi(x)]$, and if we replace $\psi(x)$ by x, we have the function of x that satisfies the problem. The following examples will clarify the method." Laplace now gives as an illustration the case when $\varphi(x) = x^q$, $\psi(x) = mx$ and $H_x = 1$, $X_x = p$. Thus he wishes to solve $f(x^q) = f(mx) + p$ for f. Since $u_z = mx$, $u_{z+1} = x^q$, he has $u_{z+1} = (u_z/m)^q$. To solve this he sets $u_1 = a$ and finds

$$u_2 = \frac{a^q}{m^q}, \qquad u_3 = \frac{a^{q^2}}{m^{q^2+q}}, \ldots \quad .$$

He introduces functions f_z, g_z so that $u_z = a^{g_z}/m^{f_z}$. Then

$$u_{z+1} = \frac{a^{qg_z}}{m^{qf_z+q}} = \frac{a^{g_{z+1}}}{m^{f_{z+1}}},$$

from which he infers that $g_{z+1} = qg_z$, $f_{z+1} = qf_z + q$, or $g_z = Aq^z$; but for $z = 2$, $g_z = q$. Thus

$$g_z = q^{z-1}, \qquad f_z = \frac{1}{q-1}(q^z - q),$$

and

$$u_z = \frac{a^{q^{z-1}}}{m^{(1/(q-1))(q^z-q)}}.$$

Laplace notes that this is the complete solution since it contains the arbitrary

[16] Laplace VIII [1776], pp. 103ff.
[17] Laplace VIII [1776], pp. 74–75.

constant a. Next he has $y_{z+1} = y_z + p$, whose complete solution is $y_z = C + pz = f(mx)$. Since $u_z = \psi(x)$, this means that

$$mx = \frac{a^{q^z - 1}}{m^{(1/(q-1))(q^z - q)}},$$

or

$$lmx = q^z \frac{la}{q} - \frac{1}{q-1}(q^z - q)lm.$$

From this Laplace concludes that

$$q^z\left(\frac{la}{q} - \frac{lm}{q-1}\right) = l\frac{mx}{m^{q/(q-1)}}.$$

[Laplace at this time wrote the logarithm as l. Later he wrote it as log.] Hence, in his notation,

$$z = \frac{ll\dfrac{mx}{m^{q/(q-1)}}}{lq} - \frac{lK}{lq}, \qquad \frac{la}{q} - \frac{lm}{q-1} = K$$

and

$$y_z = A + p\frac{ll\dfrac{mx}{m^{q/(q-1)}}}{lq},$$

where A is an arbitrary constant which can be any function of $\sin 2\pi z$ and $\cos 2\pi z$. Let $\Gamma(\sin 2\pi z, \cos 2\pi z)$ be this function. Then

$$y_z = f(mx) = \Gamma\left(\sin 2\pi\frac{ll\dfrac{mx}{m^{q/(q-1)}}}{lq}, \cos 2\pi\frac{ll\dfrac{mx}{m^{q/(q-1)}}}{lq}\right) + p\frac{ll\dfrac{mx}{m^{q/(q-1)}}}{lq};$$

and hence

$$f(x) = \Gamma\left(\sin 2\pi\frac{ll\dfrac{x}{m^{q/(q-1)}}}{lq}, \cos 2\pi\frac{ll\dfrac{x}{m^{q/(q-1)}}}{lq}\right) + p\frac{ll\dfrac{x}{m^{q/(q-1)}}}{lq}.$$

In his "Recherches sur le calcul intégral et sur le système du Monde", (Laplace VII [1772]) Laplace took up the theory of determinants or what he called *résultantes*.[18] In this paper he refers to what we now call Cramer's rule. He says, "the geometers have given general rules for this purpose" (see the *Introduction à l'analyse des lignes courbes*, of M. Cramer, and the *Mémoires de l'Académie* for the year 1764, p. 292)![19] Later he expands this

---

[18] Laplace VIII [1772], pp. 395ff.

[19] Laplace VIII [1772], p. 395. Gabriel Cramer (1704–1752) was a Swiss mathematician who followed Euler in classifying quartic curves. Etienne Bezout (1730–1783), a Frenchman, also worked on determinants, and Sylvester referred to a certain determinant as the *Bezoutiant*. Cf. Cramer [1750].

and says: "This rule is due to M. Cramer, but it can be simplified by the following procedure, that M. Bezout has given in the place cited in the *Mémoires* of the Academy."[20] Laplace gave a rather detailed discussion of the effect of interchanging rows or columns in these pages. He also developed what is sometimes called Laplace's development by columns or rows.[21] For a really complete discussion of the early history of determinants the reader is referred to a compendious work on the subject by Sir Thomas Muir (cf. Muir [1906/23]).

4.6. Laplace on Finite Sums and Integrals

In a 1777 paper Laplace reverted to his earlier interest in expressing the relations between finite differences and derivatives as well as between finite sums and integrals.[22] He considered a function u of t, t_1, t_2, ..., and supposed that these independent variables are themselves functions $t + \alpha$, $t_1 + \alpha_1$, $t_2 + \alpha_2$, ... To make things simpler notationally he considered only the case of one variable, but remarked that his analysis was equally valid for any number. He knew that (in his notation)

$$\Delta \cdot u = \alpha\left(\frac{\partial u}{\partial t}\right) + \frac{\alpha^2}{1\cdot 2}\cdot\left(\frac{\partial^2 u}{\partial t^2}\right) + \frac{\alpha^3}{1\cdot 2\cdot 3}\cdot\left(\frac{\partial^3 u}{\partial t^3}\right) + \&\text{c.},$$

and hence that

$$\Delta^2 \cdot u = \alpha\cdot\Delta\cdot\left(\frac{\partial u}{\partial t}\right) + \frac{\alpha^2}{1\cdot 2}\cdot\Delta\cdot\left(\frac{\partial^2 u}{\partial t^2}\right) + \frac{\alpha^3}{1\cdot 2\cdot 3}\cdot\Delta\cdot\left(\frac{\partial^3 u}{\partial t^3}\right) + \&\text{c.}$$

Thus by replacing u successively by $\partial u/\partial t$, $\partial^2 u/\partial t^2$, ..., he finds

$$\alpha\Delta\cdot\left(\frac{\partial u}{\partial t}\right) = \alpha^2\cdot\left(\frac{\partial^2 u}{\partial t^2}\right) + \frac{\alpha^3}{1\cdot 2}\cdot\left(\frac{\partial^3 u}{\partial t^3}\right) + \&\text{c.},$$

$$\alpha^2\Delta\cdot\left(\frac{\partial^2 u}{\partial t^2}\right) = \alpha^3\cdot\left(\frac{\partial^3 u}{\partial t^3}\right) + \frac{\alpha^4}{1\cdot 2}\cdot\left(\frac{\partial^4 u}{\partial t^4}\right) + \&\text{c.}, \&\text{c.}$$

Hence by substitution

$$\Delta^2\cdot u = \alpha^2\cdot\left(\frac{\partial^2 u}{\partial t^2}\right) + a\cdot\alpha^3\left(\frac{\partial^3 u}{\partial t^3}\right) + a^1\cdot\alpha^4\cdot\left(\frac{\partial^4 u}{\partial t^4}\right) + \&\text{c.},$$

[20] Laplace VIII [1772], p. 396.

[21] Dickson [1922], pp. 122–123. Laplace's discussion is contained in Laplace VIII [1772], pp. 395–406.

[22] Laplace IX [1777], pp. 315–335. (Note in what follows how Laplace now used the symbol ∂/∂ for partial derivatives.)

and in general

$$\Delta^i \cdot u = \alpha^i \cdot \left(\frac{\partial^i u}{\partial t^i}\right) + h \cdot \alpha^{i+1} \cdot \left(\frac{\partial^{i+1} u}{\partial t^{i+1}}\right) + h' \cdot \alpha^{i+2} \cdot \left(\frac{\partial^{i+2} u}{\partial t^{i+2}}\right) + \text{\&c.},$$

where $h, h', \ldots$ are coefficients independent of α and t, which remain to be determined. To do this he made a remark that enabled him to prove a result that Lagrange only conjectured and that we mentioned earlier on p. 165. He noted that the $h, h', \ldots$ are not only independent of α and t but of u. It therefore suffices to find them for $u = e^t$ — here Laplace used e as we do. Then

$$\begin{aligned} \Delta \cdot u &= e^{t+\alpha} - e^t = (e^\alpha - 1) \cdot e^t; \\ \Delta^2 \cdot u &= (e^\alpha - 1) \cdot (e^{t+\alpha} - e^t) = (e^\alpha - 1)^2 \cdot e^t; \qquad \text{\&c.,} \\ \Delta^i \cdot u &= (e^\alpha - 1)^i \cdot e^t; \end{aligned}$$

and we have $(e^\alpha - 1)^i = \alpha^i + h \cdot \alpha^{i+1} + h' \cdot \alpha^{i+2} + \text{\&c.}$ Now it follows from this that

$$\alpha^i \cdot \left(\frac{\partial^i u}{\partial t^i}\right) + h \cdot \alpha^{i+1} \cdot \left(\frac{\partial^{i+1} u}{\partial t^{i+1}}\right) + h' \cdot \alpha^{i+2} \cdot \left(\frac{\partial^{i+2} u}{\partial t^{i+2}}\right) + \text{\&c.} = (e^{\alpha(\partial u/\partial t)} - 1)^i,$$

provided that in the development of the right-hand member we write $\partial^n u/\partial t^n$ in place of $(\partial u/\partial t)^n$. Thus

$$\Delta^i u = (e^{\alpha(\partial u/\partial t)} - 1)^i$$

for i positive.

Now if we suppose that

$$\sum^i z = \frac{\partial^i u}{\partial t^i}$$

then clearly

$$u = \sum^i \cdot \int^i \cdot z \, \partial t^i, \qquad \Delta^i u = \int^i \cdot z \, \partial t^i;$$

and our relations above imply that

$$\sum^i z = \frac{\int^i z \, dt^i}{\alpha^i} - h\alpha \sum^i \frac{\partial z}{\partial t} - h'\alpha^2 \sum^i \frac{\partial^2 z}{\partial t^2} - \cdots \quad .$$

By repeatedly differentiating and substituting, Laplace found

$$\begin{aligned} \sum^i z = {} & \frac{\int^i z \, dt^i}{\alpha^i} + r \frac{\int^{i-1} z \, dt^{i-1}}{\alpha^{i-1}} + r' \frac{\int^{i-2} z \, dt^{i-2}}{\alpha^{i-2}} + \cdots \\ & + sz + s'\alpha \frac{\partial z}{\partial t} + s''\alpha^2 \frac{\partial^2 z}{\partial t^2} + \cdots, \end{aligned}$$

where the $r, r', \ldots, s, s', \ldots$ are independent of α and t. To determine these constants let $z = e^t$. Then this relation becomes

$$\frac{1}{(e^\alpha - 1)^i} = \frac{1}{\alpha^i} + \frac{r}{\alpha^{i-1}} + \frac{r'}{\alpha^{i-2}} + \cdots + s + \alpha s' + \alpha^2 s'' + \cdots,$$

since $\sum^i z = 1/\Delta^i z$ by definition, so that

$$\frac{\int^i z\,dt^i}{\alpha^i} + \frac{r\int^{i-1} z\,dt^{i-1}}{\alpha^{i-1}} + \cdots + sz + \alpha s' \frac{\partial z}{\partial t} + \cdots = \frac{1}{(e^{\alpha(\partial z/\partial t)} - 1)^i},$$

provided that we now write $\partial^n z/\partial t^n$ for $(\partial z/\partial t)^n$ with n positive but $\int^{-n} z\,dt^{-n}$ for n negative. Thus

$$\sum^i z = \frac{1}{(e^{\alpha(\partial z/\partial t)} - 1)^i}.$$[23]

Laplace remarked that the two expressions above could be incorporated in the following:

$$(\Delta u)^i = (e^{\alpha(\partial u/\partial t)} - 1)^i, \tag{4.23}$$

provided one understands that when i is negative, differences are replaced by sums and expressions such as $\partial^{-i}u/\partial t^{-i}$ by $\int^i u\,dt^i$. He then wrote, for positive i,

$$[\log(1 + \Delta u)]^i = [\log(1 + e^{\alpha(\partial u/\partial t)} - 1)]^i = \alpha^i \frac{\partial^i u}{\partial t^i} \tag{4.24}$$

and, for negative i, if one changes i to $-i$

$$\frac{\int^i u\,dt^i}{\alpha^i} = \frac{1}{[\log(1 + \Delta u)]^i}.$$

With the help of these relations he was able to reestablish Lagrange's symbolic interpolation results. (The fundamental point Laplace noted and Lagrange did not was that the $h, h', \ldots$ were independent of the function u.) Laplace said: "... given by Lagrange in his excellent Mémoire entitled, '*Sur une nouvelle espèce de calcul relatif à la différentiation et à l'intégration des quantités variables*', and which appears in the volume of the Berlin Academy for the year 1772. This fine analyst has there noted the very remarkable analogy between positive powers and differences, and between negative powers and sums; he was satisfied to see the analogy, but he regarded the demonstration as very difficult (cf. the Mémoire cited, pp. 186 and 195); it is this which I have undertaken to demonstrate here by a method, which, if I am not deceived, is also both as direct and simple as one can desire and which has the virtue of showing *a priori* the reason for this curious analogy."[24]

[23] Laplace IX [1777], pp. 315–320.

[24] Laplace IX [1777], p. 327. The paper of Lagrange cited appears in Lagrange III [1772], pp. 441–476.

To find the Lagrangian formulas as developed by Laplace suppose that α_1 is a value of α and that we wish to find differences and sums of u at $\alpha = \alpha_1$. Write $\Delta_1 u = u(\alpha_1/\alpha) - u(1)$ and note that

$$(1 + \Delta u)^{\alpha_1/\alpha} = (1 + e^{\alpha(\partial u/\partial t)} - 1)^{\alpha_1/\alpha} = e^{\alpha_1(\partial u/\partial t)}$$

for every α and α_1, and thus

$$[(1 + \Delta u)^{\alpha_1/\alpha} - 1]^i = (e^{\alpha_1(\partial u/\partial t)} - 1)^i.$$

(Notice that the left-hand member is the operator $E^{\alpha_1/\alpha} - 1 = (1 + \Delta)^{\alpha_1/\alpha} - 1$ acting upon u, where $Eu(x) = u(x + 1)$.) Laplace now remarks that

$$\Delta_1 u = (1 + \Delta u)^{\alpha_1/\alpha} - 1,$$

where the right-hand member is an operator acting upon u; it may be expressed in more modern terms as $[(1 + \Delta)^{\alpha_1/\alpha} - 1]\cdot u$; and he has the dual relationships

$$\Delta_1^i u = (e^{\alpha_1(\partial u/\partial t)} - 1)^i, \qquad \Sigma_1^i\, u = \frac{1}{(e^{\alpha_1(\partial u/\partial t)} - 1)^i}.$$

Hence he has

$$\Delta_1^i u = [(1 + \Delta u)^{\alpha_1/\alpha} - 1]^i, \tag{4.25}$$

$$\Sigma_1^i\, u = \frac{1}{[(1 + \Delta u)^{\alpha_1/\alpha} - 1]^i}. \tag{4.26}$$

Laplace remarks: "These two formulae encompass the theory of interpolation with all possible generality."[25]

Lagrange incidentally was particularly interested in extending these results to the case of two variables so that he could interpolate in double entry tables. To this end he wrote

$$\Delta' u = (1 + \Delta_\xi u)^{\xi'/\xi}\cdot(1 + \Delta_\psi u)^{\psi'/\psi} - 1$$

and expanded the right-hand member, finding

$$\Delta' u = \frac{\xi'}{\xi}\Delta_\xi u + \frac{\psi'}{\psi}\Delta_\psi u + \frac{\xi'(\xi' - \xi)}{2\xi^2}\Delta^2_{\xi\xi} u$$
$$+ \frac{\xi'\psi'}{\xi\psi}\Delta^2_{\xi\psi} u + \frac{\psi'(\psi' - \psi)}{2\psi^2}\Delta^2_{\psi\psi} u + \cdots,$$

where $\Delta_\xi, \Delta_\psi, \ldots$ are partial differences. He remarked that this result "... agrees with that given by M. Lambert for the same purpose in the third part of his Beiträge, etc."[26]

Lagrange also used the formula (4.26) with $i = 1$ to write out a formula "which could serve to calculate the areas of curves by the sums and differences of equidistant coordinates. Cotes, Stirling and others previously gave

[25] Laplace IX [1777], p. 327.
[26] Lagrange III [1772], p. 465.

formulae for calculating the area of a curve when a certain number of equidistant coordinates have been given; but the preceding formula is different from those of the authors. . . ."[27] His actual formula in his notation is

$$\frac{\int u\,dx}{\xi} = \sum u + \mu u + \nu\,\Delta u + \omega\Delta^2 u + \chi\Delta^3 u + \cdots,$$

where

$$\mu = \frac{1}{2}, \quad 2\nu = -\frac{1}{2\cdot 3}, \quad 3\omega = -\frac{\nu}{2} + \frac{1}{3\cdot 4}, \quad 4\chi = -\omega + \frac{\nu}{2\cdot 3} - \frac{1}{4\cdot 5}, \ldots.$$

He used this to estimate the value of $\log x = \int dx/x$. Let $\xi = 1$ in the formula above; then

$$\begin{aligned}\log x = \int \frac{dx}{x} &= \sum \frac{1}{x} + \frac{\mu}{x} + \nu\Delta\frac{1}{x} + \omega\Delta^2\frac{1}{x} + \chi\Delta^3\frac{1}{x} + \cdots \\ &= \frac{1}{x-1} + \frac{1}{x-2} + \frac{1}{x-3} + \cdots \\ &\quad + \frac{\mu}{x} - \frac{\nu}{x(x+1)} + \frac{2\omega}{x(x+1)(x+2)} \\ &\quad - \frac{2\cdot 3\cdot \chi}{x(x+1)(x+2)(x+3)} + \cdots \quad .\end{aligned}$$

From this it is quite easy to derive his final result

$$\log\left(1 + \frac{1}{x}\right) = \frac{1}{x} - \frac{\mu}{x(x+1)} + \frac{2\nu}{x(x+1)\,(x+2)} - \frac{2\cdot 3\omega}{x(x+1)(x+2)(x+3)} + \frac{2\cdot 3\cdot 4\chi}{x(x+1)(x+2)(x+3)(x+4)} - \cdots .$$ [28]

Laplace returned to the relations (4.25), (4.26) in yet another paper, on sequences.[29] He developed them this time using the generating function for the sequence. Next he replaced his original sequence y_x by $h^x y_x$. He then found, in his notation,

$$^1\Delta^n h^x y_x = h^x[h^i(1 + \Delta y_x)^i - 1]^n, \tag{4.27}$$

where $^1\Delta y_x = y(x + i) - y(x)$ and where the right-hand member is again to be viewed as an operator acting upon y_x, i.e., as $h^x[h^i(1 + \Delta)^i - 1]^n \cdot y_x$. Again he had

$$^1\sum^n (h^x y_x) = \frac{h^x}{[h^i(1 + \Delta y_x)^i - 1]^n} + ax^{n-1} + bx^{n-2} + \cdots + f, \tag{4.28}$$

---

[27] Lagrange III [1772], p. 458. Cf. p. 169 above in Chap. III.

[28] Lagrange III [1772], p. 460. (Note that $\log(1 + 1/x) = \log(1 + x) - \log x$.) This has been discussed in Chap. III above on p. 170.

[29] Laplace X [1779], pp. 1–89.

in which $a, b, \ldots, f$ are n arbitrary constants of summation. Laplace now wishes to let the increment of x approach zero while that of ${}^1\Delta y_x$ stays finite. To do this he images y_x to be a function y_1 of $x_1 = x/r$, $h^r = p$ and $i/r = \alpha$, a constant. Then $h^x = p^{x_1}$ and $h^x y_x = p^{x_1} y_1(x_1)$. He now takes his relations (4.27), (4.28) and lets both i and r approach infinity so that α remains fixed. Laplace next comments that when x changes by i, x_1 changes by α, and hence that ${}^1\Delta^n(p^{x_1}y_1)$ and ${}^1\sum^n (p^{x_1}y_1)$ will be the nth difference and sum when x_1 varies by increments α. Further he says that "... one has $\Delta y_x = dy_1$, the difference dx_1 being equal to $1/r$." On this basis he writes

$$^1\Delta^m(p^{x_1}y_1) = p^{x_1}[p^\alpha(1 + dy_1)^i - 1]^n,$$

$$^1\sum^n (p^{x_1}y_1) = \frac{p^{x_1}}{[p^\alpha(1 + dy_1)^i - 1]^n} + ax_1^{n-1} + bx_1^{n-2} + \cdots \quad .$$

But

$$(1 + dy_1)^i = e^{\alpha(dy_1/dx_1)}$$

and hence

$$^1\Delta^m(p^{x_1}y_1) = p^{x_1}(p^\alpha e^{\alpha(dy_1/dx_1)} - 1)^n, \tag{4.29}$$

$$^1\sum^m (p^{x_1}y_1) = \frac{p^{x_1}}{(p^\alpha e^{\alpha(dy_1/dx_1)} - 1)^n} + ax_1^{n-1} + bx_1^{n-2} + \cdots + f. \tag{4.30}$$

Next Laplace returns to the relations (4.27), (4.28) and supposes that $i = dx$ becomes infinitely small; in other words, he now passes to the limit as i approaches zero in both members of these relations. To do this he notes that (to first order terms) $h^i(1 + \Delta y_x)^i = 1 + dx \log [h(1 + \Delta y_x)]$, and thus he has

$$\frac{d^n(h^x y_x)}{dx^n} = h^x[\log h(1 + \Delta y_x)]^n, \tag{4.31}$$

$$\int^n h^x y_x \, dx^n = \frac{h^x}{[\log h(1 + \Delta y_x)]^n} + ax^{n-1} + bx^{n-2} + \cdots + f. \tag{4.32}$$

He again attributes much to Lagrange but says that the results (4.27)–(4.32) are his own, with the exception of (4.32), which he credits in the case $n = 1$ to Euler, in his "*Institutions de Calcul différentiel.*"

4.7. Laplace on Difference Equations

It is very noteworthy that Laplace saw he could have derived an infinity of analogous results by considering not differences or sums of y_x but other functions of y_x. He does this in a very pretty problem (p. 41):

Let $\Gamma(y_x)$ represent any linear function of $y_x, y_{x+1}, y_{x+2}, \ldots$, and ∇y_x any other linear function of these same variables; it is proposed to find the expression for $\Gamma(y_x)$ in a sequence of the quantities $\nabla y_x, \nabla^2 y_x, \nabla^3 y_x, \ldots$.

He handles the problem by means of generating functions. He then proceeds with the help of this result to interpolation formulas for two variable sequences and to the integration of linear partial difference equations. Let us now examine how Laplace used his tool of the generating function to integrate linear difference equations, both total and partial, and how he introduced two different transforms to the same end. One of these is what we now call in his honor, the Laplace transform. While some authors give credit to Oliver Heaviside as the inventor or discoverer of the Operational Calculus, it seems clear to me that a large measure of the credit for this should be conferred upon Laplace. Much of what we now discuss is in his *Théorie analytique des probabilités.*[30] Consider the linear difference equation

$$0 = L(y_x) = ay_x + by_{x+1} + cy_{x+2} + \cdots + py_{x+n-1} + qy_{n+n}, \quad (4.33)$$

and let the generating function of y_x be

$$G\{y_x\} = u = y_0 + y_1 t + \cdots + y_x t^x + \cdots \quad .$$

Laplace discovered how to find the generating function for $Z_x = L(y_x)$. Let us suppose given a sequence $\{a_x\}$ and let its generating function $G\{a_x\}$ be

$$a_0 + a_1 t + a_2 t^2 + \cdots + a_x t^x + \cdots \quad .$$

Then the function G is clearly linear since $G[\alpha\{a_x\} + \beta\{b_x\}] = \alpha G\{a_x\} + \beta G\{b_x\}$; and it is easy to see that

$$G\{a_{x+1}\} = \sum_{n=0}^{\infty} a_{n+1} t^n = \frac{G\{a_x\} - a_0}{t}.$$

Now by induction it can be shown that

$$G\{a_{x+p}\} = \frac{1}{t^p}\,[G\{a_x\} - t^{p-1}a_{p-1} - t^{p-2}a_{p-2} - \cdots - a_0].$$

Moreover we note that, formally,

$$G\{xa_x\} = \sum_{n=0}^{\infty} na_n t^n = t\frac{d}{dt}\sum_{n=0}^{\infty} a_n t^n = t\frac{d}{dt}G\{a_x\},$$

$$G\{x(x-1)a_x\} = t^2\frac{d^2G\{a_x\}}{dt^2}, \ldots, G\{(x)_m a_x\} = t^m\frac{d^mG\{a_x\}}{dt^m}.$$

Finally we notice that

$$G\{(x+1)a_{x+1}\} = \frac{dG\{a_x\}}{dt}, \qquad G\{(x+1)(x+2)a_{x+2}\} = \frac{d^2G\{a_x\}}{dt^2}, \ldots \quad .$$

---

[30] Laplace VII [1820], pp. 80ff.

With these results it is now easy to write the generating function for the operator L above in (4.33). It is evidently

$$au + \frac{bu - y_0}{t} + \frac{cu - y_1 t - y_0}{t^2} + \cdots$$

$$= u\left(a + \frac{b}{t} + \frac{c}{t^2} + \cdots + \frac{p}{t^{n-1}} + \frac{q}{t^n}\right) - \left(\frac{A}{t} + \frac{B}{t^2} + \frac{C}{t^3} + \cdots + \frac{H}{t^n}\right),$$

where the $A, B, \ldots$ are functions of the initial values $y_0, y_1, \ldots, y_n$ of the sequence. They are then, in a sense, arbitrary constants of the integration. Since the left-hand member of the equation given by (4.33) is zero, we have

$$u(at^n + bt^{n-1} + \cdots + q) = (At^{n-1} + Bt^{n-2} + \cdots + H);$$

and so y_x, the solution of (4.33), is the coefficient of t^x in the power series expansion of

$$u = \frac{At^{n-1} + Bt^{n-2} + \cdots + H}{at^n + bt^{n-1} + \cdots + p}.$$

If the equation given by (4.33) had a nonzero left-hand member v_x with the generating function $z(t)$, then we would have

$$u = \frac{At^{n-1} + Bt^{n-2} + \cdots + H - t^n z(t)}{at^n + bt^{n-1} + \cdots + p}.$$

Consider as an example the equation $0 = y_x - 2y_{x+1} + y_{x+2}$, where $a = 1$, $b = -2$, $c = 1$ and $v_x = 0$. Then the solution must be the coefficient of t^x in the expansion of

$$u = \frac{At + B}{(t-1)^2} = (At + B) \sum_{\alpha=0}^{\infty} \binom{\alpha + 1}{\alpha} t^\alpha;$$

i.e.,

$$y_x = A\binom{x}{x-1} + B\binom{x+1}{x} = Ax + B(x+1) = \alpha + \beta x,$$

where α, β are arbitrary constants. Another example is that of Stirling.[31] Suppose we write

$$\frac{1}{a_0 + a_1 t + a_2 t^2} = y_0 + y_1 t + y_2 t^2 + \cdots,$$

and multiply through by the denominator of the left-hand member. Then we clearly have $1 = a_0 y_0$, $a_0 y_1 + a_1 y_0 = 0$, and $a_2 y_x + a_1 y_{x+1} + a_0 y_{x+2} = 0$. Thus the polynomial $a_0 + a_1 t + a_2 t^2$ may be inverted by solving this difference equation.

Laplace showed how his generating function technique could be applied as easily to linear partial difference equations with constant coefficients. He

---

[31] Stirling [1749], p. 2.

then took up the case where the coefficients are polynomials in x. Thus he considered

$$\begin{aligned}0 = &ay_x + by_{x+1} + cy_{x+2} + \cdots + qy_{x+n}\\ &+ x(a'y_x + b'y_{x+1} + c'y_{x+2} + \cdots + q'y_{x+n})\\ &+ x^2(a''y_x + b''y_{x+1} + c''y_{x+2} + \cdots + q''y_{x+n})\\ &+ \cdots \quad .\end{aligned}$$

By what we saw before, the generating function for y_x will now be determined with the help of the relation

$$\begin{aligned}&u\left(a + \frac{b}{t} + \frac{c}{t^2} + \cdots + \frac{q}{t^n}\right) + t\frac{d}{dt}\left[u\left(a' + \frac{b'}{t} + \frac{c'}{t^2} + \cdots + \frac{q'}{t^n}\right)\right]\\ &\quad + t\frac{d}{dt}\left\{t\frac{d}{dt}\left[u\left(a'' + \frac{b''}{t} + \frac{c''}{t^2} + \cdots + \frac{q''}{t^n}\right)\right]\right\} + \cdots \qquad (4.34)\\ &= A + Bt + Ct^2 + \cdots + Ht^{n-1},\end{aligned}$$

in Laplace's notation.[32] (Actually Laplace should have written the right-hand member as $A + B/t + C/t^2 + \cdots + H/t^{n-1}$.) To apply this method to an example of Laplace,

$$0 = (x + 1)y_x - y_{x+1}, \text{[33]} \qquad (4.35)$$

we note that $a = 1$, $b = -1$, $a' = 1$, $b' = 0$. By what we have seen in the discussion just after (4.33) the generating function for the right-hand member is

$$t\frac{du}{dt} + u - \frac{1}{t}(u - y_0) = u\left(1 - \frac{1}{t}\right) + t\frac{du}{dt} + \frac{1}{t}y_0.$$

Thus y_x is the coefficient of t^x in the expansion of u. We shall see in a moment how Laplace handled this problem. Here he went from his generating function technique to his method of transforms.

4.8. Laplace Transforms

Laplace first replaces t in the generating function u for y_x by $\exp(\omega\sqrt{-1})$, and calls the resulting function U.[34] He then integrates $U\exp(-x\omega\sqrt{-1})$ and writes

$$\int U\,d\omega c^{-x\omega\sqrt{-1}} = \int d\omega\begin{Bmatrix}y_0c^{-x\omega\sqrt{-1}} + y_1c^{-(x-1)\omega\sqrt{-1}} + \cdots \\ + y_x + y_{x+1}c^{\omega\sqrt{-1}} + \cdots\end{Bmatrix}.$$

---

[32] Laplace VII [1820], p. 83.
[33] Laplace VII [1820], p. 128.
[34] Laplace VII [1820], pp. 83–84.

(He took $-\pi$ and $+\pi$ as his limits of integration.) Then the right-hand member reduces to $2\pi y_x$, since x is an integer, and hence

$$y_x = \frac{1}{2\pi}\int U\,d\omega(\cos x\omega - \sqrt{-1}\sin x\omega);$$

"but this formula has the disadvantage that it introduces imaginary quantities." To avoid these problems Laplace now went to another transform. He assumes that y_x is expressible as

$$y_x = \int t^{-x-1}T\,dt,$$

where $T(t)$ as well as the limits of integration are at his disposal. Consider now which is usually called Laplace's Difference equation,

$$0 = ay_x + by_{x+1} + \cdots + qy_{x+n} + x(a'y_x + b'y_{x+1} + \cdots + q'y_{x+n}).^{35} \quad (4.36)$$

If y_x is substituted into this equation, one finds that

$$0 = -Tt^{-x}\left(a' + \frac{b'}{t} + \cdots + \frac{q'}{t^n}\right)$$
$$+\int t^{-x-1}\left\{\begin{matrix} T\left(a + \frac{b}{t} + \cdots + \frac{q}{t}\right) \\ + t\frac{d}{dt}\left[T\left(a' + \frac{b'}{t} + \cdots + \frac{q'}{t^n}\right)\right]\end{matrix}\right\}, \quad (4.37)$$

since by integration by parts

$$x\int t^{-x-1}T\,dt\frac{T}{t^r} = -t^{-x}\frac{T}{t^r} + \int t^{-x}d\left(\frac{T}{t^r}\right).$$

Laplace now set the integrand in (4.37) above equal to zero; thus

$$0 = T\left(a + \frac{b}{t} + \cdots + \frac{q}{t^n}\right) + t\frac{d}{dt}\left[T\left(a' + \frac{b'}{t^i} + \cdots + \frac{q'}{t^n}\right)\right].$$

This equation determines T. Notice how it compares to the relation (4.34) for the generating function. Next Laplace needed to fix the limits of integration. He did this so that the other term in (4.37) will vanish. Thus

$$0 = Tt^{-x}\left(a' + \frac{b'}{t} + \cdots + \frac{q'}{t^n}\right).$$

This equation has $n + 1$ roots: $t = \infty$ and the n roots t_α ($\alpha = 1, 2, \ldots, n$) given by

$$0 = a' + \frac{b'}{t} + \cdots + \frac{q'}{t^n}.$$

[35] Milne–Thompson [1933], pp. 491–494.

(He has here assumed that all the roots are distinct.) Then there are n integrals

$$\int_{t_\alpha}^{\infty} t^{-x-1} T\, dt \qquad (\alpha = 1, 2, \ldots, n).$$

If each resulting integral is multiplied by an arbitrary constant and these products are summed the result is the complete integral of the equation.

Laplace at this point summed up the importance of this transform by saying (p. 86): "But the great advantage of this transformation of analytic expressions by definite integrals is to furnish a convenient approximation which converges to these expressions when they are formed by a large number of terms and factors; this is the case in the theory of probability when the number of events being considered is very large. Then the numerical calculation of the results to which one is led in solving such problems becomes impracticable, and it is essential to have a method of approximation which becomes more convergent as the results become more complicated."

Let us look again at the example (4.35). Here we have

$$0 = (x + 1)y_x - y_{x+1} = -t^{-x}T + \int t^{-x-1}\left[T\left(1 + \frac{1}{t}\right) + t\frac{dT}{dt}\right] dt,$$

and thus $tT = e^{-1/t}$ is a solution of the differential equation; the limits of integration are to be determined so that $0 = t^{-x}T = t^{-x-1}(tT) = t^{-x-1} \cdot e^{-1/t}$. Clearly $t = 0$ and $t = \infty$ are the limits and we have

$$y_x = A \int_0^{\infty} t^{-x-2} e^{-1/t}\, dt \tag{4.38}$$

as the solution.[36] Actually to handle this problem Laplace wrote $y_x = \int t^x \varphi\, dt$, i.e., he replaced t by $1/\tau$ to make the work easier. Making this change of variables in (4.38), he found

$$y_x = A \int_0^{\infty} \tau^x e^{-\tau}\, d\tau = A\Gamma(x + 1),$$

where Γ is the well-known Gamma function.

Much the same methods were used by Laplace to integrate partial difference equations. The ideas are procedures are the same but a little more complicated notationally.[37] A nice example of his method is given in the 1809 paper just cited, on pp. 188ff. He considered there the equation

$$\Delta^2 y_{x,x'} = \Delta' y_{x,x'},$$

where Δ is the partial difference relative to x and Δ' to x'. If u is the generating function for $y_{x,x'}$,

$$u(t, t') = \sum_{\alpha,\beta=0}^{\infty} y_{\alpha\beta} t^\alpha t'^\beta,$$

---

[36] Laplace VII [1820], pp. 128ff.

[37] Laplace X [1779], pp. 1–89 and XIV [1809], pp. 178–214, where these ideas are further developed.

then, Laplace says, neglecting arbitrary constants or functions which he will bring in later, the generating function for $\Delta^2 y_{x,x'} - \Delta' y_{x,x'}$ is

$$uz = u\left[\left(\frac{1}{t} - 1\right)^2 - \left(\frac{1}{t'} - 1\right)\right]$$

and hence

$$\frac{1}{t'} = 1 + \left(\frac{1}{t} - 1\right)^2 - z.$$

He then forms the generating function for $\Delta' y_{x,x'}$:

$$\begin{aligned}\frac{u}{t'^{x'}} &= u\left[1 + \left(\frac{1}{t} - 1\right)^2 - z\right]^{x'} \\ &= u\left\{1 + x'\left(\frac{1}{t} - 1\right)^2 + \frac{x'(x'-1)}{1\cdot 2}\left(\frac{1}{t} - 1\right)^4 + \cdots \right. \\ &\qquad \left. - x'z\left[1 + (x'-1)\left(\frac{1}{t} - 1\right)^2 + \cdots\right] + \cdots \right\}.\end{aligned}$$

If this is now equated to the corresponding value for $\Delta^2 y_{x,x'}$, it follows that

$$y_{x,x'} = y_{x,0} + x'\Delta^2 y_{x,0} + \frac{x'(x'-1)}{1\cdot 2}\Delta^4 y_{x,0} + \cdots,$$

where evidently $y_{x,0}$ is arbitrary. He also derived this result by going back to the original equation and setting successively $x' = 0, 1, 2, \ldots$. This gave him

$$\begin{aligned} y_{x,1} &= y_{x,0} + \Delta^2 y_{x,0}, \\ y_{x,2} &= y_{x,1} + \Delta^2 y_{x,1} = y_{x,0} + 2\Delta^2 y_{x,0} + \Delta^4 y_{x,0}, \ldots, \end{aligned}$$

which is the same result.

4.9. Method of Least Squares

One of the most useful tools of the numerical analyst as well as of the experimentalist is the Method of least squares. It was originally published by Legendre in 1805 but had been used much earlier by Gauss. Laplace summarized the matter in this way: "Legendre had the simple idea of considering the sum of the squares of the errors of the observations, and of rendering it a minimum, which furnishes directly as many final equations as there are elements to amend. This expert on geometry is the first who has published this method; but we must do Gauss the justice of remarking that he had the same idea many years before this publication and used it habitually as well as communicating it to many astronomers. Gauss in his *Théorie du mouvement elliptique* has sought to tie this method to the Theory of Probability by seeing that the same law of the errors of observations, which gives in general the rule of the arithmetic mean between many observations, also gives the rule of least squares of the errors of observations, and this is what we have seen in §23."[38]

Legendre's ideas first appeared in 1805.[39] At the very end of the volume is a "NOTE PAR M.***." It is a spirited defence of Legendre's claim that Gauss unfairly preempted his method. "At the beginning of the preceding treatise... the author, speaking of his *Method of least squares*, confined himself to recalling that he had published it for the first time in 1805 at the end of a memoir on comets, which bears the date of *March* of that year. Meanwhile, since a very celebrated geometer has not hesitated to appropriate this method to himself in a work printed in 1809, we think we must dwell for a moment on that claim, to which every impartial reader will be able to apply the proper name. See how it has been stated: after having described this method or principle as belonging to him this geometer added: 'Our *other principle which we have used since* 1795 *has recently been published by the well-known* Legendre *in a work* Nouvelles Méthodes, etc., etc.'."[40]

Gauss's method first appeared in his *Theoria motus corporum coelestium in sectionibus conicis solem ambientium.*[41] In the English translation Gauss said: "Our principle, which we have made use of since the year 1795, has lately been published by Legendre in the work *Nouvelles methodes pour la determination des orbites des cometes, Paris,* 1806, where several other properties of this principle have been explained, which for the sake of brevity, we here omit."[42]

Legendre states in the preface to his *Nouvelles Méthodes*...: "It is then necessary, when all the conditions of a problem have been expressed suitably, to determine the coefficients so that the errors are made as small as possible. To do this the method which seems to me to be the simplest and most general consists of minimizing the sum of the squares of the errors. One finds in this way as many equations as there are unknowns; which completes the determination of the elements of the orbit."[43]

Following is a description of the problem and a discussion of how various mathematicians solved it. Let us suppose given a set of m nonhomogeneous linear equations in n unknowns, with $m > n$, and define a set of m "errors" by means of the relations

$$e_i = \sum_{j=1}^{n} a_{ij}x_j - r_i \qquad (i = 1, 2, \ldots, m).$$

---

[38] Laplace VII [1820], p. 353.

[39] Legendre [1805], p. viii and pp. 72–80. This latter reference is to the *Appendix*.

[40] Legendre [1805], p. 79. The Note goes on at considerable length. Incidentally, it is of passing interest to note that a Robert Adrian of Philadelphia wrote on the subject in 1808 in a journal called *The Analyst*, published in Philadelphia by himself.

[41] Gauss, *Theoria* and *KQ*, IV *KQ*1, IV *KQ*2.

[42] *Theoria* (Eng. trans.), p. 270. (This is the quotation referred to in the prior paragraph.)

[43] Legendre [1805], p. viii.

The problem may then be stated in this way: it is desired to find, if possible, that point $(x_1, x_2, \ldots, x_n)$ in n-space which makes the e_i *in some sense* as small as possible.

The first attempts to solve the problem were made by the German astronomer Johann Tobias Mayer of Gottingen in 1748 and 1760.[44] The next attempt was due to Laplace. Let us quote Gauss on the subject: "LAPLACE made use of another principle for the solution of linear equations the number of which is greater than the number of the unknown quantities, which had been previously proposed by BOSCOVICH, namely, that the sum of the errors themselves taken positively, be made a minimum. It can be easily shown, that a system of values of unknown quantities, derived from this principle alone, must . . . exactly satisfy as many equations out of the number proposed, as there are unknown quantities, so that the remaining equations come under consideration only so far as they help to *determine the choice*: if, therefore, the equation $V = M$, for example, is of the number of those which are not satisfied, the system of values found according to this principle would in no respect be changed, even if any other value N had been observed instead of M, provided that, denoting the computed value by n, the differences $M - n$, $N - n$, were affected by the same signs. Besides, LAPLACE qualifies in some measure this principle by adding a new condition: he requires, namely, that the sum of the differences, the signs remaining unchanged, be equal to zero. Hence it follows, that the number of equations exactly represented may be less by unity than the number of unknown quantities; but what we have before said will still hold good if there are only two unknown quantities."[45]

This work of Laplace was done in 1799 and appeared in his *Mécanique Céleste*, Première parte, Livre troisième.[46] He also used his method on the problem of finding the "best value" of the period of a pendulum. The pendulum's period was measured in 15 latitudes ranging from 0°N to 74°;53N.[47]

Legendre remarked that one is generally led to a system of equations of the form $E = a + bx + cy + fz + \&c.$ in which the unknowns x, y, z, &c. are to be so determined that the errors E for each equation are to be made zero or very small. He then proposed that this be done by rendering "the sum of the squares of the errors" a minimum.[48] He notes that this sum of squares is

$$\begin{aligned} E^2 + E'^2 + E''^2 + \&c. = {} & (a + bx + cy + fz + \&c.)^2 \\ & + (a' + b'x + c'y + f'z + \&c.)^2 \\ & + (a'' + b''x + c''y + f''z + \&c.)^2 + \&c., \end{aligned}$$

---

[44] Whittaker, *WR*, pp. 258–259.

[45] Gauss, *Theoria* (Eng. trans.), pp. 270–271. Ruggiero Boscovich (1711–1787) was a Jesuit who was professor of astronomy and mathematics at Rome, Pavia, and Milan.

[46] Laplace II [1805], pp. 147–157. There he determined the size of a degree on a meridian.

[47] Laplace II [1805], pp. 171–177.

[48] Legendre [1805], *Appendice*, pp. 72–80.

and that its being a minimum implies that

$$0 = \int ab + x\int b^2 + y\int bc + z\int bf + \&c.,$$

where, e.g., he wrote $\int ab$ for the sum $ab + a'b' + a''b'' + \&c.$ In the same way he writes down the remaining equations. He then applied his method of least squares to some examples, the most detailed of which is to determine the length of a degree. He had four observations carried out by Delambre and Méchain at Dunkerque, the Panthéon in Paris, Evaux, Carcassone, and Mountjouy and used these to determine the length of the 45th degree, assuming the meridian is an ellipse.

As we shall see there is an intimate connection between probability theory and the method of least squares. If one does not discuss this connection then a certain arbitrariness arises, which Gauss remarked upon in this way:

> In conclusion, the principle that the sum of the squares of the differences between the observed and computed quantities must be a minimum may, in the following manner, be considered independently of the calculus of probabilities.
>
> When the number of unknown quantities is equal to the number of the observed quantities depending on them, the former may be so determined as exactly to satisfy the latter. But when the number of the former is less than that of the latter, an absolutely exact agreement cannot be obtained, unless the observations possess absolute accuracy. In this case care must be taken to establish the best possible agreement, or to diminish as far as practicable the differences. This idea, however, from its nature, involves something vague. For, although a system of values for the unknown quantities which makes *all* the differences respectively less than another system, is without doubt to be preferred to the latter, still the choice between two systems, one of which presents a better agreement in some observations, the other in others, is left in a measure to our judgment, and innumerable different principles can be proposed by which the former condition is satisfied. Denoting the differences between observation and calculation by Δ, Δ', Δ'', etc., the first condition will be satisfied not only if $\Delta\Delta + \Delta'\Delta' + \Delta''\Delta'' +$ etc., is a minimum (which is our principle), but also if $\Delta^4 + \Delta'^4 + \Delta''^4 +$ etc., or $\Delta^6 + \Delta'^6 + \Delta''^6 +$ etc., or in general, if the sum of any of the powers with an even exponent becomes a minimum. But of all these principles ours is the most simple; by the others we should be led into the most complicated calculations.
>
> If we were to adopt a power with an infinite even exponent, we should be led to that system in which the greatest differences become less than in any other system.[49]

4.10. Gauss on Least Squares

It was Gauss in his *Theoria Motus* who first connected probability theory to the method of least squares. By doing this he gave a rational underpinning to the method and made the choice of a Euclidean metric reasonable. The

[49] Gauss, *Theoria* (Eng. trans.), pp. 269–270.

Theoria Motus, which was written to explain how to calculate planetary positions, came into being because the methods available to the astronomers of the eighteenth century were not adequate to determine the orbit of the planet Ceres. This was first observed on January 1, 1801 at Palermo and was kept in sight until February 11, when it was lost. During this time it had described an arc of only 3 degrees, yet Gauss's method was so powerful that on the first clear night in the fall on December 7, 1801 von Zach found the planet where Gauss predicted it would be.[50]

Against this background it was natural for Gauss to concern himself with the problem of how to use redundant observations. It seems clear that the more observations available the more accurately will the orbit be known. Gauss said on the subject:

> But in such a case, if it is proposed to aim at the greatest precision, we shall take care to collect and employ the greatest possible number of accurate places. Then, of course, more data will exist than are required: but all these data will be liable to errors, however small, so that it will generally be impossible to satisfy all perfectly. Now as no reason exists, why, from among those data, we should consider any six as absolutely exact, but since we must assume, rather, upon the principles of probability, that greater or less errors are equally possible in all, promiscuously; since, moreover, generally speaking, small errors oftener occur than large ones; it is evident, that an orbit which, while it satisfies precisely six data, deviates more or less from the others, must be regarded as less consistent with the principles of the calculus of probabilities, than one which, at the same time that it differs a little from those six data, presents so much the better an agreement with the rest. The investigation of an orbit having, strictly speaking, the *maximum* probability, will depend upon a knowledge of the law according to which the probability of errors decreases as the errors increase in magnitude: but that depends upon so many vague and doubtful considerations — physiological included — which cannot be subjected to calculation, that it is scarcely, and indeed less than scarcely, possible to assign properly a law of this kind, in any case of practical astronomy. Nevertheless, an investigation of the connection between this law and the most probable orbit, which we will undertake in its utmost generality, is not to be regarded as by any means a barren speculation.[51]

Let us now examine Gauss's method in the *Theoria Motus* and then his later (1821/23) one in the *Theoria Combinationis*, which reflects a partial revision of his point of view in the light of Laplace's 1811/12 procedure in the *Théorie analytique des Probabilités*.[52] He considered μ functions, V, V', $V'', \ldots$ of ν variables $p, q, r, s, \ldots$ and supposed given values $V = M$,

---

[50] Gauss, *Theoria* (Eng. trans.), pp. xiv–xv. The first real analytical solution to the problem of determining an orbit was given by Euler in 1744, and this was improved by Lambert in 1761. Then Lagrange in 1778 and 1783 gave a mathematically beautiful solution to the problem. (Cf. Lagrange IV [1778], pp. 439–532.) This was followed in 1780 by Laplace who gave a totally different solution, parts of which are still of importance today. (Cf. Laplace X [1780], pp. 93–146.)

[51] Gauss, *Theoria* (Eng. trans.), p. 253.

[52] Gauss acknowledged this change in a letter to Bessel on February 28, 1839.

$V' = M'$, $V'' = M''$, ... obtained by direct observation. When $\mu > \nu$ then in general the problem of finding the $p, q, r, \ldots$ is overdetermined, and this is the case Gauss chose to discuss. He then supposed that the probability of an error of observation Δ is $\varphi\Delta$, in his notation; he further assumed that φ has its maximum at $\varphi = 0$ and that $\varphi(-\Delta) = \varphi(\Delta)$. Finally he supposed that the probability that Δ lies between D and D' is given by

$$\int_D^{D'} \phi\Delta \cdot d\Delta,$$

where the value of the integral for $D = -\infty$, $D' = +\infty$ is unity. He then concluded, under the assumption of independence of events, that the probability of the values $M, M', \ldots$ is

$$\varphi(M - V)\varphi(M' - V')\varphi(M'' - V'') \text{ etc.}, = \Omega$$

in which the quantities $p, q, r, \ldots$ have been substituted into $V, V', V'', \ldots$.

Gauss next invoked Bayes's theorem.[53] Since he does not mention Bayes but gives a proof he may have rediscovered the result himself. Here is his formulation of the rule: "*If, any hypothesis H being made, the probability of any determinate event E is h, and if, another hypothesis H' being made excluding the former and equally probable in itself, the probability of the same event is h': then I say, when the event E has actually occurred, that the probability that H was the true hypothesis, is to the probability that H' was the true hypothesis, as h to h'*."[54] From this Gauss concluded that after the observations $V = M$, $V' = M'$, ... have been made, the probability of the values $p, q, \ldots$ lying between p and $p + dp$, q and $q + dq, \ldots$ is

$$\frac{\Omega \, dp \, dq \cdots}{\int_{-\infty}^{+\infty} \Omega \, dp \, dq \cdots}.$$

Thus "the most probable system of values of the quantities, p, q, r, s, etc." is that which maximizes Ω subject to the hypothesis that "if any quantity has been determined by several direct observations, made under the same circumstances and with equal care, the arithmetical mean of the observed values affords the most probable value, if not rigorously, yet very nearly at least, so that it is always most safe to adhere to it."[55] Gauss therefore sets $V = V' = V''$ etc. $= p$, where

$$p = \frac{1}{\mu}(M + M' + M'' + \text{etc.}),$$

[53] This result of Thomas Bayes was published posthumously in the *Philosophical Transactions* (cf. Bayes [1763]). It served to stimulate Laplace and others to discuss the probabilities of causes which can be inferred from observations of events.

[54] Gauss, *Theoria* (Eng. trans.), p. 255. For an interesting discussion of Bayes's Rule and of conditional probabilities, cf., e.g., Feller, *Prob*, Vol. I, pp. 114–145.

[55] Gauss, *Theoria* (Eng. trans.), p. 258.

and then proceeds to determine the form of the function φ. Since φ is a maximum for this value of p, he has, by taking the logarithmic derivative,

$$\frac{\phi'(M-p)}{\phi(M-p)}+\frac{\phi'(M'-p)}{\phi(M'-p)}+\cdots=0,$$

where the parameters μ and $M, M', M'', \ldots$ are at his disposal. He chooses

$$M' = M'' = \text{etc.} = M - \mu N,$$

and he infers that

$$\frac{\phi'[(\mu-1)N]}{(1-\mu)\varphi[(\mu-1)N]}=\frac{\phi'(-N)}{\phi(-N)}$$

for any positive integer μ. From this he can conclude that $\varphi'(\Delta)/\Delta\varphi(\Delta)$ is a constant k and hence that

$$\phi(\Delta)=\kappa e^{(1/2)k\Delta\Delta},$$

where $k = -2hh$ since $\varphi(0)$ is the maximum value of φ; "and since, by the elegant theorem first discovered by Laplace, the integral

$$\int e^{-hh\Delta\Delta}\,d\Delta$$

from $\Delta = -\infty$ to $\Delta = +\infty$ is $\sqrt{\pi}/h$... our function becomes

$$\phi\Delta=\frac{h}{\sqrt{\pi}}e^{-hh\Delta\Delta}."^{56}$$

Gauss noted that the parameter h is a measure of the accuracy or precision of the observations, and deduced, as a consequence of the form of the normal or Gaussian distribution, the principle of least squares. "*Therefore, that will be the most probable system of values of the unknown quantities p, q, r, s, etc., in which the sums of the squares of the differences between the observed and expected values of the functions V, V', V", etc. is a minimum*, if the same degree of accuracy is to be presumed in all the observations. This principle, which promises to be of most frequent use in all applications of the mathematics to natural philosophy, must everywhere be considered an axiom with the same propriety as the arithmetical mean of several observed values of the same quantity is adopted as the most probable value."[57] Gauss then extended the principle to cases where the observations are of unequal accuracy. Suppose the observations $V = M$, $V' = M'$, $V'' = M''$, etc. have

[56] Gauss, *Theoria* (Eng. trans.), pp. 257–259. This is the extent of Gauss's proof. The details are not difficult to fill in. Let $f(\Delta) = \varphi'(\Delta)/\Delta\varphi(\Delta)$. Then since $f(\mu N) = f(N)$ for every positive integer μ, it follows easily that $f(rN) = f(N)$ for every positive rational r and, if one assumes f to be continuous, for every positive real r. But $f(x) = f(-x)$ and thus $f(x)$ is a constant for all x.

[57] Gauss, *Theoria* (Eng. trans.), p. 260.

the measures of precision h, h', h'', etc., then the expression $h^2v^2 + h'^2v'^2 + h''^2v''^2 + \cdots$ is a minimum where $v = V - M$, $v' = V' - M'$, $v'' = V'' - M''$, etc.

He next took up the case where V, V', V'' are linear in the unknowns p, q, r, s, etc. Thus

$$\begin{aligned} V - M &= v = -m + ap + bq + cr + ds + \text{etc.}, \\ V' - M' &= v' = -m' + a'p + b'q + c'r + ds + \text{etc.}, \\ V'' - M'' &= v'' = -m'' + a''p + b''q + c''r + d''s + \text{etc., etc.}; \end{aligned}$$

he set

$$\begin{aligned} av + a'v' + a''v'' + \text{etc.} &= P, \\ bv + b'v' + b''v'' + \text{etc.} &= Q, \\ cv + c'v' + c''v'' + \text{etc.} &= R, \\ dv + d'v' + d''v'' + \text{etc.} &= S, \text{ etc.}, \end{aligned} \tag{4.39}$$

and remarked that the ν equations which must be solved for the unknowns v, v', v'', etc. are $P = Q = R = S = 0$ etc., "provided we suppose the observations equally good; to which case we have shown . . . how to reduce the others. We here, therefore, as many linear equations as there are unknown quantities to be determined, from which the values of the latter will be obtained by common elimination."[58]

It remained for Gauss to discuss whether the system of linear equations

$$P = 0, \qquad Q = 0, \qquad R = 0, \qquad S = 0, \text{ etc.}$$

was indeed solvable. He stated that it was indeterminate or impossible "when one of the equations . . . being omitted, an equation can be formed from the rest, either identical with the omitted one or inconsistent with it, or which amounts to the same thing, when it is possible to assign a linear function

$$\alpha P + \beta Q + \gamma R + \delta S + \text{etc.} = \kappa."^{59}$$

By reasoning of this sort Gauss showed the solvability of the system. As an example he supposed that he has from observations of equal accuracy

$$\begin{aligned} p - q + 2r &= 3, \\ 3p + 2q - 5r &= 5, \\ 4p + q + 4r &= 21, \end{aligned}$$

and "from a fourth observation, to which is to be assigned one half the same accuracy only, there results $-2p + 6q + 6r = 28$." He replaces this

[58] Cf. Goldstine [1947] and [1951]. These papers by Goldstine and von Neumann were pioneering papers in numerical analysis on numerical stability. They were an attempt by the authors to give a detailed analysis of "Gaussian" elimination. Since then much has been learned.

[59] Gauss, *Theoria* (Eng. trans.), pp. 261–262 and pp. 266–268.

by $-p + 3q + 3r = 14$ and then supposes all observations to be of equal accuracy. Thus

$$P = 27p + 6q - 88,$$
$$Q = 6p + 15q + r - 70,$$
$$R = q + 54r - 107,$$

"and hence by elimination,"

$$19899p = 49154 + 809P - 324Q + 6R,$$
$$737q = 2617 - 12P + 54Q - R,$$
$$6633r = 12707 + 2P - 9Q + 123R.$$

This gives the most probable values of p, q, r, as $p = 2.470$, $q = 3.551$, $r = 1.916$ with precisions — relative to unity for the observations — of $\sqrt{19899/809} = 4.96$ for p, $\sqrt{737/54} = 3.69$ for q, and $\sqrt{2211/41} = 7.34$ for r. (We see below on p. 273 how Jacobi handled this problem.)

This more or less completes Gauss's procedure as outlined in his *Theoria Motus*. It was, as we mentioned above, abandoned by him in favor of a modification of Laplace's method.[60] This modified procedure of Gauss is essentially simpler and more elegant than that of Laplace. The proof of the latter depends upon a derivation of a form of the so-called Law of large numbers, which Gauss did not need. Let U be any function of the unknowns V, V', V'', etc., which have been observed with errors e, e', e'', etc., whose standard deviations are m, m', m'', etc. and which are independent. Then the standard deviation of E, the error in U,

$$E = \lambda e + \lambda' e' + \lambda'' e'' + \text{etc.}, \tag{4.40}$$

is given by

$$M = (\lambda^2 m^2 + \lambda'^2 m'^2 + \lambda''^2 m''^2 + \text{etc.})^{1/2}.$$

In this expression $\lambda, \lambda', \lambda''$ are the $\partial U/\partial V, \partial U/\partial V', \partial U/\partial V''$ at the true values of V, V', V'', etc.[61] The proof is quite straightforward. If the errors are assumed small or if U is linear, then the form (4.40) for the error is evident. The value for M is easily calculated since it is assumed that the expected value for any $e, e', e'', \ldots$ is 0, and since the expected value of any ee' is also 0. Moreover, if the errors $e, e', e'', \ldots$ have weights $p, p', p'', \ldots$, then the weight of E is given by

$$\frac{1}{P} = \frac{\lambda^2}{p} + \frac{\lambda'^2}{p'} + \frac{\lambda''^2}{p''} + \text{etc.}$$

(The *weight* of an observation is inversely proportional to its *dispersion*, the

---

[60] Gauss IV *KQ*1, IV *KQ*2. All this material appears together with other work by Gauss on least squares in German translation in *KQ*. In general, references to Gauss's work on least squares are to the German translation just cited. The important part of Laplace's work on the subject is in Laplace XII [1809′], pp. 301–412.

[61] Gauss, *KQ*, pp. 20–21. For the reader who wishes to compare the Latin text it is in Gauss, *Werke*, Vol. IV, pp. 19–20.

square of its standard deviation.) Gauss considers π variables V, V', V'', etc. as functions of ρ unknowns x, y, z, etc. where $\rho < \pi$. He assumes that observations $V = L$, $V' = L'$, $V'' = L''$, etc., have been made with standard deviations m, m', m'', etc. and weights p, p', p'', etc. Then he introduces the weighted variables $v = (V - L)\sqrt{p}$, $v' = (V' - L')\sqrt{p'}$, $v'' = (V'' - L'')\sqrt{p''}$, etc. This reduces the problem to the case where all weights are equal to 1.

Let

$$\left.\begin{aligned} v &= ax + by + cz + \text{etc.} + l, \\ v' &= a'x + b'y + c'z + \text{etc.} + l', \\ v'' &= a''x + b''y + c''z + \text{etc.} + l'', \text{etc.} \end{aligned}\right\}. \tag{4.41}$$

Gauss now asks for all systems of coefficients $\kappa, \kappa', \kappa''$, etc., such that

$$\kappa v + \kappa' v' + \kappa'' v'' + \text{etc.} = x - k,$$

where k is independent of x, y, z, etc.; he seeks the one which minimizes

$$\kappa^2 + \kappa'^2 + \kappa''^2 + \text{etc.}^{62}$$

To find this he sets

$$\begin{aligned} av + a'v' + a''v'' + \text{etc.} &= \xi, \\ bv + b'v' + b''v'' + \text{etc.} &= \eta, \\ cv + c'v' + c''v'' + \text{etc.} &= \zeta \end{aligned} \tag{4.42}$$

etc., and notes that ξ, η, ζ, etc. are then linear functions of x, y, z, etc. In fact

$$\begin{aligned} \xi &= x \textstyle\sum a^2 + y \sum ab + z \sum ac + \text{etc.} + \sum al, \\ \eta &= x \textstyle\sum ab + y \sum b^2 + z \sum bc + \text{etc.} + \sum bl, \\ \zeta &= x \textstyle\sum zc + y \sum bc + z \sum c^2 + \text{etc.} + \sum cl \text{ etc.}, \end{aligned} \tag{4.43}$$

where he uses $\sum a^2$ for $a^2 + a'^2 + a''^2 + \text{etc.}$, and analogously for the others Gauss now asserts that these linear equations may be solved for x in the form

$$x = A + [\alpha\alpha]\xi + [\alpha\beta]\eta + [\alpha\gamma]\zeta + \text{etc.}, \tag{4.44}$$

where the expressions $[\alpha\alpha]$, $[\alpha\beta]$, etc. are so far merely notations for coefficients. (Later he shows that $[\alpha\alpha] = \sum \alpha^2$, $[\alpha\beta] = \sum \alpha\beta$, etc., and proves that the relation (4.44) is valid. Cf. p. 219 below.) He next forms the system of equations

$$\begin{aligned} a[\alpha\alpha] + b[\alpha\beta] + c[\alpha\gamma] + \text{etc.} &= \alpha, \\ a'[\alpha\alpha] + b'[\alpha\beta] + c'[\alpha\gamma] + \text{etc.} &= \alpha', \\ a''[\alpha\alpha] + b''[\alpha\beta] + c''[\alpha\gamma] + \text{etc.} &= \alpha'', \text{etc.} \end{aligned} \tag{4.45}$$

and notes that this implies

$$\alpha v + \alpha' v' + \alpha'' v'' + \text{etc.} = x - A; \tag{4.46}$$

[62] Gauss, *KQ*, pp. 23–27; in the Latin text, pp. 21–26.

and hence

$$(\kappa - \alpha)v + (\kappa' - \alpha')v' + (\kappa'' - \alpha'')v'' + \text{etc.} = A - k,$$

which may be expressed as

$$\begin{aligned}(\kappa - \alpha)a + (\kappa' - \alpha')a' + (\kappa'' - \alpha'')a'' + \text{etc.} &= 0,\\ (\kappa - \alpha)b + (\kappa' - \alpha')b' + (\kappa'' - \alpha'')b'' + \text{etc.} &= 0,\\ (\kappa - \alpha)c + (\kappa' - \alpha')c' + (\kappa'' - \alpha'')c'' + \text{etc.} &= 0, \text{ etc.}\end{aligned}$$

With the help of the relations (4.45), these imply that

$$(\kappa - \alpha)\alpha + (\kappa' - \alpha')\alpha' + (\kappa'' - \alpha'')\alpha'' + \text{etc.} = 0,$$

so that

$$\begin{aligned}&\kappa^2 + \kappa'^2 + \kappa''^2 + \text{etc.}\\ &\quad = \alpha^2 + \alpha'^2 + \alpha''^2 + \text{etc.} + (\kappa - \alpha)^2 + (\kappa' - \alpha')^2 + (\kappa'' - \alpha'')^2 + \text{etc.};\end{aligned}$$

now we see that from this $\kappa^2 + \kappa'^2 + \kappa''^2 + \text{etc.}$ has as its minimum value $\alpha^2 + \alpha'^2 + \alpha''^2 + \text{etc.}$, and this is attained when $\kappa = \alpha$, $\kappa' = \alpha'$, $\kappa'' = \alpha''$, etc.

Let us now look at the expressions (4.41) and (4.46), which yield identities in x, y, z, etc. It is clear at once that

$$\begin{aligned}\alpha a + \alpha' a' + \alpha'' a'' + \text{etc.} &= 1,\\ \alpha b + \alpha' b' + \alpha'' b'' + \text{etc.} &= 0,\\ \alpha c + \alpha' c' + \alpha'' c'' + \text{etc.} &= 0, \text{ etc.}\end{aligned}$$

Multiplying these equations respectively by the coefficients $[\alpha\alpha]$, $[\alpha\beta]$, $[\alpha\gamma]$, etc., Gauss finds with the help of the relations (4.45) that $\alpha^2 + \alpha'^2 + \alpha''^2 + \text{etc.} = [\alpha\alpha]$, and shows that the value $x = A$ has the weight $1/[\alpha\alpha]$. In a similar manner the values of y, z, etc. can be obtained and their weights determined. (The value of $x = A$ is obtained with the help of (4.42) and (4.44) when we recall that $v = 0$, $v' = 0$, $v'' = 0$, etc. imply zero errors.)

The values of x, y, z, etc. are now found by solving the equations

$$\begin{aligned}x &= A + [\alpha\alpha]\xi + [\alpha\beta]\eta + [\alpha\gamma]\zeta + \text{etc.},\\ y &= B + [\beta\alpha]\xi + [\beta\beta]\eta + [\beta\gamma]\zeta + \text{etc.},\\ z &= C + [\gamma\alpha]\xi + [\gamma\beta]\eta + [\gamma\gamma]\zeta + \text{etc., etc.,}\end{aligned} \tag{4.47}$$

where A, B, C, etc. are the most probable — *plausibelsten* or *plausibiles* — values of the unknowns x, y, z, etc., and their weights are $1/[\alpha\alpha]$, $1/[\beta\beta]$, $1/[\gamma\gamma]$, etc. Moreover, expressions such as $[\beta\alpha]$ are symmetric; thus $[\beta\alpha] = [\alpha\beta]$. Recall the definitions of v, v', v'', etc. just before (4.41); then if $\Omega = p(V - L)^2 + p'(V' - L')^2 + p''(V'' - L'')^2 + \text{etc.} = v^2 + v'^2 + v''^2 + \text{etc.}$, Gauss remarks that

$$2\xi = \frac{\partial\Omega}{\partial x}, \qquad 2\eta = \frac{\partial\Omega}{\partial y}, \qquad 2\zeta = \frac{\partial\Omega}{\partial z}, \text{ etc.}$$

He calls the most probable values of the unknowns those that make Ω a minimum.[63] This is the same result as Gauss had obtained in his *Theoria Motus*. In the simple case of one variable x, $V = V' = V'' = x$ etc. and $a = \sqrt{p}, a' = \sqrt{p'}, a'' = \sqrt{p''}$, etc., $l = -L\sqrt{p}, l' = -L'\sqrt{p'}, l'' = -L''\sqrt{p''}$, etc., hence

$$\xi = (p + p' + p'' + \text{etc.})x - (pL + p'L' + p''L'' + \text{etc.}),$$

since

$$[\alpha\alpha] = \frac{1}{p + p' + p'' + \text{etc.}},$$

$$A = \frac{pL + p'L' + p''L'' + \text{etc.}}{p + p' + p'' + \text{etc.}}.$$

If in particular all the observations are of equal precision then the most probable value of A is

$$\frac{L + L' + L'' + \text{etc.}}{\pi},$$

where π is the number of functions V, V', V'', etc., but this is the arithmetic mean.

In what preceded Gauss assumed without proof that he could find the values of x, y, z, etc. as linear functions of ξ, η, ζ, etc. This is then done in a February 2, 1828 paper in a straightforward way.[64] Since the numbers of x, y, z, etc. and of ξ, η, ζ, etc. are equal, the equations relating them are solvable, provided the ξ, η, ζ, etc. are linearly independent. He supposes for the moment they are not but that there is a relation between them of the form $0 = F\xi + G\eta + H\zeta + \text{etc.} + K$. He therefore finds by proper choices of x, y, z, etc. that

$$\begin{aligned}
F\sum a^2 + G\sum ab + H\sum ac + \text{etc.} &= 0,\\
F\sum ab + G\sum b^2 + H\sum bc + \text{etc.} &= 0,\\
F\sum ac + G\sum bc + H\sum c^2 + \text{etc.} &= 0,\\
&\text{etc.,}\\
F\sum al + G\sum bl + H\sum cl + \text{etc.} &= -K.
\end{aligned}$$

He sets

$$\begin{aligned}
aF + bG + cH + \text{etc.} &= \Theta,\\
a'F + b'G + c'H + \text{etc.} &= \Theta',\\
a''F + b''G + c''H + \text{etc.} &= \Theta'', \text{ etc.,}
\end{aligned} \tag{4.48}$$

and notes that

$$\begin{aligned}
a\Theta + a'\Theta' + a''\Theta'' + \text{etc.} &= 0,\\
b\Theta + b'\Theta' + b''\Theta'' + \text{etc.} &= 0,\\
c\Theta + c'\Theta' + c''\Theta'' + \text{etc.} &= 0,\\
&\text{etc.,}\\
l\Theta + l'\Theta' + l''\Theta'' + \text{etc.} &= -K.
\end{aligned}$$

---

[63] Gauss, *KQ*, pp. 26ff. These are $\xi = 0$, $\eta = 0$, $\xi = 0$, etc.
[64] Gauss, *KQ*, pp. 28–53.

Then he multiplies the first of the equations (4.48) by Θ, the second by Θ', etc. and adds. This gives him

$$0 = \Theta^2 + \Theta'^2 + \Theta''^2 + \text{etc.},$$

or $0 = \Theta = \Theta' = \Theta''$ etc. Hence $K = 0$ and the functions v, v', v'', etc. are unchanged when x, y, z, etc. are replaced by kF, kG, kH, etc. The same then obtains for the V, V', V'', etc. But this is a contradiction of Gauss's assumption that the errors e, e', e'', etc. of the observations of V, V', V'', etc. are statistically independent.

Consider now the values $\lambda, \lambda', \lambda''$, etc. of v, v', v'', etc. when x, y, z, etc. are given their most probable values A, B, C, etc. Then, by the equations given by (4.41),

$$\begin{aligned} aA + bB + cC + \text{etc.} + l &= \lambda, \\ a'A + b'B + c'C + \text{etc.} + l' &= \lambda', \\ a''A + b''B + c''C + \text{etc.} + l'' &= \lambda'', \text{etc.} \end{aligned}$$

We notice that $a\lambda + a'\lambda' + a''\lambda'' + \text{etc.}$ is the value of ξ corresponding to $x = A$, $y = B$, $z = C$, etc., and hence $\xi = 0$; similarly $0 = \eta = \zeta$ etc. and $\sum a\lambda = 0$, $\sum b\lambda = 0$, $\sum c\lambda = 0$, etc. and therefore $\sum \alpha\lambda = 0$, $\sum \beta\lambda = 0$, $\sum \gamma\lambda = 0$, etc. If we set $M = \lambda^2 + \lambda'^2 + \lambda''^2 + \text{etc.}$, it follows from these last relations that

$$l\lambda + l'\lambda' + l''\lambda'' + \text{etc.} = \lambda^2 + \lambda'^2 + \lambda''^2 + \text{etc.} = M.$$

Next Gauss writes

$$\begin{aligned} \Omega = x^2 \sum a^2 + y^2 \sum b^2 + z^2 \sum c^2 + \text{etc.} + 2xy \sum ab + 2xz \sum ac + 2yz \sum bc \\ + \text{etc.} + 2x \sum al + 2y \sum bl + 2z \sum cl + \text{etc.} + \sum l^2, \end{aligned} \tag{4.49}$$

and calls this his first form. By simple manipulations he writes Ω in the form

$$\begin{aligned} \Omega &= \xi x + \eta y + \zeta z + \text{etc.} + lv + l'v' + l''v'' + \text{etc.} \\ &= \xi(x - A) + \eta(y - B) + \zeta(z - C) + \text{etc.} + M, \end{aligned}$$

which he calls his second form. Then with the help of (4.47) he has his third form

$$\begin{aligned} \Omega = [\alpha\alpha]\xi^2 + [\beta\beta]\eta^2 + [\gamma\gamma]\zeta^2 + \text{etc.} + 2[\alpha\beta]\xi\eta \\ + 2[\alpha\gamma]\xi\zeta + 2[\beta\gamma]\eta\zeta + \text{etc.} + M, \end{aligned}$$

and his fourth form

$$\Omega = M + \sum (v - \lambda)^2.$$

Gauss was leading up to a diagonalization of the quadratic form Ω, which he carries out both in the *Theoria Motus* and the *Combination*.[65] Before embarking on this, however, let us look at Gauss's work in somewhat more modern terms.

---

[65] For the latter cf. Gauss, *KQ*, pp. 66ff., and for the former 107ff.

Let $\tilde{A}$ be the matrix of coefficients in (4.41); i.e.,

$$\tilde{A} = \begin{pmatrix} a & b & c & \cdots \\ a' & b' & c' & \cdots \\ a'' & b'' & c'' & \cdots \\ \cdots & \cdots & \cdots & \cdots \end{pmatrix}.$$

Notice that the column of $l, l', l'', \ldots$ is not included in $\tilde{A}$. Then $\tilde{A}$ is a ρ by ρ matrix. If $\nu = (v, v', v'', \ldots)$, $\omega = (x, y, z, \ldots)$, $\rho = (\xi, \eta, \zeta, \ldots)$, $\lambda = (l, l', l'', \ldots)$ and $\alpha = (a, a', a'', \ldots)$, $\beta = (b, b', b'', \ldots), \ldots$, then Gauss's relations (4.41), (4.42), and (4.43) become

$$\nu = \tilde{A}\omega + \lambda, \qquad \rho = \tilde{A}^*\nu = \tilde{A}^*\tilde{A}\omega + \tilde{A}^*\lambda = \tilde{B}\omega + \tilde{A}^*\lambda, \tag{4.50}$$

where $\tilde{B} = \tilde{A}^*\tilde{A}$. (Note that $\tilde{A}^*$ is the transpose of $\tilde{A}$.) The relations (4.47) are then clearly $\omega = -\tilde{A}^{-1}\lambda + \tilde{B}^{-1}\rho$. (We see from this, incidentally, that Gauss's vector $\kappa = (A, B, C, \ldots)$ is $-\tilde{A}^{-1}\lambda$, where the A, B, C, etc. appear in the equations given by (4.47).)

Next we define another matrix

$$\mathfrak{A} = \begin{pmatrix} \alpha & \beta & \gamma & \cdots \\ \alpha' & \beta' & \gamma' & \cdots \\ \alpha'' & \beta'' & \gamma'' & \cdots \\ \cdots & \cdots & \cdots & \cdots \end{pmatrix},$$

and note that the equations (4.45) and their analogs for $\beta, \gamma, \ldots$ are then expressible as $\tilde{A}\mathfrak{A}\kappa = I$, where I is the identity matrix. Thus $\mathfrak{A}^*$ is the inverse of $\tilde{A}$. We also see this by an inspection of the relations (4.47), which are expressible as

$$\omega = \kappa + \mathfrak{A}^*\mathfrak{A}\rho = -\tilde{A}^{-1}\lambda + \tilde{A}^{-1}\mathfrak{A}^{*-1}\rho = \kappa + \tilde{B}^{-1}\rho. \tag{4.51}$$

The quadratic form Ω in (4.49) is evidently of the form

$$\begin{aligned} \Omega &= (\tilde{A}\omega + \lambda, \tilde{A}\omega + \lambda) = (\tilde{A}\omega, \tilde{A}\omega) + (\tilde{A}\omega, \lambda) + (\lambda, \tilde{A}\omega) + (\lambda, \lambda) \\ &= (\omega, \tilde{B}\omega) + 2(\omega, \tilde{A}^*\lambda) + (\lambda, \lambda), \end{aligned}$$

where (,) is the usual inner product symbol; thus, e.g., $(\omega, \rho) = \sum x_i\xi_i$. Now the relations (4.51) tell us that Ω is representable as $(\tilde{B}\rho, \tilde{B}\rho) = (\rho, \tilde{B}^2\rho)$. Gauss now in effect showed how to factor the positive definite matrix $\tilde{B}^2$ into a product of the form

$$T^*DT = \tilde{B}^2,$$

where T is a lower triangular matrix and D is a diagonal one.[66] I.e., each element above the main diagonal is equal to zero and each one on that

[66] This factoring first appears in Art. 182 of the *Theoria Motus*. Cf. *Theoria* (Eng. trans.), pp. 264–265 or *KQ*, pp. 66–69 and pp. 107–111. It also is basic to Gauss's paper, "Pallas," in *KQ*, pp. 124–126. Of course the discussion above is in modern matricial terms, which Gauss did not use.

diagonal to unity; moreover each diagonal element of D is positive and Ω is positive definite. This decomposition is the basis for the so-called Gaussian elimination method.[67] Gauss exhibits the elements of T and D in the form of transformations. Thus he writes

$$\eta' = \eta - \frac{[ab]}{[aa]}\xi, \qquad \xi'' = \xi - \frac{[ac]}{[aa]}\xi - \frac{[bc, 1]}{[bb, 1]}\eta', \text{ etc.},$$

where

$$[bb, 1] = [bb] - \frac{[ab]^2}{[aa]}, \qquad [bc, 1] = [bc] - \frac{[ab][ac]}{[aa]}, \text{ etc.},$$

$$[cc, 2] = [cc] - \frac{[ac]^2}{[aa]} - \frac{[bc, 1]^2}{[bb, 1]},$$

$$[cd, 2] = [cd] - \frac{[ac][ad]}{[aa]} - \frac{[bc, 1][bd, 1]}{[bb, 1]}, \text{ etc.},$$

$$[dd, 3] = [dd] - \frac{[ad]^2}{[aa]} - \frac{[bd, 1]^2}{[bb, 1]} - \frac{[cd, 2]^2}{[cc, 2]}, \text{ etc. etc.},$$

and notes that in terms of these variables

$$\Omega = \frac{\xi^2}{[aa]} + \frac{\xi'^2}{[bb, 1]} + \frac{\xi''^2}{[cc, 2]} + \frac{\xi'''^2}{[dd, 3]} = \text{etc.}$$

This gives the elements of D. They are obtained from the matrix

$$\begin{pmatrix} [aa] & [ab] & [ac] & \cdots \\ [ba] & [bb] & [bc] & \cdots \\ [ca] & [cb] & [cc] & \cdots \\ \cdots & \cdots & \cdots & \cdots \end{pmatrix}.$$

Clearly $[aa]$ is a principal minor of order 1, $[aa][bb, 1]$ of order 2, $[aa][bb, 1][cc, 2]$ of order 3, etc. Moreover, we know that all principal minors are nonnegative since the matrix is nonnegative. Hence all elements of D are nonnegative.

Laplace's method for handling the problem of least squares was to start with a large number of errors, all of which are equidistributed, and to form a linear combination of them. He then derived a form of the law of large numbers to show that this sum $\sum \lambda_i e_i$ is normally distributed. He then proceeded much as we saw above.

In case the number of normal equations becomes too large for easy calculation, Gauss already in his *Supplementum* had an iterative scheme for solving linear equations.[68] He suggested dividing up the system into two or more systems by temporarily neglecting unwanted variables. He proposed solving these subsystems and then substituting these approximations back in to get

[67] Goldstine [1947].

[68] Gauss, *KQ*, pp. 76–77.

improved subsystems. These yield still new solutions, and this procedure may be iterated as often as desired. Later Gauss described to Gerling an approximation technique that anticipated the procedure now often called the Gauss–Seidel process.[69]

4.11. Gauss on Numerical Integration

On September 16, 1814 Gauss presented to the Göttingen Society a paper entitled *Methodus nova integralium valores per approximationem inveniendi.*[70] Here Gauss set out his method of numerical integration. To describe it let us first reconsider with Gauss the so-called Newton–Cotes schemes as explained by Cotes in his *Harmonia Mensurarum.* (Cotes, *HAR.*) (We saw on p. 114 how Stirling handled the same problem.) Let the interval of integration $h - g = \Delta$ of y be divided into n equal parts — g, $g + \Delta/n$, $g + 2\Delta/n$, $g + 3\Delta/n$ etc. up to $g + \Delta = h$. The weights or coefficients R, R', R'', R''' etc. are found such that

$$\int_0^1 y\, dt \sim AR + A'R' + A''R'' + A'''R''' + \text{etc.}, \tag{4.52}$$

where A, A', A'', A''' etc. are the ordinates $y(g)$, $y(g + \Delta/n)$, $y(g + 2\Delta/n)$, $y(g + 3\Delta/n)$ etc. Gauss remarks that in the case $n = 4$ (there are five ordinates; this is the one we worked out in our study of Stirling) the error is about $-K^{\text{VI}}/2688$, in the case $n = 5$ it is about $-11K^{\text{VII}}/52500$, in the case $n = 6$ it is about $-K^{\text{VIII}}/38880$, where Gauss assumes y to be of the form

$$y = K + K't + K''t^2 + K'''t^3 + \text{etc.}^{71} \tag{4.53}$$

Evidently the parameters in the Newton–Cotes schemes are the quantities R, R', R'', R''' etc. In this paper, Gauss posed the following question and gave a complete answer: what could be accomplished if in addition the partition points of the interval Δ a, a', a'', a''' etc. are also available as parameters to be chosen? As we shall see, if these numbers are taken as the $n + 1$ roots of the Legendre polynomial P_{n+1} on the interval 0, 1 then the integration formula (4.52) above is correct for y any polynomial (4.53) of degree $2n + 1$. Gauss did not make explicit use of Legendre's polynomials.[72] Instead he used his hypergeometric function $F(\alpha, \beta, \gamma; x)$ which contains Legendre's polynomial

---

[69] Gerling published the method in a work on practical geometry. (Gerling [1845].) See also Seidel [1874], pp. 81–108. Jacobi brought out a similar one. Jacobi III [1845], pp. 468–478. Seidel was a student of Jacobi.

[70] Gauss III *NI*, pp. 165–196; cf. also, pp. 202–206.

[71] Gauss III *NI*, p. 172. Steffensen [1950], pp. 169–170; Markoff [1896], pp. 61–72 or Runge, *RK*, pp. 275–285.

[72] They were introduced by Legendre [1785].

P_{n+1} as a special case. There is nothing to indicate he was aware of Legendre's paper.[73] Gauss's method is not only elegant but it differs from the more modern approaches in that he arrives at his result from first principles, and not, as is now done, by showing how the result follows from properties of Legendre's polynomials. That is to say, Gauss shows us his discovery of the optimal location of the a, a', a'' etc., whereas modern proofs start from a knowledge of their location.

Consider the function

$$\begin{aligned} T(t) &= (t-a)(t-a')(t-a'')(t-a''')\cdots(t-a^{(n)}) \\ &= t^{n+1} + \alpha t^n + \alpha' t^{n-1} + \alpha'' t^{n-2} + \text{etc.} + \alpha^{(n)}. \end{aligned} \tag{4.54}$$

It is fundamental to the Newton–Cotes scheme. In fact if A, A', A'', A''' etc. are the ordinates of the curve y at the points a, a', a'', a''' etc., then an approximation to y is given by

$$Y = \frac{AT}{M(t-a)} + \frac{A'T}{M'(t-a')} + \frac{A''T}{M''(t-a'')} + \text{etc.} + \frac{A^{(n)}T}{M^{(n)}(t-a^{(n)})}, \tag{4.55}$$

where M, M', M'' etc. are

$$\left.\frac{T}{t-a}\right|^{t=a}, \quad \left.\frac{T}{t-a'}\right|^{t=a'}, \quad \left.\frac{T}{t-a''}\right|^{t=a''} \text{etc.}$$

The Newton–Cotes coefficients R, R', R'' etc. are now easy to derive in terms of T. If we integrate the equation given by (4.55) above, we find

$$\int_0^1 Y\,dt = RA + R'A' + R''A'' + \text{etc.} + R^{(n)}A^{(n)},$$

where

$$R = \int_0^1 \frac{T}{M(t-a)}\,dt, \quad R' = \int_0^1 \frac{T}{M'(t-a')}\,dt, \quad R'' = \int_0^1 \frac{T}{M''(t-a'')}\,dt \text{ etc.}$$

Thus for $n = 5$ we have $T/(t-a'') = 5^5t^5 - 13\cdot 5^4t^4 + 59\cdot 5^3t^3 - 107\cdot 5^2tt + 60\cdot 5\cdot t$, $M'' = 2 \times 1 \times (-1) \times (-2) \times (-3) = -12$, and so $R'' = 25/144$.[74]

---

[73] Whittaker, *WW*, pp. 311ff. and the 1812 paper, Gauss *III HGF*, pp. 126–162. Gauss's paper on this series was first presented to the Göttingen Society on January 30, 1812. The series is

$$F(\alpha, \beta, \gamma, x) + 1 + \frac{\alpha\beta}{1\cdot\gamma}x + \frac{\alpha(\alpha+1)\beta(\beta+1)}{1\cdot 2\cdot\gamma(\gamma+1)}x^2 + \frac{\alpha(\alpha+1)(\alpha+2)\beta(\beta+1)(\beta+2)}{1\cdot 2\cdot 3\gamma(\gamma+1)(\gamma+2)}x^3 + \cdots.$$

[74] Gauss III, *NI*, p. 167. On pp. 168–169 Gauss works out the Newton–Cotes cases for $n = 1, 2, \ldots, 10$ and gives all the values of R, R', R'' etc. in each case.

Now if we go back to (4.54) above and note that for $t = a$ we have

$$0 = T(a) = a^{n+1} + \alpha a^n + \alpha' a^{n-1} + \alpha'' a^{n-2} + \text{etc.} + \alpha^{(n)},$$

then

$$\begin{aligned} T = t^{n+1} - a^{n+1} + \alpha(t^n - a^n) + \alpha'(t^{n-1} - a^{n-1}) \\ + \alpha''(t^{n-2} - a^{n-2}) + \text{etc.} + \alpha^{(n-1)}(t - a). \end{aligned}$$

Therefore $T/(t - a)$ can be given a simple representation as a polynomial, and we can express its integral as

$$\begin{aligned} \int_0^1 \frac{T\,dt}{t-a} = {} & a^n + \alpha a^{n-1} + \alpha' a^{n-2} + \alpha'' a^{n-3} + \text{etc.} + \alpha^{(n-1)} \\ & + \frac{1}{2}(a^{n-1} + \alpha a^{n-2} + \alpha' a^{n-3} + \text{etc.} + \alpha^{(n-2)}) \\ & + \frac{1}{3}(a^{n-2} + \alpha a^{n-3} + \alpha' a^{n-4} + \text{etc.} + \alpha^{(n-3)}) \\ & + \frac{1}{4}(a^{n-3} + \alpha a^{n-4} + \alpha' a^{n-5} + \text{etc.} + \alpha^{(n-4)}) \\ & + \text{etc.} \\ & + \frac{1}{n-1}(aa + \alpha a + \alpha') \\ & + \frac{1}{n}(a + \alpha) \\ & + \frac{1}{n+1}. \end{aligned}$$

Gauss saw that this was closely related to the analytic part T' of $-T \log(1 - t^{-1})$. In fact he wrote

$$T\left(t^{-1} + \frac{1}{2}t^{-2} + \frac{1}{3}t^{-3} + \frac{1}{4}t^{-4} + \text{etc.}\right) = T' + T'',$$

where T'' is the principal part of $-T \log(1 - t^{-1})$, i.e., it is the part containing negative exponents of t, and T' is the part containing the nonnegative ones. Then it is easy to show from the above that

$$\int_0^1 \frac{T\,dt}{t-a} = T'(a).$$

He further remarked that the expressions

$$\frac{T'}{(dT/dt)},$$

evaluated successively for $t = a, a', a'', a''', \ldots, a^{(n)}$, yield the values $R, R', R'', R''', \ldots R^{(n)}$.[75]

[75] Gauss III *NI*, p. 175.

Gauss was led to the Legendre polynomials by remarking that the a, a', a'' etc. need not be equally spaced and could in fact be any numbers, rational or irrational; all the above would still hold just as before.

Let $k^{(m)}$ be the difference between the true value of $\int_0^1 t^m\,dt$ and its approximation $RA + R'A' + R''A'' + \text{etc.} + R^{(n)}A^{(n)}$ where the A, A', A'' etc. are now the ordinates of t^m. Thus

$$Ra^m + R'a'^m + R''a''^m + \text{etc.} + R^{(n)}a^{(n)m} = \frac{1}{m+1} - k^{(m)}.$$

Now, with the help of these relations, Gauss was able to write for $m \geq 0$,

$$\frac{R}{t-a} + \frac{R'}{t-a'} + \frac{R''}{t-a''} + \text{etc.} + \frac{R^{(n)}}{t-a^{(n)}}$$

$$= (1-k)t^{-1} + \left(\frac{1}{2} - k'\right)t^{-2} + \left(\frac{1}{3} - k''\right)t^{-3} + \left(\frac{1}{4} - k'''\right)t^{-4} + \text{etc.} \tag{4.56}$$

$$= t^{-1} + \frac{1}{2}t^{-2} + \frac{1}{3}t^{-3} + \frac{1}{4}t^{-4} + \text{etc.} - \Theta$$

where he had expressed by Θ the series

$$\Theta = kt^{-1} + k't^{-2} + k''t^{-3} + k'''t^{-4} + \text{etc.}$$

But k, k', k'', k''' etc., up to and including $k^{(n)}$, all vanish since the functions $t, t^2, t^3, \ldots, t^n$ are fitted exactly by the procedure outlined above. Thus

$$\Theta = k^{(n+1)}t^{-(n+2)} + k^{(n+2)}t^{-(n+3)} + k^{(n+3)}t^{-(n+4)} + \text{etc.},$$

and we see that

$$-T\log(1 - t^{-1}) = T\left(t^{-1} + \frac{1}{2}t^{-2} + \frac{1}{3}t^{-3} + \frac{1}{4}t^{-4} + \text{etc.}\right)$$

$$= T' + T''$$

$$= T\left(\frac{R}{t-a} + \frac{R'}{t-a'} + \frac{R''}{t-a''} + \text{etc.} + \frac{R^{(n)}}{t-a^{(n)}}\right) + T\Theta.$$

We know that the expressions $T/(t-a)$, $T/(t-a')$ etc. are polynomials in T whose values for $t = a, a', a''$ etc. are MR, $M'R'$, $M''R''$ etc. But these are also the values of T' for $t = a, a', a''$ etc. Hence Gauss is able to conclude that $T'' = T\Theta$ or $\Theta = T''/T$.

If the function y is expressible as $y = K + K't + K''tt + K'''t^3 + \text{etc.}$, then Gauss's approximation to the integral is

$$k^{(n+1)}K^{(n+1)} + k^{(n+2)}K^{(n+2)} + k^{(n+3)}K^{(n+3)}\ \text{etc.}, \tag{4.57}$$

where clearly

$$K^{(m)} = \frac{1}{1\cdot 2\cdot 3\cdots m}\cdot\frac{d^m y}{dt^m}\bigg|^{t=0} = \frac{\Delta^m}{1\cdot 2\cdot 3\cdots m}\cdot\frac{d^m y}{dx^m}\bigg|^{x=g},$$

when $t = \Delta \cdot x$. Gauss now proceeds to improve the degree of approximation by considering the function T''. What he wishes to do is to determine the coefficients $\alpha, \alpha', \alpha''$ etc. appearing in T so that the coefficients of $t^{-1}, t^{-2}, \ldots, t^{-(n+1)}$ in T'' vanish.[76] Thus for $n = 0$ the coefficient of t^{-1} is $\alpha + 1/2$, and therefore $T = t - 1/2$ is such that the coefficient of t^{-1} in T'' is zero; similarly for $n = 1$, $T = tt - t + 1/6$; for $n = 2$, $T = t^3 - 3tt/2 + 3t/5 - 1/20$.

For symmetry Gauss made a change of variables. Set $u = 2t - 1$, $b = 2a - 1$, $b' = 2a' - 1$, $b'' = 2a'' - 1$ etc., and $U = (u - b)(u - b')(u - b'') \cdots (u - b^{(n)})$. Then Gauss considers

$$U\left(u^{-1} + \frac{1}{3}u^{-3} + \frac{1}{5}u^{-5} + \frac{1}{7}u^{-7} + \text{etc.}\right) = U' + U'',$$

and shows that

$$\frac{U'}{(dU/du)}$$

for $u = b, b', b'', b'''$ etc. is R, R', R'', R''' etc. In terms of these variables Gauss writes

$$y = L + \frac{L'}{2}u + \frac{L''}{4}u^2 + \frac{L'''}{8}u^3 + \text{etc.},$$

and the counterpart of (4.57) above is now

$$l^{(n+1)}L^{(n+1)} + l^{(n+2)}L^{(n+2)} + l^{(n+3)}L^{(n+3)} + \text{etc.},$$

where the $l^{(m)}$ are the true values of the integrals $\int (t - \frac{1}{2})^m\, dt$. Thus

$$l^{(m)} = k^{(m)} - \frac{1}{2}mk^{(m-1)} + \frac{1}{4}\cdot\frac{m\cdot m - 1}{1\cdot 2}k^{(m-2)}$$
$$- \frac{1}{8}\cdot\frac{m\cdot m - 1\cdot m - 2}{1\cdot 2\cdot 3}k^{(m-3)} + \text{etc.}$$

If we set $U''/U = \Omega$ — this is the analog of Θ above — then

$$\Omega = 2^{n+1}l^{(n+1)}u^{-(n+2)} + 2^{n+2}l^{(n+2)}u^{-(n+3)} + 2^{n+3}l^{(n+3)}u^{-(n+4)} + \text{etc.}$$

In terms of the variable u we find that the cases $n = 0, 1, 2$ we just worked out for T give for U the values $U = u$, $uu - 1/3$ and $u^3 - 3u/5$. We see that these are proportional to P_1, P_2, P_3, the first three Legendre polynomials.[77]

---

[76] Gauss III *NI*, p. 184.
[77] Courant, *Methoden*, Vol. I, pp. 70–74.

Gauss now expressed the function $\varphi = u^{-1} + \frac{1}{3}u^{-3} + \frac{1}{5}u^{-5} + \frac{1}{7}u^{-7} +$ etc. as a continued fraction in the form

$$\cfrac{1}{u - \cfrac{\frac{1}{3}}{u - \cfrac{\frac{2\cdot 2}{3\cdot 5}}{u - \cfrac{\frac{3\cdot 3}{5\cdot 7}}{u - \cfrac{\frac{4\cdot 4}{7\cdot 9}}{u - \text{etc.}}}}}}\,{}^{78}$$

Consider now with Gauss the expression of this continued fraction when the fractions are "unwrapped," i.e., in the form of a series. To do this he first shows that the continued fraction

$$\varphi = \cfrac{v}{w + \cfrac{v'}{w' + \cfrac{v''}{w'' + \cfrac{v'''}{w''' + \text{etc.}}}}}$$

may be written as

$$\varphi = \frac{v}{WW'} - \frac{vv'}{W'W''} + \frac{vv'v''}{W''W'''} - \frac{vv'v''v'''}{W'''W''''} + \text{etc.},$$

where $W = 1$, $W' = wW$, $W'' = w'W' + v'W$, $W''' = w''W'' + v''W'$ etc. In our case, clearly, $v = 1$, $v' = -1/3$, $v'' = -4/15$, $v''' = -9/35$, $v'''' = -16/63$ etc., and $W = 1$, $W' = u$, $W'' = uu - 1/3$, $W''' = u^3 - 3u/5$, $W'''' = u^4 - 6uu/7 + 3/35$ etc.

It is not difficult to show that

$$\varphi - \frac{V^{(m)}}{W^{(m)}} = \frac{2\cdot 2\cdot 3\cdot 3\cdots m\cdot m}{3\cdot 3\cdot 5\cdot 5\cdots(2m-1)(2m+1)W^{(m)}W^{(m+1)}}$$
$$+ \frac{2\cdot 2\cdot 3\cdot 3\cdots(m+1)(m+1)}{3\cdot 3\cdot 5\cdot 5\cdots(2m+1)(2m+3)W^{(m+1)}W^{(m+2)}}$$
$$+ \text{etc.},$$

[78] Gauss III *NI*, p. 186. This is shown in his paper Gauss III *HGF*, pp. 123–162. This paper is the one referred to on hypergeometric series. On p. 136 of that paper he gives the continued fraction development of $\log(1 + t)/(1 - t)$. With $u = 1/t$ this is the one above. Furthermore, on p. 127 he points out that this is $2tF(1/2, 1, 3/2, t^2)$, where F is the hypergeometric series.

where the V, V', V'' etc. are formed in this way: $V = 0$, $V' = v$, $V'' = w'V' + v'V$, $V''' = w''V'' + v''V'$ etc.; thus in our case $V = 0$, $V' = 1$, $V'' = u$, $V''' = uu - 4/15$ etc. Gauss then remarks that if $\varphi - V^{(m)}/W^{(m)}$ is expressed in a descending series, its first term will be

$$\frac{2\cdot 2\cdot 3\cdot 3\cdots m\cdot mu^{-(2m+1)}}{3\cdot 3\cdot 5\cdot 5\cdots (2m-1)(2m+1)}$$

and consequently $\varphi W^{(m)}$ will consist of the polynomial $V^{(m)}$ and an infinite series whose first term is

$$\frac{2\cdot 2\cdot 3\cdot 3\cdots mmu^{-(m+1)}}{3\cdot 3\cdot 5\cdot 5\cdots (2m-1)(2m+1)}.$$

Recall that U is a polynomial of degree $n + 1$ and that $\varphi U = U' + U''$, where U' is a polynomial in u and U'' is the principal part; moreover, that what Gauss wants is to choose U so that φU does not contain $u^{-1}, u^{-2}, u^{-3}, \ldots, u^{(-n+1)}$. He notes that if he chooses U to be $W^{(n+1)}$, U' will be $V^{(m+1)}$, and the first term of U'' will be the expression above with $m = n + 1$. Call the $n + 1$ roots of $W^{(n+1)}$ $0, b, b', b'', \ldots, b^{(n)}$. Then in terms of these roots, the coefficients $R, R', R'', \ldots, R^{(n)}$ can be found as before. Thus Gauss can say that his formula for the integral is precise through order $2n + 1$ with an error of approximately

$$\frac{1}{2^{2n+2}}\cdot\frac{2\cdot 2\cdot 3\cdot 3\cdots (n+1)(n+1)}{3\cdot 3\cdot 5\cdot 5\cdots (2n+1)(2n+3)}L^{(2n+2)}$$
$$= \frac{1\cdot 1\cdot 2\cdot 2\cdot 3\cdot 3\cdots (n+1)(n+1)}{2\cdot 6\cdot 6\cdot 10\cdot 10\cdot 14\cdots (4n+2)(4n+6)}L^{(2n+2)}.$$

Let us now examine with Gauss the form of $U = W^{(n+1)}$; it is

$$u^{n+1} - \frac{(n+1)n}{2\cdot(2n+1)}u^{n-1} + \frac{(n+1)n(n-1)(n-2)}{2\cdot 4(2n+1)(2n-1)}u^{n-3}$$
$$- \frac{(n+1)n(n-1)(n-2)(n-3)(n-4)}{2\cdot 4\cdot 6\cdot(2n+1)(2n-1)(2n-3)}u^{n-5} + \text{etc.},$$

and this, he points out, is $u^{n+1}F(-\tfrac{1}{2}n, -\tfrac{1}{2}(n+1), -(n+\tfrac{1}{2}), u^{-2})$, where F is the well-known hypergeometric function of Gauss. He next found the form of $U'' = \varphi W^{(n+1)} - V^{(n+1)}$ and noted that

$$\varphi W'' - V'' = w'(\varphi W' - V') + v'(\varphi W - V),$$
$$\varphi W''' - V''' = w''(\varphi W'' - V'') + v''(\varphi W' - V'),$$
$$\varphi W'''' - V'''' = w'''(\varphi W''' - V''') + v'''(\varphi W'' - V''), \text{ etc.}$$

Thus he could conclude that

$$U'' = \frac{2\cdot 2\cdot 3\cdot 3\cdot 4\cdot 4\cdots(n+1)(n+1)}{3\cdot 3\cdot 5\cdot 5\cdot 7\cdot 7\cdot 9\cdots(2n+1)(2n+3)}$$
$$\times\, u^{-(n+2)}\cdot F\left(\frac{1}{2}n+1, \frac{1}{2}n+\frac{3}{2}, n+\frac{5}{2}, u^{-2}\right).$$

At the conclusion of this paper Gauss shows his love of numerical calculation. He works out the functions U, U', T, T' for $n = 0, 1, \ldots, 6$; he finds the roots a, a', a'' etc. for each case and also the coefficients R, R', R'' etc. — all to 16 decimal places. Thus for $n = 6$ he has

$$U = u^7 - \frac{21}{13}u^5 + \frac{105}{143}u^3 - \frac{35}{429}u,$$

$$U' = u^6 - \frac{50}{39}u^4 + \frac{283}{715}uu - \frac{256}{15015},$$

$$T = t^7 - \frac{7}{2}t^6 + \frac{63}{13}t^5 - \frac{175}{52}t^4 + \frac{175}{143}t^3 - \frac{63}{286}tt + \frac{7}{429}t - \frac{1}{3432},$$

$$T' = t^6 - 3t^5 - \frac{535}{156}t^4 - \frac{145}{78}t^3 + \frac{1377}{2860}tt - \frac{223}{4290}t + \frac{323}{240240},$$

$$\begin{aligned}
a &= 0.0254460438\,286202,\\
a' &= 0.1292344072\,003028,\\
a'' &= 0.2970774243\,113015,\\
a''' &= 0.5,\\
a'''' &= 0.7029225756\,886985,\\
a''''' &= 0.8707655927\,996972,\\
a'''''' &= 0.9745539561\,713798,
\end{aligned}$$

$$\begin{aligned}
R &= R'''''' = 0.0647424830\,844348, && \log. = 8.8111893529,\\
R' &= R''''' = 0.1398526957\,446384, && 9.1456708421,\\
R'' &= R'''' = 0.1909150252\,525595, && 9.2809401093,\\
R''' &= \frac{256}{1225} = 0.2089795918\,367347, && 9.3201038766.
\end{aligned}$$

The error in the integration formula is about $L^{xiv}/176679360$. Gauss finished the paper by calculating $\int dx/\log x$ from $x = 100{,}000$ to $200{,}000$ using successively $n = 0, 1, 2, \ldots, 6$. He kept 10 decimals and found in the last case of 7 ordinates, that

$$\left\{\begin{aligned}
\Delta RA &= 561.1213804\\
\Delta R'A' &= 1202.0551998\\
\Delta R''A'' &= 1621.6290819\\
\Delta R'''A''' &= 1753.4212406\\
\Delta R''''A'''' &= 1584.9790252\\
\Delta R^{\mathrm{V}}A^{\mathrm{V}} &= 1152.0681116\\
\Delta R^{\mathrm{VI}}A^{\mathrm{VI}} &= 530.9690816\\
\hline
\text{Summa} &= 8406.2431211.
\end{aligned}\right.$$

He closes with the observation: "The value of the same integral has been found by the calculations of the famous Bessel to be = 8406,24312."[79]

It is in the paper on the hypergeometric series that Gauss defined what we now call the Gamma function and established some of its properties. He introduced a function

$$\Pi(k, z) = \frac{1 \cdot 2 \cdot 3 \cdots k}{(z+1)(z+2)(z+3)\cdots(z+k)} k^z,$$

where k is a positive integer, and $\Pi(z) = \Pi(\infty, z)$, which is Gauss's notation for the limit of $\Pi(k, z)$ as k approaches ∞. He was now able to write

$$F(\alpha, \beta, \gamma, 1) = \frac{\Pi(\gamma - 1) \cdot \Pi(\gamma - \alpha - \beta - 1)}{\Pi(\gamma - \alpha - 1) \cdot \Pi(\gamma - \beta - 1)}. \tag{4.57'}$$

Next he observes that (note that Gauss wrote $\sin^2 t$ as $\sin t^2$) $t = \sin t \cdot F(\frac{1}{2}, \frac{1}{2}, \frac{3}{2}, \sin t^2)$ and hence

$$\frac{1}{2}\pi = 1 + \frac{1 \cdot 1}{2 \cdot 3} + \frac{1 \cdot 1 \cdot 3}{2 \cdot 4 \cdot 5} + \frac{1 \cdot 1 \cdot 3 \cdot 5}{2 \cdot 4 \cdot 6 \cdot 7} + \text{etc.}$$

But this implies, by (4.57′) above, that $F(\frac{1}{2}, \frac{1}{2}, \frac{3}{2}, 1) = \Pi\frac{1}{2} \cdot \Pi(-\frac{1}{2})/\Pi 0 \cdot \Pi 0$; but $\Pi 0 = 1$, $\Pi\frac{1}{2} = (\frac{1}{2})\Pi(-\frac{1}{2})$, and hence $\Pi(-\frac{1}{2}) = \sqrt{\pi}$, $\Pi\frac{1}{2} = \frac{1}{2}\sqrt{\pi}$.

By another result of Gauss, $\sin nt = n \sin t \cdot F(\frac{1}{2}n + \frac{1}{2}, -\frac{1}{2}n + \frac{1}{2}, \frac{3}{2}, \sin t^2)$, and hence

$$\sin\frac{n\pi}{2} = \frac{n\Pi\frac{1}{2} \cdot \Pi(-\frac{1}{2})}{\Pi(-\frac{1}{2}n) \cdot \Pi\frac{1}{2}n};$$

from this he deduces an "elegant formula":

$$\Pi\frac{1}{2}n \cdot \Pi\left(-\frac{1}{2}n\right) = \frac{\frac{1}{2}n\pi}{\sin\frac{1}{2}n\pi}.$$

He now sets $n = 2z$ and has

$$\Pi(-z) \cdot \Pi(+z) = \frac{z\pi}{\sin z\pi}, \qquad \Pi(-z) \cdot \Pi(z - 1) = \frac{\pi}{\sin z\pi}.$$

This can be used to express

$$\sin z\pi = z\pi(1 - zz)\left(1 - \frac{zz}{4}\right)\left(1 - \frac{zz}{9}\right) \text{ etc. in inf.,}$$

$$\cos z\pi = (1 - 4zz)\left(1 - \frac{4zz}{9}\right)\left(1 - \frac{4zz}{25}\right) \text{ etc. in inf.}$$

There are many other elegant results which one could record, but which are not strictly relevant.

---

[79] Gauss III *NI*, p. 196. "E calculis clar. BESSEL valor eiusdem integralis inventus est = 8406,24312."

4.12. Gauss on Interpolation

In Gauss's *Nachlass* (Volume III) is a paper entitled *Theoria interpolationis methodo nova tractata* (pp. 265ff.). The first problem Gauss posed in this paper was to find the sum of the series

$$S^n = \alpha a^n + \beta b^n + \gamma c^n + \delta d^n + \text{etc.},$$

where a, b, c, d, etc. are m quantities and where

$$\frac{1}{(a-b)(a-c)(a-d)(a-c)\cdots} = \alpha,$$

$$\frac{1}{(b-a)(b-c)(b-d)(b-e)\cdots} = \beta,$$

$$\frac{1}{(c-a)(c-b)(c-d)(c-e)\cdots} = \gamma,$$

$$\frac{1}{(d-a)(d-b)(d-c)(d-e)\cdots} = \delta, \text{ etc.}$$ [80]

He first writes

$$P = \frac{\alpha}{1-ax} + \frac{\beta}{1-bx} + \frac{\gamma}{1-cx} + \frac{\delta}{1-dx} + \text{etc.},$$

and notes that, if each denominator is expanded in a power series and the resulting series are rearranged in ascending powers of x, P becomes

$$S^0 + S^1x + S^2xx + S^3x^3 + \text{etc. in infin.}$$

Next he defines Q to be the product $(1-ax)(1-bx)(1-cx)(1-dx)\ldots$. Then PQ is a polynomial of degree $m-1$; by the definition of α, β, etc. for $x = 1/a$, $PQ = 1/a^{m-1} = \alpha(1-b/a)(1-c/a)(1-d/a)\ldots$ and similarly for $x = 1/b$, $PQ = 1/b^{m-1}$, etc. Thus $PQ - x^{m-1}$ vanishes for $x = 1/a, 1/b, 1/c, 1/d, \ldots$; i.e., for m values. Hence it is identically zero and $PQ = x^{m-1}$. Gauss next notes that, by definition, the function Q is expressible as $Q = 1 - Ax + Bxx - Cx^3 + Dx^4 - \text{etc.}$, where A is the sum of the $a, b, c, d, \ldots$, B is the sum of their products taken two at a time, etc. Moreover we have $x^{m-1}/Q = P = S^0 + S^1x + S^2xx + S^3x^3 + \text{etc.}$, and hence $S^0 = 0$, $S^1 = 0$, $S^2 = 0$, etc. through $S^{m-2} = 0$; and clearly $S^{m-1} = 1$, $S^m = A$. In fact

$$\begin{aligned} S^m &= A, \\ S^{m+1} &= AS^m - BS^{m-1} &&= AA - B, \\ S^{m+2} &= AS^{m+1} - BS^m + CS^{m-1} &&= A^3 - 2BA + C, \text{ etc.} \end{aligned}$$

---

[80] Gauss III *TI*, pp. 265–327. The problem just mentioned was first treated by Euler in his *Inst. Calc. Integr.*, *Opera*, Vol. II, p. 432. He showed there that $S^0 = S^1 = \cdots = S^{m-2} = 0$. (We see it directly below.)

It is clear that $S^{m+1} = aa + bb + cc + \cdots + B$. Furthermore, Gauss remarked that the function $1/Q$ is of the form $\sum a^\lambda b^\mu c^\nu \cdots \times x^{\lambda+\mu+\nu\ldots}$. From this expansion he was able to estimate the value of S^n. For n negative he found, e.g.,

$$S^{-1} = \pm\frac{1}{abcd\cdots}, \qquad S^{-2} = \pm\frac{1/a + 1/b + 1/c + 1/d + \cdots}{abcd\cdots}, \text{ etc.}$$

Gauss went on to apply these results to the complex numbers E^{ia}, E^{ib}, E^{ic}, $E^{id}, \ldots$ and E^{-ia}, E^{-ib}, E^{-ic}, $E^{-id}, \ldots$ where E is the base of the natural logarithms and i is $\sqrt{-1}$. He forms two expressions S^n and T^n:

$$\begin{aligned}
&\frac{E^{ina}}{(E^{ia} - E^{ib})(E^{ia} - E^{ic})(E^{ia} - E^{id})\cdots}\\
&+ \frac{E^{inb}}{(E^{ib} - E^{ia})(E^{ib} - E^{ic})(E^{ib} - E^{id})\cdots}\\
&+ \frac{E^{inc}}{(E^{ic} - E^{ia})(E^{ic} - E^{ib})(E^{ic} - E^{id})\cdots}\\
&+ \frac{E^{ind}}{(E^{id} - E^{ia})(E^{id} - E^{ib})(E^{id} - E^{ic})\cdots} + \text{etc.} = S^n,\\
&+ \frac{E^{-ina}}{(E^{-ia} - E^{-ib})(E^{-ia} - E^{-ic})(E^{-ia} - E^{-id})\cdots}\\
&+ \frac{E^{-inb}}{(E^{-ib} - E^{-ia})(E^{-ib} - E^{-ic})(E^{-ib} - E^{-id})\cdots}\\
&+ \frac{E^{-inc}}{(E^{-ic} - E^{-ia})(E^{-ic} - E^{-ib})(E^{-ic} - E^{-id})\cdots}\\
&+ \frac{E^{-ind}}{(E^{-id} - E^{-ia})(E^{-id} - E^{-ib})(E^{-id} - E^{-ic})\cdots} + \text{etc.} = T^n.
\end{aligned}$$

Then as before $S^0, S^1, S^2, \ldots, S^{m-2}$, and also $T^0, T^1, T^2, \ldots, T^{m-2}$, are all zero; $S^{m-1} = T^{m-1} = 1$; S^m is the sum of the quantities E^{ia}, E^{ib}, E^{ic}, $E^{id}, \ldots$; S^{m+1} is the sum of their products taken two at a time, etc. The T^m, T^{m+1}, T^{m+2}, etc. are formed similarly out of E^{-ia}, E^{-ib}, E^{-ic}, $E^{-id}, \ldots$. It is not hard to see that

$$\frac{1}{2}(S^n + T^n) = \begin{cases} 0 & (n = 0, 1, 2, \ldots, m-2)\\ 1 & (n = m-1)\\ \cos a + \cos b + \cos c + \cos d + \cdots & (n = m). \end{cases}$$

Similarly $\frac{1}{2}(S^{m+1} + T^{m+1})$ is the sum of the cosines of all angles which can be formed by adding in pairs the $a, b, c, d, \ldots$ in all possible ways, etc. while

$$\frac{S^n - T^n}{2i} = 0 \qquad \text{for } n = 0, 1, 2, \ldots, m-1;$$

and

$$\frac{S^n - T^n}{2i} = \sin a + \sin b + \sin c + \sin d + \cdots \qquad \text{for } n = m, \text{ etc.}$$

Gauss now examined the functions S^n, T^n in trigonometric form. He found that the first term in S^n, e.g., is expressible as

$$\frac{E^{i((n+1-\frac{1}{2}m)a-\frac{1}{2}s)}}{(2i)^{m-1}\sin\frac{1}{2}(a-b)\sin\frac{1}{2}(a-c)\sin\frac{1}{2}(a-d)\cdots},$$

where $a + b + c + d + \cdots = s$; a quite similar expression holds for T^n, so that the first terms in $\frac{1}{2}(S^n + T^n)$, $\frac{1}{2}(S^n - T^n)$ are alternately

$$\frac{\cos((n+1-\frac{1}{2}m)a-\frac{1}{2}s)}{(2i)^{m-1}\sin\frac{1}{2}(a-b)\sin\frac{1}{2}(a-c)\sin\frac{1}{2}(a-d)\cdots},$$

$$\frac{\sin((n+1-\frac{1}{2}m)a-\frac{1}{2}s)}{(2i)^{m-1}\sin\frac{1}{2}(a-b)\sin\frac{1}{2}(a-c)\sin\frac{1}{2}(a-d)\cdots},$$

according to whether m is odd or even. The other terms in $\frac{1}{2}(S^n \pm T^n)$ are also easy to express.

Gauss now forms two new functions U^λ, V^λ, where $\lambda = n + 1 - \frac{1}{2}m$, as

$$\begin{aligned}
&\frac{\cos(\lambda a + k)}{\sin\frac{1}{2}(a-b)\sin\frac{1}{2}(a-c)\sin\frac{1}{2}(a-d)\cdots} \\
&+ \frac{\cos(\lambda b + k)}{\sin\frac{1}{2}(b-a)\sin\frac{1}{2}(b-c)\sin\frac{1}{2}(b-d)\cdots} \\
&+ \frac{\cos(\lambda c + k)}{\sin\frac{1}{2}(c-a)\sin\frac{1}{2}(c-b)\sin\frac{1}{2}(c-d)\cdots} \\
&+ \frac{\cos(\lambda d + k)}{\sin\frac{1}{2}(d-a)\sin\frac{1}{2}(d-b)\sin\frac{1}{2}(d-c)\cdots} + \text{etc.} = U^\lambda, \\
&\frac{\sin(\lambda a + k)}{\sin\frac{1}{2}(a-b)\sin\frac{1}{2}(a-c)\sin\frac{1}{2}(a-d)\cdots} \\
&+ \frac{\sin(\lambda b + k)}{\sin\frac{1}{2}(b-a)\sin\frac{1}{2}(b-c)\sin\frac{1}{2}(b-d)\cdots} \\
&+ \frac{\sin(\lambda c + k)}{\sin\frac{1}{2}(c-a)\sin\frac{1}{2}(c-b)\sin\frac{1}{2}(c-d)\cdots} \\
&+ \frac{\sin(\lambda d + k)}{\sin\frac{1}{2}(d-a)\sin\frac{1}{2}(d-b)\sin\frac{1}{2}(d-c)\cdots} + \text{etc.} = V^\lambda,
\end{aligned} \tag{4.58}$$

where k is arbitrary. These formulas also hold for λ replaced by $-\lambda$, since $\cos(-\lambda a + k) = \cos(\lambda a - k)$, $\sin(-\lambda a + k) = -\sin(\lambda a - k)$, and we need only replace k by $-k$. Consider first the case where m is odd. Then λ

is an integer plus a half. Moreover, for $n = 0, 1, \ldots, m - 2$, or $\lambda = -\frac{1}{2}m + 1$, $-\frac{1}{2}m + 2, \ldots, \frac{1}{2}m - 1$, or (if we omit negative values of λ), for $\lambda = \frac{1}{2}, \frac{3}{2}, \frac{5}{2}$, $\ldots, \frac{1}{2}m - 1$, $U^{\lambda} = 0 = V^{\lambda}$ and

$$U^{\frac{1}{2}m} = (2i)^{m-1} \cos\left(\frac{1}{2}s + k\right), \qquad V^{\frac{1}{2}m} = (2i)^{m-1} \sin\left(\frac{1}{2}s + k\right),$$

$$U^{\frac{1}{2}m+1} = (2i)^{m-1}\left\{\cos\left(\frac{1}{2}s + k + a\right) + \cos\left(\frac{1}{2}s + k + b\right)\right.$$
$$\left. + \cos\left(\frac{1}{2}s + k + c\right) + \cos\left(\frac{1}{2}s + k + d\right) + \text{etc.}\right\},$$

$$V^{\frac{1}{2}m+1} = (2i)^{m-1}\left\{\sin\left(\frac{1}{2}s + k + a\right) + \sin\left(\frac{1}{2}s + k + b\right)\right.$$
$$\left. + \sin\left(\frac{1}{2}s + k + c\right) + \sin\left(\frac{1}{2}s + k + d\right) + \text{etc.}\right\}.$$

Now consider the case where m is even; λ is, of course, an integer and $U^{\lambda} = V^{\lambda} = 0$ for $\lambda = 0, 1, 2, \ldots, \frac{1}{2}m - 1$. Moreover,

$$U^{\frac{1}{2}m} = 2^{m-1} i^{m-2} \sin\left(\frac{1}{2}s + k\right),$$

$$V^{\frac{1}{2}m} = -2^{m-1} i^{m-2} \cos\left(\frac{1}{2}s + k\right),$$

$$U^{\frac{1}{2}m+1} = 2^{m-1} i^{m-2}\left\{\sin\left(\frac{1}{2}s + k + a\right) + \sin\left(\frac{1}{2}s + k + b\right)\right.$$
$$\left. + \sin\left(\frac{1}{2}s + k + c\right) + \sin\left(\frac{1}{2}s + k + d\right) + \text{etc.}\right\},$$

$$V^{\frac{1}{2}m+1} = -2^{m-1} i^{m-2}\left\{\cos\left(\frac{1}{2}s + k + a\right) + \cos\left(\frac{1}{2}s + k + b\right)\right.$$
$$\left. + \cos\left(\frac{1}{2}s + k + c\right) + \cos\left(\frac{1}{2}s + k + d\right) + \text{etc.}\right\}.$$

Gauss gave expressions for higher superscripts as well.

Consider now a polynomial X of the form $\alpha + \beta x + \gamma xx + \delta x^3 + \text{etc.}$, of degree m, and suppose that for m different values $a, b, c, d, \ldots$ of x the values of X are $A, B, C, D \ldots$. Finally, if T corresponds to $x = t$, then

$$\begin{aligned} A &= \alpha + \beta a + \gamma aa + \delta a^3 + \cdots, \\ B &= \alpha + \beta b + \gamma bb + \delta b^3 + \cdots, \\ C &= \alpha + \beta c + \gamma cc + \delta c^3 + \cdots, \\ D &= \alpha + \beta d + \gamma dd + \delta d^3 + \cdots, \\ &\text{etc.} \\ T &= \alpha + \beta t + \gamma tt + \delta t^3 + \cdots \quad . \end{aligned}$$

Gauss multiplies the first equation by $1/(a - b)(a - c)(a - d)\cdots(a - t)$, the second by $1/(b - a)(b - c)(b - d)\cdots(b - t)$, etc., and forms

$$\begin{aligned} &\frac{A}{(a - b)(a - c)(a - d)\cdots(a - t)} \\ &+ \frac{B}{(b - a)(b - c)(b - d)\cdots(b - t)} \\ &+ \frac{C}{(c - a)(b - c)(b - d)\cdots(b - t)} \\ &+ \frac{D}{(d - a)(d - b)(d - c)\cdots(d - t)} \\ &+ \text{etc.} \\ &+ \frac{T}{(t - a)(t - b)(t - c)(t - d)\cdots} = W. \end{aligned}$$

Then, by the above, it follows immediately that $W = 0$ and

$$\begin{aligned} T = &\frac{(t - b)(t - c)(t - d)\cdots}{(a - b)(a - c)(a - d)\cdots} A \\ &+ \frac{(t - a)(t - c)(t - d)\cdots}{(b - a)(b - c)(b - d)\cdots} B \\ &+ \frac{(t - a)(t - b)(t - d)\cdots}{(c - a)(c - b)(c - d)\cdots} C \\ &+ \frac{(t - a)(t - b)(t - c)\cdots}{(d - a)(d - b)(d - c)\cdots} D \\ &+ \text{etc.} \end{aligned} \qquad (4.59)$$

(This is the Lagrange formula.) Now Gauss rewrites this in the form

$$T = A + A'(t - a) + A''(t - a)(t - b) + A'''(t - a)(t - b)(t - c) + \cdots,$$

and gives expressions for the coefficients. In the case where $b = a + 1$, $c = a + 2, \ldots$ he gives the familiar result and exhibits the error term involving the mth difference.

To illustrate the methods of sections 1–8 of his paper he gives two examples. In the first he assumes given the logarithms (base 10) of the sines of 23°, 24°, 25°, 26°. Thus $A = 9.5918780$; $B = 9.6093133$; $C = 9.6259483$; $D = 9.6418420$ and

$$\log \sin 24°30' = -A/16 + 9B/16 + 9C/16 - D/16 = 9.6177270.$$

(The reader should note that Gauss writes, e.g., $A = 9.5918780$, omitting -10.) Moreover, if instead of $\log \sin 23°$, he had $\log \sin 27° = 9.6570468$ then $\log \sin 24°30' = 5B/16 + 15C/16 - 5D/16 + E/16 = 9.6177267375$.

The error in the former case is $3\Delta/128$, in the latter $-5\Delta/128$, where Δ is the fourth difference. In this example $\Delta = -0.0000066$. In his second example Gauss gives the longitudes of the moon and sun for five days around the time of a new moon and interpolates for the precise time of that event.

In the succeeding sections of the paper he goes from this sort of interpolation to trigonometric interpolation. Here he considers a finite series X of the form

$$\begin{aligned}\alpha + \alpha' \cos x + \alpha'' \cos 2x + \alpha''' \cos 3x + \text{etc.}\\ + \beta' \sin x + \beta'' \sin 2x + \beta''' \sin 3x + \text{etc.},\end{aligned}$$

which contains $2m + 1$ coefficients.[81] As before he assumes that at the $2m + 1$ points $x = a, b, c, d$ etc. the function X takes on the known values $A, B, C, D \ldots$; and for $x = t$ it has the value T. This gives $2m + 2$ equations:

$$\begin{aligned}A &= \alpha + \alpha' \cos a + \alpha'' \cos 2a + \alpha''' \cos 3a + \text{etc.}\\ &\quad + \beta' \sin a + \beta'' \sin 2a + \beta''' \sin 3a + \text{etc.}\\ &\text{etc.,}\\ T &= \alpha + \alpha' \cos t + \alpha'' \cos 2t + \alpha''' \cos 3t + \text{etc.}\\ &\quad + \beta' \sin t + \beta'' \sin 2t + \beta''' \sin 3t + \text{etc.}\end{aligned}$$

Gauss now multiplies these equations by the expressions

$$\begin{aligned}&\frac{1}{\sin \frac{1}{2}(a-b) \sin \frac{1}{2}(a-c) \sin \frac{1}{2}(a-d) \cdots \sin \frac{1}{2}(a-t)},\\ &\text{etc.,} \qquad\qquad (4.60)\\ &\frac{1}{\sin \frac{1}{2}(t-a) \sin \frac{1}{2}(t-b) \sin \frac{1}{2}(t-c) \sin \frac{1}{2}(t-d) \cdots},\end{aligned}$$

and adds, thereby forming an expression he calls W. Then, by what preceded, he knows that $W = 0$ and hence

$$\begin{aligned}T = {} & \frac{\sin \frac{1}{2}(t-b) \sin \frac{1}{2}(t-c) \sin \frac{1}{2}(t-d) \cdots}{\sin \frac{1}{2}(a-b) \sin \frac{1}{4}(a-c) \sin \frac{1}{2}(a-d) \cdots} A\\ & + \frac{\sin \frac{1}{2}(t-a) \sin \frac{1}{2}(t-c) \sin \frac{1}{2}(t-d) \cdots}{\sin \frac{1}{2}(b-a) \sin \frac{1}{2}(b-c) \sin \frac{1}{2}(b-d) \cdots} B\\ & + \frac{\sin \frac{1}{2}(t-a) \sin \frac{1}{2}(t-b) \sin \frac{1}{2}(t-d) \cdots}{\sin \frac{1}{2}(c-a) \sin \frac{1}{2}(c-b) \sin \frac{1}{2}(c-d) \cdots} C \qquad (4.61)\\ & + \frac{\sin \frac{1}{2}(t-a) \sin \frac{1}{2}(t-b) \sin \frac{1}{2}(t-c) \cdots}{\sin \frac{1}{2}(d-a) \sin \frac{1}{2}(d-b) \sin \frac{1}{2}(d-c) \cdots} D\\ & + \text{etc.}\end{aligned}$$

This is Gauss's formula for trigonometric interpolation; it is generally considered the one to use in interpolating a periodic function. There is a similar result attributed to Hermite which differs from that of Gauss in having all

---

[81] Gauss III *TI*, p. 279.

angles twice as large, i.e., no "half-angles" appear in it. It is however the same as Gauss's result (4.61′) below.

In the case where we only have $2m$ known values A, B, C, D etc. instead of $2m + 1$, it is no longer possible to conclude that $W = 0$, as above. In this case Gauss multiplies X by $\cos(\frac{1}{2}x + k)$ and finds

$$\begin{aligned}
&\alpha \cos\left(\frac{1}{2}x + k\right) + \frac{1}{2}\alpha' \cos\left(\frac{3}{2}x + k\right) + \frac{1}{2}\alpha'' \cos\left(\frac{5}{2}x + k\right) + \cdots \\
&\qquad + \frac{1}{2}\alpha^m \cos\left(\left(m + \frac{1}{2}\right)x + k\right) \\
&+ \frac{1}{2}\alpha' \cos\left(\frac{1}{2}x - k\right) + \frac{1}{2}\alpha'' \cos\left(\frac{3}{2}x - k\right) + \frac{1}{2}\alpha''' \cos\left(\frac{5}{2}x - k\right) + \cdots \\
&\qquad + \frac{1}{2}\alpha^m \cos\left(\left(m - \frac{1}{2}\right)x - k\right) \\
&\qquad + \frac{1}{2}\beta' \sin\left(\frac{3}{2}x - k\right) + \frac{1}{2}\beta'' \sin\left(\frac{5}{2}x + k\right) + \cdots \\
&\qquad + \frac{1}{2}\beta^m \sin\left(\left(m + \frac{1}{2}\right)x + k\right) \\
&+ \frac{1}{2}\beta' \sin\left(\frac{1}{2}x - k\right) + \frac{1}{2}\beta'' \sin\left(\frac{3}{2}x - k\right) + \frac{1}{2}\beta''' \sin\left(\frac{5}{2}x - k\right) + \cdots \\
&\qquad + \frac{1}{2}\beta^m \sin\left(\left(m - \frac{1}{2}\right)x - k\right),
\end{aligned}$$

where the last term in X is $\alpha^m \cos mx + \beta^m \sin mx$. In effect Gauss then forms the function

$$\begin{aligned}
W &= \frac{A \cos(\frac{1}{2}a + k)}{\sin\frac{1}{2}(a - b)\sin\frac{1}{2}(a - c)\sin\frac{1}{2}(a - d)\cdots\sin\frac{1}{2}(a - t)} \\
&+ \frac{B \cos\frac{1}{2}(b + k)}{\sin\frac{1}{2}(b - a)\sin\frac{1}{2}(b - c)\sin\frac{1}{2}(b - d)\cdots\sin\frac{1}{2}(b - t)} \\
&+ \frac{C \cos(\frac{1}{2}c + k)}{\sin\frac{1}{2}(c - a)\sin\frac{1}{2}(c - b)\sin\frac{1}{2}(c - d)\cdots\sin\frac{1}{2}(c - t)} \\
&+ \frac{D \cos(\frac{1}{2}d + k)}{\sin\frac{1}{2}(d - a)\sin\frac{1}{2}(d - b)\sin\frac{1}{2}(d - c)\cdots\sin\frac{1}{2}(d - t)} \\
&+ \text{etc.} \\
&+ \frac{T \cos(\frac{1}{2}t + k)}{\sin\frac{1}{2}(t - a)\sin\frac{1}{2}(t - b)\sin\frac{1}{2}(t - c)\sin\frac{1}{2}(t - d)\cdots}.
\end{aligned}$$

He is now able to use his earlier results to show that in the present case (see (4.58) and following discussion)

$$W = \frac{1}{2}(2i)^{2m}\alpha^m \cos\left(\frac{1}{2}(a + b + c + d + \cdots + t) + k\right)$$
$$+ \frac{1}{2}(2i)^{2m}\beta^m \sin\left(\frac{1}{2}(a + b + c + d + \cdots + t) + k\right).$$

Next he chooses the parameter k, which is still arbitrary, as $-\frac{1}{2}t$ and sets $s = a + b + c + d + \cdots$, omitting t from the sum. (This gives $W = 2^{2m-1}i^{2m}[\alpha^m \cos\frac{1}{2}s + \beta^m \sin\frac{1}{2}s]$.) He then solves for T in the expression above and finds

$$T = \frac{\sin\frac{1}{2}(t-b)\sin\frac{1}{2}(t-c)\sin\frac{1}{2}(t-d)\cdots}{\sin\frac{1}{2}(a-b)\sin\frac{1}{2}(a-c)\sin\frac{1}{2}(a-d)\cdots} A\cos\frac{1}{2}(t-a)$$
$$+ \frac{\sin\frac{1}{2}(t-a)\sin\frac{1}{2}(t-c)\sin\frac{1}{2}(t-d)\cdots}{\sin\frac{1}{2}(b-a)\sin\frac{1}{2}(b-c)\sin\frac{1}{2}(b-d)\cdots} B\cos\frac{1}{2}(t-b)$$
$$+ \frac{\sin\frac{1}{2}(t-a)\sin\frac{1}{2}(t-b)\sin\frac{1}{2}(t-d)\cdots}{\sin\frac{1}{2}(c-a)\sin\frac{1}{2}(c-b)\sin\frac{1}{2}(c-d)\cdots} C\cos\frac{1}{2}(t-c)$$
$$+ \frac{\sin\frac{1}{2}(t-a)\sin\frac{1}{2}(t-b)\sin\frac{1}{2}(t-c)\cdots}{\sin\frac{1}{2}(d-a)\sin\frac{1}{2}(d-b)\sin\frac{1}{2}(d-c)\cdots} D\cos\frac{1}{2}(t-d)$$
$$+ \text{etc.}$$
$$+ 2^{2m-1}i^{2m}\sin\frac{1}{2}(t-a)\sin\frac{1}{2}(t-b)\sin\frac{1}{2}(t-c)\sin\frac{1}{2}(t-d)$$
$$\times\left(\alpha^m\cos\frac{1}{2}s + \beta^m\sin\frac{1}{2}s\right).$$

If we wish to find the $2m + 1$ values $\alpha, \alpha', \beta', \alpha'', \beta''$ etc. from this formula for T, given $2m$ values $A, B, C, D\ldots$, it is clear that some quantity, say, $\alpha^m\cos\frac{1}{2}s + \beta^m\sin\frac{1}{2}s$, must be arbitrary; it can then be set, e.g., to zero. Gauss accordingly sets $\alpha^m\cos\frac{1}{2}s + \beta^m\sin\frac{1}{2}s = 0$ and by differencing replaces the previous relation for T by

$$T = A + A' \sin\frac{1}{2}(t-a)\cos\frac{1}{2}(t-b)$$
$$+ A''\sin\frac{1}{2}(t-a)\sin\frac{1}{2}(t-b)$$
$$+ A'''\sin\frac{1}{2}(t-a)\sin\frac{1}{2}(t-b)\sin\frac{1}{2}(t-c)\cos\frac{1}{2}(t-d)$$
$$+ A''''\sin\frac{1}{2}(t-a)\sin\frac{1}{2}(t-b)\sin\frac{1}{2}(t-c)\sin\frac{1}{2}(t-d)$$

$$+ A^{\mathrm{V}} \sin\frac{1}{2}(t-a)\sin\frac{1}{2}(t-b)\sin\frac{1}{2}(t-c)\sin\frac{1}{2}(t-d)$$
$$\times \sin\frac{1}{2}(t-e)\cos\frac{1}{2}(t-f)$$
$$+ A^{\mathrm{VI}} \sin\frac{1}{2}(t-a)\sin\frac{1}{2}(t-b)\sin\frac{1}{2}(t-c)\sin\frac{1}{2}(t-d)$$
$$\times \sin\frac{1}{2}(t-e)\sin\frac{1}{2}(t-f)$$
$$+ \text{etc.},$$

where the coefficients A', A'', A''', A'''' etc. are expressible as

$$A' = \frac{A}{\sin\frac{1}{2}(a-b)} + \frac{B}{\sin\frac{1}{2}(b-a)},$$

$$A'' = \frac{A\cos\frac{1}{2}(a-c)}{\sin\frac{1}{2}(a-b)\sin\frac{1}{2}(a-c)} + \frac{B\cos\frac{1}{2}(b-c)}{\sin\frac{1}{2}(b-a)\sin\frac{1}{2}(b-c)}$$
$$+ \frac{C}{\sin\frac{1}{2}(c-a)\sin\frac{1}{2}(c-b)},$$

$$A''' = \frac{A}{\sin\frac{1}{2}(a-b)\sin\frac{1}{2}(a-c)\sin\frac{1}{2}(a-d)}$$
$$+ \frac{B}{\sin\frac{1}{2}(b-a)\sin\frac{1}{2}(b-c)\sin\frac{1}{2}(b-d)}$$
$$+ \frac{C}{\sin\frac{1}{2}(c-a)\sin\frac{1}{2}(c-b)\sin\frac{1}{2}(c-d)}$$
$$+ \frac{D}{\sin\frac{1}{2}(d-a)\sin\frac{1}{2}(d-b)\sin\frac{1}{2}(d-c)},$$

etc.

To find a new and more elegant representation for these coefficients Gauss introduces some ancillary variables and writes

$$A' = \frac{A-B}{\sin\frac{1}{2}(a-b)}, \qquad B' = \frac{B-C}{\sin\frac{1}{2}(b-c)},$$

$$A'' = \frac{A'\cos\frac{1}{2}(a-c) - B'}{\sin\frac{1}{2}(a-c)},$$

$$A''' = \frac{A''\cos\frac{1}{2}(a-c) + A'\sin\frac{1}{2}(a-c) - B''}{\sin\frac{1}{2}(a-d)},$$

$$A'''' = \frac{A'''\cos\frac{1}{2}(a-e) - B'''}{\sin\frac{1}{2}(a-e)},$$

$$A^{\mathrm{V}} = \frac{A'''' \cos \frac{1}{2}(a-e) + A''' \sin \frac{1}{2}(a-e) - B''''}{\sin \frac{1}{2}(a-f)},$$

$$A^{\mathrm{VI}} = \frac{A^{\mathrm{V}} \cos \frac{1}{2}(a-g) - B^{\mathrm{V}}}{\sin \frac{1}{2}(a-g)},$$

$$C' = \frac{C - D}{\sin \frac{1}{2}(c-d)} \text{ etc.}$$

$$B'' = \frac{B' \cos \frac{1}{2}(b-d) - C'}{\sin \frac{1}{2}(b-d)} \text{ etc.}$$

$$B''' = \frac{B'' \cos \frac{1}{2}(b-d) + B'' \sin \frac{1}{2}(b-d) - C''}{\sin \frac{1}{2}(b-e)} \text{ etc.}$$

$$B'''' = \frac{B''' \cos \frac{1}{2}(b-f) - C'''}{\sin \frac{1}{2}(b-f)} \text{ etc.}$$

$$B^{\mathrm{V}} = \frac{B'''' \cos \frac{1}{2}(b-f) + B''' \sin \frac{1}{2}(b-f) - C''''}{\sin \frac{1}{2}(b-g)} \text{ etc.}$$

$$B^{\mathrm{VI}} = \frac{B^{\mathrm{V}} \cos \frac{1}{2}(b-h) - C^{\mathrm{V}}}{\sin \frac{1}{2}(a-h)} \text{ etc.,}$$

etc.

(In this case Gauss's values for the Fourier coefficients appear in (4.64) and earlier.) Gauss next turns in sections 13–14 to some related topics and proves several theorems he needs later. First, if X is of the form

$$\begin{aligned} &\alpha + \alpha' \cos x + \alpha'' \cos 2x + \alpha''' \cos 3x + \text{etc.} \\ &\quad + \beta' \sin x + \beta'' \sin 2x + \beta''' \sin 3x + \text{etc.,} \end{aligned} \tag{4.62}$$

or of the form

$$\begin{aligned} &\gamma \cos \frac{1}{2} x + \gamma' \cos \frac{3}{2} x + \gamma'' \cos \frac{5}{2} x + \text{etc.} \\ &+ \delta \sin \frac{1}{2} x + \delta' \sin \frac{3}{2} x + \delta'' \sin \frac{5}{2} x + \text{etc.,} \end{aligned} \tag{4.63}$$

and if for $x = a$, $X = 0$, then X is divisible by $b \sin \frac{1}{2}(x - a)$; in the prior case the quotient is of the posterior form and in the posterior case it is of the prior form. Suppose that X is representable in the first form. Then

$$\begin{aligned} X &= \alpha + \alpha' \cos x + \alpha'' \cos 2x + \cdots + \beta' \sin x + \beta'' \sin 2x + \cdots, \\ 0 &= \alpha + \alpha' \cos a + \alpha'' \cos 2a + \cdots + \beta' \sin a + \beta'' \sin 2a + \cdots, \end{aligned}$$

and by subtraction

$$\begin{aligned} X = {} &\alpha'(\cos x - \cos a) + \alpha''(\cos 2x - \cos 2a) + \cdots \\ &+ \beta'(\sin x - \sin a) + \beta''(\sin 2x - \sin 2a) + \cdots \quad . \end{aligned}$$

But

$$\begin{aligned}\frac{\cos nx - \cos na}{\sin \frac{1}{2}(x-a)} = &-2\sin\left(\left(n-\frac{1}{2}\right)x + \frac{1}{2}a\right)\\ &-2\sin\left(\left(n-\frac{3}{2}\right)x + \frac{3}{2}a\right)\\ &-2\sin\left(\left(n-\frac{5}{2}\right)x + \frac{5}{2}a\right)\\ &-\text{etc.}\\ &-2\sin\left(\frac{1}{2}x + \left(n-\frac{1}{2}\right)a\right),\end{aligned}$$

$$\begin{aligned}\frac{\sin nx - \sin na}{\sin \frac{1}{2}(x-a)} = &+2\cos\left(\left(n-\frac{1}{2}\right)x + \frac{1}{2}a\right)\\ &+2\cos\left(\left(n-\frac{3}{2}\right)x + \frac{3}{2}a\right)\\ &+2\cos\left(\left(n-\frac{5}{2}\right)x + \frac{5}{2}a\right)\\ &+\text{etc.}\\ &+2\cos\left(\frac{1}{2}x + \left(n-\frac{1}{2}\right)a\right).\end{aligned}$$

This shows the first part of the result. Suppose next that X is representable in the second form. Gauss then writes

$$X = \gamma\cos\frac{1}{2}x + \gamma'\cos\frac{3}{2}x + \text{etc.} + \delta\sin\frac{1}{2}x + \delta'\sin\frac{3}{2}x + \text{etc.},$$

$$\begin{aligned}0 = \Big(&\gamma\cos\frac{1}{2}a + \gamma'\cos\frac{3}{2}a + \text{etc.} + \delta\sin\frac{1}{2}a\\ &+ \delta'\sin\frac{3}{2}a + \text{etc.}\Big)\cos\frac{1}{2}(x-a);\end{aligned}$$

and hence the general terms of the difference are expressible as

$$\begin{aligned}\frac{\cos nx - \cos\frac{1}{2}(x-a)\cos na}{\sin \frac{1}{2}(x-a)} = &-2\sin\left(\left(n-\frac{1}{2}\right)x + \frac{1}{2}a\right)\\ &-2\sin\left(\left(n-\frac{3}{2}\right)x + \frac{3}{2}a\right)\\ &-2\sin\left(\left(n-\frac{5}{2}\right)x + \frac{5}{2}a\right)\\ &-\text{etc.}\\ &-2\sin(x + (n-1)a)\\ &-\sin na,\end{aligned}$$

$$\begin{aligned}\frac{\sin nx - \cos\frac{1}{2}(x-a)\sin na}{\sin\frac{1}{2}(x-a)} = &+2\cos\left(\left(n-\frac{1}{2}\right)x+\frac{1}{2}a\right)\\ &+2\cos\left(\left(n-\frac{3}{2}\right)x+\frac{3}{2}a\right)\\ &+2\cos\left(\left(n-\frac{5}{2}\right)x+\frac{5}{2}a\right)\\ &+\text{etc.}\\ &+2\cos(x+(n-1)a)\\ &+\cos na\end{aligned}$$

and the second part of the result follows.

In the same way he shows divisibility by $\sin\frac{1}{2}(x-b)$. Now, if all the coefficients $\beta', \beta'', \beta'''$ etc. vanish, i.e., if X is an even function, then there are only $m+1$ unknown coefficients to be determined. Suppose for $x = a, b, c, d$ etc. X takes on the values A, B, C, D etc. Gauss first expresses X as a trigonometric polynomial of the form

$$\gamma + \gamma'\cos x + \gamma''(\cos x)^2 + \gamma'''(\cos x)^3 + \text{etc.} + \gamma^m(\cos x)^m$$

and second, by his previous results (4.59), finds

$$\begin{aligned}T = &\frac{(\cos t - \cos b)(\cos t - \cos c)(\cos t - \cos d)\cdots}{(\cos a - \cos b)(\cos a - \cos c)(\cos a - \cos d)\cdots}A\\ &+\frac{(\cos t - \cos a)(\cos t - \cos c)(\cos t - \cos d)\cdots}{(\cos b - \cos a)(\cos b - \cos c)(\cos b - \cos d)\cdots}B\\ &+\frac{(\cos t - \cos a)(\cos t - \cos b)(\cos t - \cos d)\cdots}{(\cos c - \cos a)(\cos c - \cos b)(\cos c - \cos d)\cdots}C\\ &+\frac{(\cos t - \cos a)(\cos t - \cos b)(\cos t - \cos c)\cdots}{(\cos d - \cos a)(\cos d - \cos b)(\cos d - \cos c)\cdots}D\\ &+\text{etc.}\end{aligned} \qquad (4.61')$$

If now all the coefficients $\alpha, \alpha', \alpha'', \alpha'''$ etc. vanish in (4.60), i.e., if X is an odd function, X vanishes for all multiples of 180°. For n even he has

$$\frac{\sin nx}{\sin x} = 2\cos(n-1)x + 2\cos(n-3)x + 2\cos(n-5)x + \cdots + 2\cos x,$$

and for n odd

$$= 2\cos(n-1)x + 2\cos(n-3)x + 2\cos(n-5)x + \cdots + 2\cos 2x + 1.$$

Thus $X/\sin x$ is expressible as $\delta + \delta'\cos x + \delta''(\cos x)^2 + \delta'''(\cos x)^3 +$ etc. $+ \delta^{m-1}(\cos x)^{m-1}$. Suppose that for $x = a, b, c, d\ldots$

$$\frac{X}{\sin x}$$

takes on the values

$$\frac{A}{\sin \alpha}, \frac{B}{\sin b}, \frac{C}{\sin c}, \frac{D}{\sin d}, \ldots \quad .$$

Then Gauss concludes that

$$T = \frac{\sin t(\cos t - \cos b)(\cos t - \cos c)(\cos t - \cos d)\cdots}{\sin a(\cos a - \cos b)(\cos a - \cos c)(\cos a - \cos d)\cdots} A$$

$$+ \frac{\sin t(\cos t - \cos a)(\cos t - \cos c)(\cos t - \cos d)\cdots}{\sin b(\cos b - \cos a)(\cos b - \cos c)(\cos b - \cos d)\cdots} B$$

$$+ \frac{\sin t(\cos t - \cos a)(\cos t - \cos b)(\cos t - \cos d)\cdots}{\sin c(\cos c - \cos a)(\cos c - \cos b)(\cos c - \cos d)\cdots} C$$

$$+ \frac{\sin t(\cos t - \cos a)(\cos t - \cos b)(\cos t - \cos c)\cdots}{\sin d(\cos d - \cos a)(\cos d - \cos b)(\cos d - \cos c)\cdots} D$$

$$+ \text{etc.}$$

Consider now, with Gauss, two lemmas. Let a, b, c, d etc. be μ arcs in arithmetic progression, $b - a = c - b = d - c$ etc. $= 360°/\mu$, and let P be $\sin \frac{1}{2}(t - a) \sin \frac{1}{2}(t - b) \sin \frac{1}{2}(t - c) \sin \frac{1}{2}(t - d)$ etc. Then

$$P = \mp \frac{\sin \frac{1}{2}\mu(t - a)}{2^{\mu - 1}},$$

where the upper sign goes with μ even and the lower with μ odd. Gauss remarks that this follows from a related result of Euler (*Introd. in Anal. Inf.* I Section 240, *Opera Omnia*, Vol. VIII) which says that

$$\sin nz = 2^{n-1} \sin z \sin\left(\frac{\pi}{n} - z\right) \sin\left(\frac{\pi}{n} + z\right) \sin\left(\frac{2\pi}{n} - z\right) \sin\left(\frac{2\pi}{n} + z\right) \cdots \quad .$$

His second lemma tells us that if $a, b, c, d\ldots$ are as above, then

$$(\cos t - \cos a)(\cos t - \cos b)(\cos t - \cos c)(\cos t - \cos d)\cdots = \frac{\cos \mu t - \cos \mu a}{2^{\mu - 1}}.$$

To show this he writes

$$\cos t - \cos a = 2 \sin \frac{1}{2}(t - a) \sin \frac{1}{2}(-t - a),$$

$$\cos t - \cos b = 2 \sin \frac{1}{2}(t - b) \sin \frac{1}{2}(-t - b),$$

$$\cos t - \cos c = 2 \sin \frac{2}{2}(t - c) \sin \frac{1}{2}(-t - c),$$

$$\cos t - \cos d = 2 \sin \frac{1}{2}(t - d) \sin \frac{1}{2}(-t - d),$$

etc.;

and hence their product is, by the first lemma,

$$= 2^\mu \times \mp\frac{\sin\frac{1}{2}\mu(t-a)}{2^{\mu-1}} \times \mp\frac{\sin\frac{1}{2}\mu(-t-a)}{2^{\mu-1}}$$

$$= \frac{\sin\frac{1}{2}\mu(t-a)\sin\frac{1}{2}\mu(-t-a)}{2^{\mu-2}} = \frac{\cos\mu t - \cos\mu a}{2^{\mu-1}}.$$

This is all preliminary to Gauss's finding the values of the Fourier coefficients.

Consider the series

$$\begin{aligned}&\alpha + \alpha'\cos x + \alpha''\cos 2x + \alpha'''\cos 3x + \cdots + \alpha^{(m)}\cos mx\\ &\quad + \beta'\sin x + \beta''\sin 2x + \beta'''\sin 3x + \cdots + \beta^{(m)}\sin mx,\end{aligned}$$

and set $\mu = 2m + 1$. Now, by what we saw before,

$$\begin{aligned}&\sin\frac{1}{2}(t-b)\sin\frac{1}{2}(t-c)\sin\frac{1}{2}(t-d)\cdots\\ &= \frac{\sin\frac{1}{2}\mu(t-a)}{2^{\mu-1}\sin\frac{1}{2}(t-a)}\\ &= \frac{1}{2^{\mu-1}}(1 + 2\cos(t-a) + 2\cos 2(t-a) + 2\cos 3(t-a) + \cdots\\ &\qquad + 2\cos m(t-a)).\end{aligned}$$

Hence when $t = a$ we have

$$\sin\frac{1}{2}(a-b)\sin\frac{1}{2}(a-c)\sin\frac{1}{2}(a-d)\cdots = \frac{\mu}{2^{\mu-1}}.$$

Similarly

$$\sin\frac{1}{2}(b-a)\sin\frac{1}{2}(b-c)\sin\frac{1}{2}(b-c)\cdots = -\frac{\mu}{2^{\mu-1}}, \text{ etc.}$$

In the expression (4.61) for T the coefficient of A becomes, in this case,

$$\frac{1}{\mu}(1 + 2\cos(t-a) + 2\cos 2(t-a) + 2\cos 3(t-a) + \cdots + 2\cos m(t-a)),$$

with comparable expressions for the coefficients of B, C, D etc. Hence we may write T in the form

$$\begin{aligned}T &= \frac{1}{\mu}(A + B + C + D + \cdots)\\ &+ \frac{2}{\mu}(A\cos a + B\cos b + C\cos c + D\cos d + \cdots)\cos t\\ &+ \frac{2}{\mu}(A\sin a + B\sin b + C\sin c + D\sin d + \cdots)\sin t\end{aligned}$$

$$+ \frac{2}{\mu}(A \cos 2a + B \cos 2b + C \cos 2c + D \cos 2d + \cdots) \cos 2t$$

$$+ \frac{2}{\mu}(A \sin 2a + B \sin 2b + C \sin 2c + D \sin 2d + \cdots) \sin 2t$$

$$+ \text{etc.}$$

$$+ \frac{2}{\mu}(A \cos ma + B \cos mb + C \cos mc + D \cos md + \cdots) \cos mt$$

$$+ \frac{2}{\mu}(A \sin ma + B \sin mb + C \sin mc + D \sin md + \cdots) \sin mt.$$

This gives the values of α, α', β', β'' etc. in their well-known form. We next consider with Gauss the case when $\mu \neq 2m + 1$. First let us suppose that X does not terminate with $\cos mx$ and $\sin mx$.[82] Gauss remarks that

$$\cos na + \cos nb + \cos nc + \cos nd + \cdots = \mu \cos na,$$
$$\sin na + \sin nb + \sin nc + \sin nd + \cdots = \mu \sin na,$$

in the case where n is divisible by μ; and in the contrary case these two series are each equal to 0. To see why this is true Gauss lets the sums be P and Q, respectively. Then it is easy to see that

$$P \cos n(b - a) - Q \sin n(b - a) = P,$$
$$P \sin n(b - a) + Q \cos n(b - a) = Q.$$

From these relations we note that $P = Q = 0$, $\cos n(b - a) = 1$. Now with Gauss, let us write the terms in X after $\cos mx$, $\sin mx$ in the form

$$\alpha^{m+1} \cos (m + 1)x + \alpha^{m+2} \cos (m + 2)x + \text{etc.}$$
$$+ \beta^{m+1} \sin (m + 1)x + \beta^{m+2} \sin (m + 2)x + \text{etc.},$$

and let us consider the approximate expression for X

$$\gamma + \gamma' \cos t + \gamma'' \cos 2t + \gamma''' \cos 3t + \cdots + \gamma^m \cos mt$$
$$+ \delta' \sin t + \delta'' \sin 2t + \delta''' \sin 3t + \cdots + \delta^m \sin mt,$$

where γ, γ', δ' etc. are the finite Fourier coefficients, i.e.,

$$\gamma = \frac{1}{\mu}(A + B + C + D + \text{etc.})$$

etc., and where A, B, C, D etc. have the values

$$A = \alpha + \alpha' \cos a + \alpha'' \cos 2a + \cdots + \alpha^m \cos ma + \alpha^{m+1} \cos (m + 1)a + \cdots$$
$$+ \beta' \sin a + \beta'' \sin 2a + \cdots + \beta^m \sin ma + \beta^{m+1} \sin (m + 1)a + \cdots$$

etc.

---

[82] Gauss III *TI*, pp. 296ff. This is Sections 21ff.

By the results above it is easy to see that

$$\begin{aligned}
\gamma &= \alpha + \alpha^{\mu}\cos\mu a + \alpha^{2\mu}\cos 2\mu a + \alpha^{3\mu}\cos 3\mu a + \text{etc.}\\
&\quad + \beta^{\mu}\sin\mu a + \beta^{2\mu}\sin 2\mu a + \beta^{3\mu}\sin 3\mu a + \text{etc.},\\
\gamma' &= \alpha' + (\alpha^{\mu-1} + \alpha^{\mu+1})\cos\mu a + (\alpha^{2\mu-1} + \alpha^{2\mu+1})\cos 2\mu a + \text{etc.}\\
&\quad + (\beta^{\mu-1} + \beta^{\mu+1})\sin\mu a + (\beta^{2\mu-1} + \beta^{2\mu+1})\sin 2\mu a + \text{etc.},\\
\delta' &= \beta' - (\beta^{\mu-1} - \beta^{\mu+1})\cos\mu a - (\beta^{2\mu-1} - \beta^{2\mu+1})\cos 2\mu a - \text{etc.}\\
&\quad - (\alpha^{\mu-1} - \alpha^{\mu+1})\sin\mu a - (\alpha^{2\mu-1} - \alpha^{2\mu+1})\sin 2\mu a - \text{etc.},\\
\gamma'' &= \alpha'' + (\alpha^{\mu-2} + \alpha^{\mu+2})\cos\mu a + (\alpha^{2\mu-2} + \alpha^{2\mu+2})\cos 2\mu a + \text{etc.}\\
&\quad + (\beta^{\mu-2} + \beta^{\mu+2})\sin\mu a + (\beta^{2\mu-2} + \beta^{2\mu+2})\sin 2\mu a + \text{etc.},\\
\delta'' &= \beta'' - (\beta^{\mu-2} - \beta^{\mu+2})\cos\mu a - (\beta^{2\mu-2} - \beta^{2\mu+2})\cos 2\mu a - \text{etc.},\\
&\quad - (\alpha^{\mu-2} - \alpha^{\mu+2})\sin\mu a - (\alpha^{2\mu-2} - \alpha^{2\mu+2})\sin 2\mu a - \text{etc.},
\end{aligned}$$

etc.

$$\begin{aligned}
\gamma^{m} &= \alpha^{m} + (\alpha^{\mu-m} + \alpha^{\mu+m})\cos\mu a + (\alpha^{2\mu-m} + \alpha^{2\mu+m})\cos 2\mu a + \text{etc.}\\
&\quad + (\beta^{\mu-m} + \beta^{\mu+m})\sin\mu a + (\beta^{2\mu-m} + \beta^{2\mu+m})\sin 2\mu a + \text{etc.},\\
\delta^{m} &= \beta^{m} - (\beta^{\mu-m} - \beta^{\mu+m})\cos\mu a - (\beta^{2\mu-m} - \beta^{2\mu+m})\cos 2\mu a - \text{etc.}\\
&\quad - (\beta^{\mu-m} - \alpha^{\mu+m})\sin\mu a - (\alpha^{2\mu-m} - \alpha^{2\mu+m})\sin 2\mu a - \text{etc.}
\end{aligned}$$

If the series converges and if α^{m+1}, β^{m+1}, α^{m+2}, β^{m+2} etc. are negligible, then the values γ, γ', δ' etc. are good approximations to α, α', β' etc.

Consider once again the case where $\mu = 2m$. Gauss expresses T in the form

$$\begin{aligned}
T = {} & \frac{1}{\mu}(A + B + C + D + \text{etc.})\\
& + \frac{2}{\mu}(A\cos a + B\cos b + C\cos c + D\cos d + \text{etc.})\cos t\\
& + \frac{2}{\mu}(A\sin a + B\sin b + C\sin c + D\sin d + \text{etc.})\sin t\\
& + \frac{2}{\mu}(A\cos 2a + B\cos 2b + C\cos 2c + D\cos 2d + \text{etc.})\cos 2t\\
& + \frac{2}{\mu}(A\sin 2a + B\sin 2b + C\sin 2c + D\sin 2d + \text{etc.})\sin 2t\\
& + \text{etc.}\\
& + \frac{1}{\mu}(A\cos ma + B\cos mb + C\cos mc + D\cos md + \text{etc.})\cos mt\\
& + \frac{1}{\mu}(A\sin ma + B\sin mb + C\sin mc + D\sin md + \text{etc.})\sin mt\\
& + (\alpha^{m}\sin ma - \beta^{m}\cos ma)(\sin ma\cos mt - \cos ma\sin mt).
\end{aligned}$$

When this expression with the last terms omitted, is compared to

$$\begin{aligned}
&\gamma + \gamma'\cos t + \gamma''\cos 2t + \text{etc.} + \gamma^{m}\cos mt\\
&\quad + \delta'\sin t + \delta''\sin 2t + \text{etc.} + \delta^{m}\sin mt,
\end{aligned}$$

he finds that all coefficients γ, γ', δ' etc., except for γ^m, δ^m, are determined by the same formulas as for the case $\mu = 2m + 1$. For γ^m, δ^m, however, he has

$$\alpha^m = \gamma^m + (\alpha^m \sin ma - \beta^m \cos ma) \sin ma,$$
$$\beta^m = \delta^m - (\alpha^m \sin ma - \beta^m \cos ma) \cos ma.$$

Notice, moreover, that since $b - a = c - b = \text{etc.} = 360°/\mu$,

$$\cos ma = -\cos mb = \cos mc = -\cos md \text{ etc.},$$
$$\sin ma = -\sin mb = \sin mc = -\sin md \text{ etc.}$$

Now from these he infers that

$$\gamma^m = \frac{\cos ma}{\mu}(A - B + C - D + \text{etc.}),$$
$$\delta^m = \frac{\sin ma}{\mu}(A - B + C - D + \text{etc.}), \qquad (4.64)$$

and that

$$\alpha^m \cos ma + \beta^m \sin ma = \frac{1}{\mu}(A - B + C - D + \text{etc.}).$$

If the series for X does not terminate with $\cos mx$ and $\sin mx$, then the above series for T is incomplete and the γ, γ', δ' etc. differ from the α, α', β' etc. This is comparable to the case studied above for $\mu \neq 2m + 1$.

Let us now turn, with Gauss, to the case where once again the series for X does not necessarily terminate with $\cos mx$ and $\sin mx$, and where now the subsequent Fourier coefficients are not negligible. Suppose further that there are $\pi = \mu\nu$ values of X corresponding to the points $x = a, a', a'', \ldots, b, b', b'', \ldots, c, c', c'', \ldots, d, d', d'', \ldots$ etc. with

$$a' = a + \frac{1}{\pi} 360°, \qquad a'' = a + \frac{2}{\pi} 360° \text{ etc.}$$

$$b = a + \frac{\nu}{\pi} 360° = a + \frac{1}{\mu} 360°, \qquad b' = a + \frac{\nu + 1}{\pi} 360° = a' + \frac{1}{\mu} 360° \text{ etc.}$$

$$c = a + \frac{2\nu}{\pi} 360° = a + \frac{2}{\mu} 360° \text{ etc.}$$

(This material is discussed in Section 25.) Then the total set of π values of x can be divided into ν groups of μ terms each:

$$x = a, b, c, d \text{ etc.};$$
$$x = a', b', c', d' \text{ etc.};$$
$$x = a'', b'', c'', d'' \text{ etc.};$$
etc.[83]

---

[83] Gauss III *TI*, pp. 303ff. This fascinating work of Gauss was neglected and was rediscovered by Cooley and Tukey in an important paper in 1965. Cf. Cooley, *CTA*, pp. 297–301.

To handle this problem Gauss imagines that the coefficients γ, γ', δ', etc. in

$$\begin{aligned} X' = \gamma + \gamma' \cos x + \gamma'' \cos 2x + \cdots \gamma^m \cos mx \\ + \delta' \sin x + \delta'' \sin 2x + \cdots \delta^m \sin mx \end{aligned}$$

are themselves variable, as we shall see. Consider the quantities

$$y = \mu a, \qquad y = \mu a' = \mu a + \frac{1}{\nu} 360°, \qquad y = \mu a'' = \mu a + \frac{2}{\nu} 360°, \text{ etc.};$$

these form a complete set of ν values which we may use to determine the coefficients γ, γ', δ' etc. of series each of the form

$$\begin{aligned} \varepsilon + \varepsilon' \cos y + \varepsilon'' \cos 2y + \cdots + \varepsilon^n \cos ny \\ + \zeta' \sin y + \zeta'' \sin 2y + \cdots + \zeta^n \sin ny, \end{aligned}$$

where $n = \frac{1}{2}\nu - \frac{1}{2}$ or $\frac{1}{2}\nu$, according as ν is odd or even. Gauss remarks that if these series in y are substituted into X' there results a function Z of two variables x and y. Moreover if y is replaced by μx, then Gauss calls the resulting function X''. He further calls X''' the function that one would find by proceeding to apply his previous methods directly to the π values of x above. He then notes that X''' has $\pi/2 - 1/2$ or $\pi/2$ ordinates, depending upon whether π is odd or even. He compares X'' and X''', and remarks that the former has $\mu n + m$ ordinates, distinguishing three cases:

(I) μ and ν are both odd. Then $\mu n + m = \frac{1}{2}\mu\nu - \frac{1}{2} = \frac{1}{2}\pi - \frac{1}{2}$; X'' and X''' will be identical;

(II) μ is even and ν is odd. Then $\mu n + m = \frac{1}{2}\mu\nu = \frac{1}{2}\pi$; the functions X'' and X''' agree up to this point. It suffices to compare their final terms. Suppose the last term in X'' is $\gamma^m \cos mx + \delta^m \sin mx$, the last term in the series for γ^m is $k \cos \mu nx + l \sin \mu nx$, and in that for δ^m is $k' \cos \mu nx + l' \sin \mu nx$. Then in X'' we have

$$(k \cos \mu nx + l \sin \mu nx) \cos mx + (k' \cos \mu nx + l' \sin \mu nx) \sin mx,$$

where (in Section 15) Gauss showed that

$$k = \pm \cos (s/2)/2^{\mu-1}, l = \pm \sin (s/2)/2^{\mu-1},$$

with the upper sign for $\mu/2$ even, and the lower one for $\mu/2$ odd. This leads to the $(\mu n + m)$th term $\frac{1}{2}(k - l') \cos (\mu n + m)x + \frac{1}{2}(l + k') \sin (\mu n + m)x$. Moreover, Gauss goes on to show that

$$(k - l') \sin (\mu n + m)a = (l + k') \cos (\mu n + m)a;$$

(III) ν is even. Then $\mu n + m = \frac{1}{2}\pi + m$; X'' and X''' have differing terms. Suppose that in the coefficients γ^λ or δ^λ we have $k \cos \mu nx + l \sin \mu nx$. Then γ^λ leads in X'' to $\frac{1}{2}k \cos (\mu n + \lambda)x + \frac{1}{2}l \sin (\mu n + \lambda)x + \frac{1}{2}k \cos (\mu n - \lambda)x + \frac{1}{2}l \sin (\mu n - \lambda)x$, and δ^λ to $-\frac{1}{2}l \cos (\mu n + \lambda)x + \frac{1}{2}k \sin (\mu n + \lambda)x + \frac{1}{2}l \cos (\mu n - \lambda)x - \frac{1}{2}k \sin (\mu n - \lambda)x$. Now by a simple manipulation the former expression becomes $\frac{1}{2}(k + k \cos \pi a + l \sin \pi a) \cos (\frac{1}{2}\pi - \lambda)x + \frac{1}{2}(l + k \sin \pi a - l \cos \pi a) \sin (\frac{1}{2}\pi - \lambda)x$, and the latter becomes $\frac{1}{2}(l - l \cos \pi a + k \sin \pi a) \cos (\frac{1}{2}\pi - \lambda)x - \frac{1}{2}(k + l \sin \pi a + k \cos \pi a) \times$

$\sin(\frac{1}{2}\pi - \lambda)x$. (These follow since, e.g., $\cos(\mu n + \lambda)x = \cos(\pi/2 + \lambda)x = \cos \pi a \cdot \cos(\pi/2 - \lambda)x + \sin \pi a \cdot \sin(\pi/2 - \lambda)x$ for the π values of x we are considering.) But Gauss notes that by calculation $k \sin \pi a/2 = l \cos \pi a/2$, and hence

$$\frac{1}{2}(k + k \cos \pi a + l \sin \pi a) = k,$$

$$\frac{1}{2}(l + k \sin \pi a - l \cos \pi a) = l.$$

Thus in X'' he finds that γ^{λ} leads to the term $k \cos(\pi/2 - \lambda)x + l \sin(\pi/2 - \lambda)x$, and that δ^{λ} leads to $l \cos(\pi/2 - \lambda)x - k \sin(\pi/2 - \lambda)x$.

To illustrate these ideas Gauss took data on the position of the asteroid Pallas from Baron von Zach's tables. The data tabulated below from Gauss's paper give the declination X of Pallas as a function of the right ascension x. (The abbreviations *Austr*. and *Bor*. stand for the adjectives *australis*, southern, and *borealis*, northern.) We notice there are twelve values of x which span the period of 360°. Thus $\pi = 12$.

x	X	
0°	6°48′ Bor....	= + 408′
30	1 29	+ 89
60	1 6 Austr.....	− 66
90	0 10 Bor	+ 10
120	5 38	+ 338
150	13 27	+ 807
180	20 38	+1238
210	25 11	+1511
240	26 23	+1583
270	24 22	+1462
300	19 43	+1183
330	13 24	+ 804

Now, to form the finite Fourier series for this function, Gauss broke up the values of x into three groups of four each as shown below.

$a = 0°$	$A = +408$	$a' = 30°$	$A' = +89$	$a'' = 60°$	$A'' = -66$
$b = 90°$	$B = +10$	$b' = 120°$	$B' = +338$	$b'' = 150°$	$B'' = +807$
$c = 180°$	$C = +1238$	$c' = 210°$	$C' = +1511$	$c'' = 240°$	$C'' = +1583$
$d = 270°$	$D = +1462$	$d' = 300°$	$D' = +1183$	$d'' = 330°$	$D'' = +804$

and then formed the functions

$$X' = \gamma + \gamma' \cos x + \gamma'' \cos 2x + \delta' \sin x + \delta'' \sin 2x \tag{4.65}$$

for each group. ($\mu = 4$, $\nu = 3$; this is case (II) above where $n = 1$, $m = 2$.)

Thus he found three Fourier expansions:

Pro periodo	ubi $y = 4x$	γ	γ'	δ'	γ''	δ''
prima	0°	+779.5	−415.0	−726.0	+43.5	0
secunda	120°	+780.2	−404.5	−721.4	+ 9.9	+17.1
tertia	240°	+782.0	−413.5	−713.3	+11.7	−20.3

— one for each group.

Gauss now viewed each coefficient $\gamma, \gamma', \delta', \gamma'', \delta''$ as a periodic function of the variable $y(= 4x)$ and formed its Fourier series. This gave him the expansion for γ,

$$\frac{1}{3}(779\rlap{.}''5 + 780.2 + 782.0)$$

$$+ \frac{2}{3}(779.5 + 780.2 \cos 120° + 782.0 \cos 240°) \cos 4x$$

$$+ \frac{2}{3}(780.2 \sin 120° + 782.0 \sin 240°) \sin 4x,$$

and after simplification he found the expansions

$$\begin{aligned} \gamma &= 780.6 - 1.1 \cos 4x - 1.0 \sin 4x, \\ \gamma' &= -411.0 - 4.0 \cos 4x + 5.2 \sin 4x, \\ \delta' &= -720.2 - 5.8 \cos 4x - 4.7 \sin 4x, \\ \gamma'' &= +\ 21.7 + 21.8 \cos 4x - 1.1 \sin 4x, \\ \delta'' &= -\ 1.1 + 1.1 \cos 4x + 21.6 \sin 4x. \end{aligned} \tag{4.66}$$

Now when these are substituted into the formula (4.65) above for X', and a little manipulation is performed, there results the expansion

$$\begin{aligned} &780.6 - 411.0 \cos x - 720.2 \sin x + 43.4 \cos 2x - 2.2 \sin 2x \\ &\quad - 4.3 \cos 3x + 5.5 \sin 3x - 1.1 \cos 4x - 1.0 \sin 4x + 0.3 \cos 5x - 0.3 \sin 5x \\ &\quad + 0.1 \cos 6x.\end{aligned}$$ [84]

Recall from our previous discussion that in the present case (II) Gauss showed that the $(\mu n + m)$th term in X'' was $\frac{1}{2}(k - l') \cos (\mu n + m)x + \frac{1}{2}(l + k') \sin (\mu n + m)x$, where γ^m, δ^m contained the terms $k \cos \mu nx + l \sin \mu nx$, $k' \cos \mu nx + l' \sin \mu nx$, respectively. Hence in this case, where $\mu = 4, \nu = 3, m = 2, n = 1$, we have $k = 21.8, l = -1.1, k' = 1.1, l' = 21.6$; and we see that the expression above reduces to $0.1 \cos 6x$, since $l + k' = 0$.

Next Gauss chooses $\mu = 3$, $\nu = 4$. Then we have his case (III), and $3n + m = 6 + m$, and $n = 2$. In this situation he divides up his 12 values into four groups of three each and writes $X' = \gamma + \gamma' \cos x + \delta' \sin x$. He finds now

Pro periodo	ubi $y = 3x$	γ	γ'	δ'
prima	0°	+776.3	−368.3	−718.8
secunda	90°	+786.0	−414.5	−676.0
tertia	180°	+785.0	−453.0	−721.1
quarta	270°	+775.0	−408.2	−765.0,

[84] Gauss III *TI*, p. 310. (The *signum* of 411.0 is erroneously given there as +.)

and deduces the values

$$\begin{aligned}\gamma \ldots\ldots &+ 780.6 - 4.3\cos 3x + 5.5\sin 3x + 0.1\cos 6x,\\ \gamma' \ldots\ldots &- 411.0 + 42.3\cos 3x - 3.2\sin 3x + 0.3\cos 6x,\\ \delta' \ldots\ldots &- 720.2 + 1.2\cos 3x + 44.5\sin 3x + 0.3\cos 6x.\end{aligned}$$

When he substitutes these into X' he finds

$$\begin{aligned}&+ 780.6 - 411.0\cos x - 720.2\sin x + 43.4\cos 2x - 2.2\sin 2x\\ &- 4.3\cos 2x + 5.5\sin 3x - 1.1\cos 4x - 1.0\sin 4x + 0.15\cos 5x\\ &- 0.15\sin 5x + 0.1\cos 6x + 0.15\cos 7x + 0.15\sin 7x.\end{aligned}$$

But we saw above that he could replace $\cos(6 + \lambda)x$, $\sin(6 + \lambda)x$ by $\cos\pi a\cos(6 - \lambda)x + \sin\pi a\cdot\cos(6 - \lambda)x = \cos(6 - \lambda)x$ and $\sin\pi a\cos(6 - \lambda)x - \cos\pi a\sin(6 - \lambda)x = -\sin(6 - \lambda)x$. Thus in this case, where $\lambda = 1$, we have $\cos 7x = \cos 5x$, $\sin 7x = -\sin 5x$, and the form above becomes identical to the previous one for all 12 values of x given in von Zach's table.

This completes the first 28 sections of Gauss's paper. He now proceeds in Sections 29–40 to discuss various important special cases in order to abbreviate the amount of calculation needed, and then illustrates by an example: the so-called Equation of center (cf. p. 171, Section 3.12 above) for the newly discovered asteroid Juno (the year of discovery was 1804) given the eccentricity $e = 0.254236$ and the values of this equation by 10° intervals from 0° to 360°.[85] (This "equation" is a Fourier series in a quantity proportional to time.)

In Section 29 Gauss takes a finite Fourier series X given μ values A, B, C, D etc. for X at $x = a, b, c, d$ etc. Recall that $b - a = c - b = d - c$ etc. $= 360°/\mu$, and hence that $\sin\mu a = \sin\mu b = \sin\mu c = \sin\mu d$ etc. There are then two cases: either this common value is zero or not. In Section 30 Gauss assumes the latter and sets $\mu = n + 1$. He further assumes that his series X is even and thus expressible as

$$X = \alpha + \alpha'\cos x + \alpha''\cos 2x + \alpha'''\cos 3x + \cdots + \alpha''\cos nx.$$

Gauss now wishes to find the coefficients in T, the value of X when $x = t$. By what we saw earlier we have

$$\begin{aligned}&(\cos t - \cos b)(\cos t - \cos c)(\cos t - \cos d)\text{ etc.}\\ &= \frac{1}{2^{\mu-1}} \times \frac{\cos\mu t - \cos\mu a}{\cos t - \cos a}\\ &= \frac{1}{2^{\mu-1}\sin a}\{\sin\mu a + 2\sin(\mu - 1)a\cos t + 2\sin(\mu - 2)a\cos 2t + \text{etc.}\\ &\qquad + 2\sin a\cos(\mu - 1)t\},\end{aligned}$$

[85] Since Gauss writes "pro novo planeta *Iunone*" we may reasonably date this paper as having been written somewhere near to 1804. Moreover, in some remarks by Schering appearing on pp. 328–330 of this volume (III) of Gauss's *Werke* it is noted that the paper is an improved version of an investigation begun by Gauss in October, 1805.

and hence

$$(\cos a - \cos b)(\cos a - \cos c)(\cos a - \cos d) \text{ etc.}$$
$$= \frac{1}{2^{\mu-1}\sin a}\{\sin \mu a + 2\sin(\mu - 1)a\cos a$$
$$+ 2\sin(\mu - 2)a\cos 2a + \text{etc.} + 2\sin a\cos(\mu - 1)a\}.$$

Now the second term in this last series may be written as $\sin \mu a + \sin(\mu - 2)a$; the last term as $\sin \mu a - \sin(\mu - 2)a$; the third term as $\sin \mu a + \sin(\mu - 3)a$; the penultimate term as $\sin \mu a - \sin(\mu - 3)a$; etc. Hence the sum of that series is $\mu \sin \mu a$. Look now at the coefficient of A in the relation (4.61′). It is the quotient of $(\cos t - \cos b)(\cos t - \cos c)\cdots$ divided by $(\cos a - \cos b)\cdot(\cos a - \cos c)\cdots$ and hence is

$$\frac{1}{\mu\sin\mu a}\{\sin \mu a + 2\sin(\mu - 1)a\cos t + 2\sin(\mu - 2)a\cos 2t + \text{etc.}$$
$$+ 2\sin a\sin(\mu - 1)t\}.$$

Exactly similar expressions obtain for the coefficients of B, C, D etc. Thus the coefficients of

$$T'' = \varepsilon + \varepsilon'\cos t + \varepsilon''\cos 2t + \varepsilon'''\cos 3t + \text{etc.} + \varepsilon^n\cos nt$$

are given by the relations — recall $n = \mu + 1$ —

$$\varepsilon = \frac{1}{\mu}(A + B + C + D + \text{etc.}),$$

$$\varepsilon' = \frac{1}{\mu\sin\mu a}\{A\sin(\mu - 1)a + B\sin(\mu - 1)b + C\sin(\mu - 1)c$$
$$+ D\sin(\mu - 1)d + \text{etc.}\},$$

$$\varepsilon'' = \frac{2}{\mu\sin\mu a}\{A\sin(\mu - 2)a + B\sin(\mu - 2)b + C\sin(\mu - 2)c$$
$$+ D\sin(\mu - 2)d + \text{etc.}\},$$

$$\varepsilon''' = \frac{2}{\mu\sin\mu a}\{A\sin(\mu - 3)a + B\sin(\mu - 3)b + C\sin(\mu - 3)c$$
$$+ D\sin(\mu - 3)d + \text{etc.}\},$$

etc.

$$\varepsilon^n = \frac{2}{\mu\sin\mu a}(A\sin a + B\sin b + C\sin c + D\sin d + \text{etc.}).$$

In Section 31 Gauss considers the relation of the $\varepsilon, \varepsilon', \varepsilon''$ etc. to the coefficients $\gamma, \gamma', \delta', \gamma'', \delta''$ etc. of

$$X' = \gamma + \gamma'\cos x + \gamma''\cos 2x + \cdots + \gamma^m\cos mx$$
$$+ \delta'\sin x + \delta''\sin 2x + \cdots + \delta^m\sin mx, \tag{4.67}$$

where $m = \frac{1}{2}\mu - \frac{1}{2}$ or $\frac{1}{2}\mu$ according as μ is odd or even. He shows that

$$\begin{aligned} \varepsilon &= \gamma, & & \\ \varepsilon' &= \gamma' - \delta' \operatorname{cotan} \mu a, & \varepsilon^{\mu-1} &= \delta' \operatorname{cosec} \mu a, \\ \varepsilon'' &= \gamma'' - \delta'' \operatorname{cotan} \mu a, & \varepsilon^{\mu-2} &= \delta'' \operatorname{cosec} \mu a, \\ \varepsilon''' &= \gamma''' - \delta''' \operatorname{cotan} \mu a, & \varepsilon^{\mu-3} &= \delta''' \operatorname{cosec} \mu a, \\ &\text{etc.} & & \\ \left(\frac{1}{2}\right)\varepsilon^m &= \gamma^m - \delta^m \operatorname{cotan} \mu a, & \left(\frac{1}{2}\right)\varepsilon^{\mu-m} &= \delta^m \operatorname{cosec} \mu a, \end{aligned}$$

where the factor $(\frac{1}{2})$ above is to be included for μ even but deleted for μ odd. The case μ odd follows at once by an examination of comparable coefficients. The case μ even ($m = \mu - m$) is a little more complex. For one thing, we have two values for $\varepsilon^{(1/2)\mu}$, namely: $2\gamma^{(1/2)\mu} - 2\delta^{(1/2)\mu} \operatorname{cotan} \mu a$ and $2\delta^{(1/2)\mu} \operatorname{cosec} \mu a$. We have already seen that $\gamma^{(1/2)\mu} \sin \frac{1}{2}\mu a = \delta^{(1/2)\mu} \cos \frac{1}{2}\mu a$; and we can readily infer from this that the two values of $\varepsilon^{(1/2)\mu}$ coincide. Gauss wrote the value of $\varepsilon^{(1/2)\mu}$ as $\gamma^{(1/2)\mu} - \delta^{(1/2)\mu} \operatorname{cotan} \mu a + \delta^{(1/2)\mu} \operatorname{cosec} \mu a$. (Recall that $b - a = c - b = \cdots = 360°/\mu = 180°/m$, and thus we have $\sin ma = -\sin mb = \sin mc = \cdots$.) From this it follows that

$$\varepsilon^m = \frac{2 \sin ma}{4m \sin ma \cos ma}(A - B + C - D + \text{etc.}) = 2\gamma^m - 2\delta^m \operatorname{cotan} \mu a.$$

If we call X'' the series $\varepsilon + \varepsilon' \cos x + \varepsilon'' \cos 2x + \varepsilon''' \cos 3x + \cdots + \varepsilon^{\mu-1} \cos(\mu - 1)x$, then clearly

$$\begin{aligned} &X'' - (X' - \delta' \sin x - \delta'' \sin 2x - \delta''' \sin 3x - \text{etc.} - \delta^m \sin mx) \\ &= -\operatorname{cotan} \mu a(\delta' \cos x + \delta'' \cos 2x + \delta''' \cos 3x + \cdots + \delta^m \cos mx) \\ &\quad + \operatorname{cosec} \mu a(\delta' \cos(\mu - 1)x + \delta'' \cos(\mu - 2)x + \cdots + \delta^m \cos(\mu - m)x). \end{aligned}$$

In Section 32 Gauss takes up the case where X does not stop with $\cos nx$ but contains the additional terms $\alpha^\mu \cos \mu x + \alpha^{\mu+1} \cos(\mu + 1)x + \text{etc.}$ Then

$$\begin{aligned} \varepsilon &= \alpha + \alpha^\mu \cos \mu a + \alpha^{2\mu} \cos 2\mu a + \text{etc.}, \\ \varepsilon' &= \frac{1}{\sin \mu a}\{\alpha' \sin \mu a + \alpha^{\mu+1} \sin 2\mu a + \alpha^{2\mu+1} \sin 3\mu a + \text{etc.} \\ &\qquad - \alpha^{2\mu-1} \sin \mu a - \alpha^{3\mu-1} \sin 2\mu a - \alpha^{4\mu-1} \sin 3\mu a - \text{etc.}\}, \\ \varepsilon'' &= \frac{1}{\sin \mu a}\{\alpha'' \sin \mu a + \alpha^{\mu+2} \sin 2\mu a + \alpha^{2\mu+2} \sin 3\mu + \text{etc.} \\ &\qquad - \alpha^{2\mu-2} \sin \mu a - \alpha^{3\mu-2} \sin 2\mu a - \alpha^{4\mu-2} \sin 3\mu a - \text{etc.}\}, \\ &\text{etc.}, \\ \varepsilon^{\mu-1} &= \frac{1}{\sin \mu a}\{\alpha^{\mu-1} \sin \mu a + \alpha^{2\mu-1} \sin 2\mu a + \alpha^{3\mu-1} \sin 3\mu a + \text{etc.} \\ &\qquad - \alpha^{\mu+1} \sin \mu a - \alpha^{2\mu+1} \sin 2\mu a - \alpha^{3\mu+1} \sin 3\mu a - \text{etc.}\}. \end{aligned}$$

In Section 33 Gauss shifts to the case where X is representable as

$$\beta' \sin x + \beta'' \sin 2x + \beta''' \sin 3x + \text{etc.} + \beta^n \sin nx \qquad (\mu = n).$$

Then the coefficients ζ', ζ'', etc. $\zeta^{\mu-1}$, ζ^μ of

$$T'' = \zeta' \sin t + \zeta'' \sin 2t + \zeta''' \sin 3t + \text{etc.} + \zeta^\mu \sin \mu t$$

are given by the relations

$$\zeta' = \frac{2}{\mu \sin \mu a}\{A \cos(\mu - 1)a + B \cos(\mu - 1)b + C \cos(\mu - 1)c + D \cos(\mu - 1)d + \text{etc.}\}$$

$$\zeta'' = \frac{2}{\mu \sin \mu a}\{A \cos(\mu - 2)a + B \cos(\mu - 2)b + C \cos(\mu - 2)c + D \cos(\mu - 2)d + \text{etc.}\},$$

$$\zeta''' = \frac{2}{\mu \sin \mu a}\{A \cos(\mu - 3)a + B \cos(\mu - 3)b + C \cos(\mu - 3)c + D \cos(\mu - 3)d + \text{etc.}\},$$

etc.,

$$\zeta^{\mu-1} = \frac{2}{\mu \sin \mu a}\{A \cos a + B \cos b + C \cos c + D \cos d + \text{etc.}\},$$

$$\zeta^{\mu} = \frac{2}{\mu \sin \mu a}\{A + B + C + D + \text{etc.}\}.$$

Now consider the function X' in the relation (4.67), where $m = \frac{1}{2}\mu - \frac{1}{2}$ or $\frac{1}{2}\mu$ according as μ is odd or even. Then in Section 34 Gauss shows that

$$\begin{aligned} & & \zeta^\mu &= \gamma \operatorname{cosec} \mu a,\\ \zeta' &= \delta' + \gamma' \operatorname{cotan} \mu a, & \zeta^{\mu-1} &= \gamma' \operatorname{cosec} \mu a,\\ \zeta'' &= \delta'' + \gamma'' \operatorname{cotan} \mu a, & \zeta^{\mu-2} &= \gamma'' \operatorname{cosec} \mu a,\\ \zeta''' &= \delta''' + \gamma''' \operatorname{cotan} \mu a, & \zeta^{\mu-3} &= \gamma''' \operatorname{cosec} \mu a,\\ &\text{etc.} & &\\ \left(\frac{1}{2}\right)\zeta^m &= \delta^m + \gamma^m \operatorname{cotan} \mu a, & \left(\frac{1}{2}\right)\zeta^{\mu-m} &= \gamma^m \operatorname{cosec} \mu a, \end{aligned}$$

where, as before, the factor $(\frac{1}{2})$ is present for μ even and absent for odd, and in the former case $\zeta^{(1/2)\mu}$ is chosen to be $\delta^{(1/2)\mu} + \gamma^{(1/2)\mu} \operatorname{cosec} \mu a + \gamma^{(1/2)\mu} \times \operatorname{cotan} \mu a$. As we saw above, if X does not terminate with $\sin nx$ but goes on, then the coefficients ζ' ζ'', ζ''' etc. ζ^μ are expressible as

$$\zeta' = \frac{1}{\sin \mu a}\{\delta' \sin \mu a + \delta^{\mu+1} \sin 2\mu a + \delta^{2\mu+1} \sin 3\mu a + \text{etc.} + \delta^{2\mu-1} \sin \mu a + \delta^{3\mu-1} \sin 2\mu a + \delta^{4\mu-1} \sin 3\mu a + \text{etc.}\},$$

$$\zeta'' = \frac{1}{\sin \mu a}\{\delta'' \sin \mu a + \delta^{\mu+2} \sin 2\mu a + \delta^{2\mu+2} \sin 3\mu a + \text{etc.} + \delta^{2\mu-2} \sin \mu a + \delta^{3\mu-2} \sin 2\mu a + \delta^{4\mu-2} \sin 3\mu a + \text{etc.}\},$$

$$\zeta''' = \frac{1}{\sin \mu a}\{\delta''' \sin \mu a + \delta^{\mu+3} \sin 2\mu a + \delta^{2\mu+3} \sin 3\mu a + \text{etc.}$$
$$+ \delta^{2\mu-3} \sin \mu a + \delta^{3\mu-3} \sin 2\mu a + \delta^{4\mu-3} \sin 3\mu a + \text{etc.}\},$$

etc.,

$$\zeta^{\mu} = \frac{1}{\sin \mu a}(\delta^{\mu} \sin \mu a + \delta^{2\mu} \sin 2\mu a + \delta^{3\mu} \sin 3\mu a + \text{etc.}).$$

In Section 36 Gauss proves the following theorem. Let λ, μ be arbitrary integers and let $\lambda' = \lambda - k\mu$, $\lambda'' = \lambda - (k+1)\mu$. Then for any a the functions

$$P = \sin(\lambda' - \lambda'')a \cos \lambda x + \sin(\lambda'' - \lambda)a \cos \lambda' x + \sin(\lambda - \lambda')a \cos \lambda'' x,$$
$$Q = \sin(\lambda' - \lambda'')a \sin \lambda x + \sin(\lambda'' - \lambda)a \sin \lambda' x + \sin(\lambda - \lambda')a \sin \lambda'' x$$

are divisible by $\cos \mu x - \cos \mu a$ for every x. Let us follow his proof. He first discusses the case where λ is divisible by μ. First we have

$$\begin{aligned} P(a) &= \sin \mu a \cos \lambda a - \sin(k+1)\mu a \cos(\lambda - k\mu)a + \sin k\mu a \cos(\lambda - (k+1)\mu)a \\ &= \sin \mu a \cos \lambda a - \sin(k+1)\mu a[\cos \lambda a \cos k\mu a + \sin \lambda a \sin k\mu a] \\ &\quad + \sin k\mu a[\cos \lambda a \cos(k+1)\mu a + \sin \lambda a \sin(k+1)\mu a] \\ &= \cos \lambda a[\sin \mu a - (\sin(k+1)\mu a \cos k\mu a + \cos(k+1)\mu a \sin k\mu a] \\ &= \cos \lambda a[\sin \mu a - \sin \mu a] \\ &= 0, \end{aligned}$$

and similarly $Q(a) = 0$. Hence we may rewrite $P(x)$, $Q(x)$ as $P(x) - P(a)$, $Q(x) - Q(a) \cdot \sin \mu x / \sin \mu a$. It is not hard now to see that $P(x)$ is divisible by $\cos \mu x - \cos \mu a$. To this end, notice, e.g., that for λ a positive multiple of μ the term

$$\begin{aligned} &\sin \mu a(\cos \lambda x - \cos \lambda a) \\ &= (\cos \mu x - \cos \mu a)\{2 \sin \mu a \cos(\lambda - \mu)x + 2 \sin 2\mu a \cos(\lambda - 2\mu)x \\ &\qquad + 2 \sin 3\mu a \cos(\lambda - 3\mu)x + \text{etc.} \\ &\qquad + 2 \sin(\lambda - \mu)a \cos \mu x + \sin \lambda a\} \end{aligned}$$

is so divisible, as are the terms in $\cos \lambda' x$ and $\cos \lambda'' x$. Further,

$$\begin{aligned} &\sin \mu a \sin \lambda x - \sin \lambda a \sin \mu x \\ &= (\cos \mu x - \cos \mu a)\{2 \sin \mu a \sin(\lambda - \mu)x + 2 \sin 2\mu a \sin(\lambda - 2\mu)x \\ &\qquad + 2 \sin 3\mu a \sin(\lambda - 3\mu)x + \text{etc.} + 2 \sin(\lambda - \mu)a \sin \mu x\}. \end{aligned}$$

(Perhaps the easiest way to see these results is to carry out the indicated multiplications and group terms appropriately.) If λ is negative then everything still holds. If λ is not divisible by μ, Gauss chooses l to be any integer divisible by μ and writes $\lambda = l + \theta$, $\lambda' = l' + \theta$, $\lambda'' = l'' + \theta$, where now l', l'' are chosen divisible by μ. Now let

$$\sin(l' - l'')a \cos lx + \sin(l'' - l)a \cos l'x + \sin(l - l')a \cos l''x = P',$$
$$\sin(l' - l'')a \sin lx + \sin(l'' - l)a \sin l'x + \sin(l - l')a \sin l''x = Q'.$$

Then clearly $P = P' \cos \theta x - Q' \sin \theta x$, $Q = P' \sin \theta x + Q' \cos \theta x$, and P', Q' are now of the form we just discussed. They are then divisible by $(\cos \mu x - \cos \mu a)$ and hence P and Q are also.

Gauss now returned to his study of even or odd functions under the assumption that $\sin \mu a$ is zero. He therefore supposes that $a = 180°/\mu$ and proceeds in Sections 38–39 to find the coefficients of the series. Then in the last section, Section 41, he illustrates his methods by finding the Equation of center mentioned above.

4.13. Gauss on Rounding Errors

Gauss in the *Theoria Motus* discusses the problem of rounding errors as they affect the accuracy of a numerical calculation. As far as I can tell from the works of authors I have examined, he appears to have been the first to do this systematically: he also seems to have noted the phenomenon we now call numerical stability, as may be seen in the following passage:

> Since none of the numbers which we take out from logarithmic and trigonometric tables admit of absolute precision, but are all to a certain extent approximate only, the results of all calculations performed by the aid of these numbers can only be approximately true. In most cases, indeed, the common tables which are exact to the seventh place of decimals, that is, never deviate from the truth either in excess or defect beyond half of an unit in the seventh figure, furnish more than the requisite accuracy, so that the unavoidable errors are evidently of no consequence: nevertheless it may happen, that in special cases the effect of the errors of the tables is so augmented that we may be obliged to reject a method, otherwise the best, and substitute another in its place. Cases of this kind can occur in those computations which we have just explained; on which account, it will not be foreign to our purpose to introduce here some inquiries concerning the degree of precision allowed in these computations by the common tables. Although this is not the place for a thorough examination of this subject, which is of the greatest importance to the practical computer, yet we will conduct the investigation sufficiently far for our own object, from which point it may be further perfected and extended to other operations by any one requiring it.[86]

He now considers the relation between the so-called true and eccentric "anomalies"

$$\tan \frac{1}{2} E = \tan \frac{1}{2} v \tan \left(45° - \frac{1}{2} \varphi\right). \tag{4.68}$$

Here E is the eccentric and v the true anomaly while $\cos \varphi = \sqrt{(1 - ee)}$, e being the eccentricity of the orbit. He supposes that φ and E are given exactly, and further supposes that the error due to forming $\log \tan E/2$ and

[86] Gauss, *Theoria* (Eng. trans.), p. 31.

log tan $(45° - \varphi/2)$ is ω. He concludes that the error in forming the difference, log tan $v/2$, will then be 2ω and that "the greatest error in the determination of the angle $v/2$ will be

$$\frac{3\omega d\frac{1}{2}v}{d \log \tan \frac{1}{2}v} = \frac{3\omega \sin v}{2\lambda},$$

λ denoting the modulus of the logarithms used in this calculation." From this he shows that the error of v in seconds is $0''.0712 \sin v$ if seven-place Briggsian logarithms are used; "if smaller tables to five places only, are used, the error may amount to $7''.12$." Next he estimates the error in computing $e \cos E$ by the use of logarithms, and finds an error of $(3\omega e \cos E)/\lambda$; hence $1 - e \cos E = r/a$ will have this same error (r is the radius vector of the orbit and a is the semimajor axis of the elliptical orbit). Now the logarithm of this quantity $1 - e \cos E$ will clearly be in error by $(1 + \delta)\omega$, where $\delta = (3e \cos E)/(1 - e \cos E)$. If we suppose log a given correctly, then the possible error in the radius vector is acceptable for e very small, but for e near 1 the denominator $1 - e \cos E$ can be quite dangerously small. He therefore concludes that the formula $r = a(1 - e \cos E)$ "is less suitable in this case." It is easy to see that δ is expressible as $3(a - r)/r = 3e(\cos v + e)/(1 - e^2)$, since another relation involving E and v is $\cos E = (\cos v + e)/(1 + e \cos v)$. This shows the danger of using the apparently innocuous formula $r = a(1 - e \cos E)$ for values of e near 1.

Next Gauss considers the relation

$$\sin \frac{1}{2}(v - E) = \sin \frac{1}{2}\varphi \sin v\sqrt{\frac{r}{p}} = \sin \frac{1}{2}\varphi \sin E\sqrt{\frac{a}{r}}.$$

He remarks that log $\sqrt{a/r}$ is liable to the error $\frac{1}{2}(1 + \delta)\omega$, and thus the expression log $\sin \frac{1}{2}\varphi \sin E\sqrt{a/r}$ to that of $\frac{1}{2}\cdot 5(1 + \delta)\omega$.

Let us look briefly at his analysis of Kepler's equation $M = E - e \sin E$. He notes that, given the so-called eccentric anomaly E, the error in finding the so-called mean anomaly M will be $3\omega e \sin E/\lambda$, "which limit is to be multiplied by 206265″ if wanted expressed in seconds." He then infers that the inverse problem of finding E given M may be in error by as much as

$$\frac{3\omega e \sin E}{\lambda}\cdot\frac{dE}{dM}\cdot 206265'' = \frac{3\omega e a \sin E}{\lambda r}\cdot 206265'', \tag{4.69}$$

"even if the equation $E - e \sin E = M$ should be satisfied with all the accuracy which the tables admit."[87] Suppose now we are given the mean anomaly M exactly. Then the true anomaly v will be in error from two sources. First, the error in finding v given E, and second, the error in finding E itself from M. The former error causes little problem but the latter, as we just saw, can cause very serious difficulty. He therefore concludes that the

[87] Gauss, *Theoria* (Eng. trans.), pp. 33–35.

total error is obtained by multiplying dv/dE by the error (4.69). This gives

$$\frac{3\omega e \sin E}{\lambda} \cdot \frac{dv}{dM} \cdot 206265'' = \frac{3\omega ea \sin v}{\lambda r} \cdot 206265'' = \left(\frac{e \sin v + \frac{1}{2} ee \sin 2v}{1 - ee}\right) 0''.0712, \tag{4.70}$$

if seven-place tables are used. This relation follows from the formula (4.69) and, as we see from (4.68), the fact that

$$\frac{dv}{dE} = \frac{\sin v}{\sin E},$$

it also depends upon the relation $a/r = (1 - e \cos v)/(1 - e^2)$. Gauss then tabulated this error (4.70); it is given below:

e	maximum error.	e	maximum error.	e	maximum error.
0.90	0″.42	0.94	0″.73	0.98	2″.28
0.91	0 .48	0.95	0 .89	0.99	4 .59
0.92	0 .54	0.96	1 .12	0.999	46 .23
0.93	0 .62	0.97	1 .50		

(This table is on p. 35.) He then goes on to consider the case of objects moving in hyperbolic orbits and finds the comparable results. He proceeds to note the essential instability of the equations we were considering and says:

> The methods above treated, both for the determination of the true anomaly from the time and for the determination of the time from the true anomaly, do not admit of all the precision that might be required in those conic sections of which the eccentricity differs but little from unity, that is, in ellipses and hyperbolas which approach very near to the parabola; indeed, unavoidable errors, increasing as the orbit tends to resemble the parabola, may at length exceed all limits. Larger tables, constructed to more than seven figures, would undoubtedly diminish this uncertainty, but they would not remove it, nor would they prevent its surpassing all limits as soon as the orbit approached too near the parabola. Moreover, the methods given above become in this case very troublesome, since a part of them require the use of indirect trials frequently repeated, of which the tediousness is even greater if we work with the larger tables. It certainly, therefore, will not be superfluous, to furnish a peculiar method by means of which the uncertainty in this case may be avoided, and sufficient precision may be obtained with the help of the common tables.[88]

He shows how to develop new relations which are stable but this is not germane to our account.

---

[88] Gauss, *Theoria* (Eng. trans.), p. 38.

5. Other Nineteenth Century Figures

5.1. Introduction

After Gauss there were a considerable number of excellent mathematicians who came along more or less in his footsteps and either continued his ideas on numerical analysis or who utilized their own discoveries to make more elegant what earlier mathematicians had done. Thus, for example, we find on the one hand Jacobi reconsidering some of Gauss's work and on the other Cauchy using his Residue theorem to establish interpolation formulas.

5.2. Jacobi on Numerical Integration

There is an interesting discussion of Gauss's method of numerical integration by Jacobi.[1] Consider a polynomial function $y = f(x)$ defined on $0 \leq x \leq 1$, and n points $\alpha', \alpha'', \alpha''', \ldots, \alpha^{(n)}$ on this interval. Then he defines a function φ by the relation

$$\varphi(x) = (x - \alpha')(x - \alpha'')(x - \alpha''')\cdots(x - \alpha^{(n)}),$$

and forms

$$G(x) = \frac{f(\alpha')}{\varphi'(\alpha')(x - \alpha')} + \frac{f(\alpha'')}{\varphi'(\alpha'')(x - \alpha'')} + \frac{f(\alpha''')}{\varphi'(\alpha''')(x - \alpha''')} + \cdots$$
$$+ \frac{f(\alpha^{(n)})}{\varphi'(\alpha^{(n)})(x - \alpha^{(n)})}, \qquad (5.1)$$

where φ' is the derivative of φ. Now if f is of degree $n - 1$, then $f = G\varphi = U$; if f is of degree $n + p$, $f = U + V\varphi$, with $U = G\varphi$, U of degree $n - 1$ or less and V of degree p. Jacobi now considers the error

$$\Delta = \int y\,dx - \int U\,dx = \int \varphi(x)V\,dx,$$

[1] Jacobi VI [1826], pp. 1–11. The paper is entitled "Über Gauss' neue Methode die Werthe der Integrale näherungsweise zu finden."

and poses the problem of choosing the $\alpha', \alpha'', \alpha''', \ldots, \alpha^{(n)}$ so that this error is as small as possible. If f/φ is expressed as a power series, then G contains the negative powers of x and V the nonnegative powers of f/φ. Thus if

$$f(x) = a + a'x + a''x^2 + \cdots + a^{(n)}x^n + a^{(n+1)}x^{n+1} + \cdots + a^{(2n)}x^{2n} + \cdots,$$

$$\frac{1}{\varphi(x)} = \frac{A'}{x^n} + \frac{A''}{x^{n+1}} + \frac{A'''}{x^{n+2}} + \cdots + \frac{A^{(n+1)}}{x^{2n}} + \cdots,$$

$$V = a^{(n)}A' + a^{(n+1)}(A'x + A'') + a^{(n+2)}(A'x^2 + A''x + A''') + \cdots$$
$$+ a^{(2n-1)}(A'x^{n-1} + A''x^{n-2} + \cdots + A^{(n)}) + \cdots \quad .$$

Jacobi notes that the first coefficients $a, a', a'', \ldots, a^{(n-1)}$ do not appear in V, and he comments that the next n can also be removed by choosing the $\alpha', \alpha'', \alpha''', \ldots, \alpha^{(n)}$ so that the integrals

$$\int \varphi(x)\,dx, \quad \int x\varphi(x)\,dx, \quad \int x^2\varphi(x)\,dx, \ldots, \int x^{n-1}\varphi(x)\,dx \tag{5.2}$$

when evaluated between the limits $x = 0$ and $x = 1$ all vanish. (In 1828 limits are not yet attached to integrals but by 1835 he was using our present scheme.)

Jacobi now writes

$$\int uv\,dx = u\int v\,dx - \int\left(\frac{du}{dx}\int v\,dx\right)dx,$$

$$\int\left(\frac{du}{dx}\int v\,dx\right)dx = \frac{du}{dx}\int^2 v\,dx^2 - \int\left(\frac{d^2u}{dx^2}\int^2 v\,dx^2\right)dx,$$

$$\int\left(\frac{d^2u}{dx^2}\int^2 v\,dx^2\right)dx = \frac{d^2u}{dx^2}\int^3 v\,dx^3 - \int\left(\frac{d^3u}{dx^3}\int^3 v\,dx^3\right)dx,$$

$$\vdots$$

$$\int\left(\frac{d^mu}{dx^m}\int^m v\,dx^m\right)dx = \frac{d^mu}{dx^m}\int^{m+1} v\,dx^{m+1} - \int\left(\frac{d^{m+1}u}{dx^{m+1}}\int^{m+1} v\,dx^{m+1}\right)dx,$$

where he uses the symbol $\int^m v\,dx^m$ to mean the m-fold indefinite integral of v. The result is of course nothing other than a repetition of an integration by parts. This gives him the relation

$$\int uv\,dx = u\int v\,dx - \frac{du}{dx}\int^2 v\,dx^2 + \frac{d^2u}{dx^2}\int^3 v\,dx^3 - \cdots$$
$$+ (-1)^m\frac{d^mu}{dx_m}\int^{m+1} v\,dx^{m+1}$$
$$+ (-1)^{m+1}\int\left(\frac{d^{m+1}u}{dx^{m+1}}\int^{m+1} v\,dx^{m+1}\right)dx. \tag{5.3}$$

He now sets $u = x^m$, $v = \varphi(x)$ and *seriatim* $m = 0, 1, 2, 3, \ldots, n - 1$. This gives him expressions for the n integrals (5.2) above, from which it is clear that their vanishing is equivalent to the vanishing of the definite integrals

$$\int \varphi(x)\,dx, \qquad \int^2 \varphi(x)\,dx^2, \qquad \int^3 \varphi(x)\,dx^3, \ldots, \int^n \varphi(x)\,dx^n.$$

Let $\pi(x) = \int^n \varphi(x)\,dx^n$. Then Jacobi's problem is equivalent to finding a function π which together with its 1st, 2d, 3d, ..., $(n-1)$th derivative vanishes at $x = 0$ and $x = 1$. Hence $\pi(x)$ must contain x^n and $(x-1)^n$ as factors. Furthermore, any function with the factor $x^n(x-1)^n$ satisfies Jacobi's criterion. Jacobi therefore sets

$$\pi(x) = x^n(x-1)^n M.$$

Now since φ is a polynomial of degree n, π must be a polynomial of degree $2n$ and M a constant. Hence

$$\begin{aligned}\varphi(x) &= M\frac{d^n x^n(x-1)^n}{dx^n}\\ &= x^n - \frac{n^2}{2n}x^{n-1} + \frac{n^2(n-1)^2}{1\cdot 2\cdot 2n(2n-1)}x^{n-2} - \frac{n^2(n-1)^2(n-2)^2}{1\cdot 2\cdot 3\cdot 2n(2n-1)(2n-2)}x^{n-3}\\ &\quad + \cdots + (-1)^n\frac{n(n-1)(n-2)\cdots 1}{2n(2n-1)(2n-2)\cdots(n+1)},\end{aligned}$$

where he has chosen

$$M = \frac{1}{2n(2n-1)(2n-2)\cdots(n+1)}.$$

(This is, apart from a constant factor, the Jacobi polynomial $G_n(1, 1, x) = F(n+1, -n, 1, x)$. Cf. Courant, *Methoden*, Vol. I, pp. 76ff. Moreover, this φ is also, apart from a constant factor, $P_n(1-2x)$, where P_n is the nth Legendre polynomial.)

He now has the desired expressions for φ. The roots of $\varphi(x) = 0$ are the $\alpha', \alpha'', \alpha''', \ldots, \alpha^{(n)}$ of Gauss's method of numerical integration (cf. Section 4.3 above). They are all real and lie between 0 and 1 since the roots of $\pi(x)$ are 0 and 1. (This can be shown by the use of Rolle's theorem.) Jacobi now sets $u = V$, $v = \varphi$, and $m = n - 1$ in the relation (5.3). Then, since the n first definite integrals of φ vanish, and since the indefinite integral

$$\int^n \varphi(x)\,dx^n = \frac{x^n(x-1)^n}{2n(2n-1)\cdots(n+1)},$$

he has the following relation between definite integrals:

$$\Delta = \int \varphi(x)V\,dx = \frac{(-1)^n}{2n(2n-1)(2n-2)\cdots(n+1)}\int x^n(x-1)^n\frac{d^nV}{dx^n}\,dx.$$

To conclude, Jacobi proceeds to simplify the integral in the right-hand side of this expression. He considers a function $t(x)$ such that t vanishes at some as yet arbitrary point $x = l$. Then if $u = t^{m+1}(x)$, we know that u, du/dx, $d^2u/dx^2, \ldots, d^mu/dx^m$ all vanish for $x = l$. Now the expressions

$$u \int v\,dx, \quad \frac{du}{dx}\int^2 v\,dx^2, \quad \frac{d^2u}{dx^2}\int^3 v\,dx^3, \ldots, \frac{d^mu}{dx^m}\int^{m+1} v\,dx^{m+1},$$

(with arbitrary v), where the integrals are formed between the limits $x = 0$ and $x = l$, also vanish. Therefore the integral from $x = 0$ to $x = l$ of uv becomes

$$\int uv\,dx = \int t^{m+1}v\,dx = (-1)^{m+1}\int \left(\frac{d^{m+1}t^{m+1}}{dx^{m+1}}\int^{m+1} v\,dx^{m+1}\right) dx.$$

Jacobi now chooses $t = 1 - x$, $l = 1$, $m = n - 1$, $v = x^n d^nV/dx^n$. Then between the limits $x = 0$ and $x = 1$

$$\int (1 - x)^n x^n \frac{d^nV}{dx^n}\,dx = 1\cdot 2\cdot 3\cdots n \int \left(\int^n x^n \frac{d^nV}{dx^n}\,dx^n\right) dx$$

$$= 1\cdot 2\cdot 3\cdots n \int^{n+1} x^n \frac{d^nV}{dx^n}\,dx^{n+1},$$

and therefore

$$\Delta = \frac{1\cdot 2\cdot 3\cdots n}{2n(2n-1)\cdots(n+1)}\int^{n+1} x^n \frac{d^nV}{dx^n}\,dx^{n+1}.$$

He says all integrals are to be chosen so that they vanish at $x = 0$, and the last one has the limits $x = 0$ and $x = 1$.[2]

It is perhaps worth noting that choices other than the zeros of the Legendre polynomials have been made for the abscissae. If, e.g., they are taken to be the roots of the appropriate Chebyshev polynomial

$$T_n(x) = \frac{1}{2^{n-1}}\cos n(\text{arc}\cos x) \qquad (n = 1, 2, \ldots),$$

or of

$$U_n(x) = \frac{1}{2^{n-1}}\frac{\sin(n+1)(\text{arc}\cos x)}{\sin(\text{arc}\cos x)} \qquad (n = 1, 2, \ldots),$$

then Fejér has shown that the R, R', R'', etc. are positive and that the corresponding sums will converge to the integral of f from $x = -1$ to $+1$.[3]

Chebyshev himself approached the problem from a probabilistic point of view. He imagined that $y(a)$, $y(a')$, etc. are measured quantities which are in error by amounts e, e', etc. He then considered the error in the integral $E = Re + R'e' +$ etc., and evaluated the probability that $|E| < \lambda$ under the

---

[2] Recall that the functions φ of Jacobi are proportional to the Legendre polynomials on the interval 0, 1. Cf., e.g., Courant, *Methoden*, Vol. I, pp. 70–74.

[3] L. Fejér [1933], pp. 287–309.

assumption that the probabilities of the $e, e', \ldots$ are equal. He said that

$$P(|E| < \lambda) = \left(\frac{2}{\pi}\right)^{1/2} \int_0^{\lambda/\sqrt{R_n^2 + R'^2 + \text{etc.}}} e^{-t^2/2}\,dt,$$

and he sought to choose the R, R', R'', etc. so that $R^2 + R'^2 + R''^2 + \text{etc.}$ is a minimum. This leads to a curious set of polynomials whose roots are not always real.[4] Note that what Chebyshev sought was to maximize the probability $P(|E| < \lambda)$. It is easy to see that $R = R' = R'' = \text{etc.}$ in this case; thus if the abscissae are $x_1, x_2, \ldots, x_n$, then the integral $\int_{-1}^{+1} f(x)\,dx$ is

$$I = \frac{2}{n}\,[f(x_1) + f(x_2) + \cdots + f(x_n)].$$

To find the x_1, x_2, etc. Chebyshev noted that they are the roots of the polynomial part of

$$x^n \exp\left(-\sum_{\alpha=1}^{\infty} \frac{n}{2\alpha(2\alpha + 1)x^{2\alpha}}\right).$$

These functions and their roots for $n = 1, 2, \ldots, 7$ have been tabulated by Chebyshev.[5] (Note that they are not the well-known Chebyshev polynomials.)

If the function to be integrated is $f(x)/\sqrt{1 - x^2}$, then Bronwin showed that

$$\int_{-1}^{+1} \frac{f(x)\,dx}{\sqrt{1 - x^2}} = \int_0^{\pi} f(\cos\varphi)\,d\varphi = \frac{\pi}{n} \sum_{\alpha=1}^{2n-1} f\left(\cos\frac{\alpha\pi}{2n}\right),$$

provided that the $2n$th differences of f vanish.[6]

There are a number of other formulas for numerical summation which are akin to those of Euler–Maclaurin and Laplace. They go by the generic names of Lubbock and Woolhouse. Their purpose is to facilitate the summing of large numbers of terms. Thus they give

$$\sum_{\alpha=0}^{mn-1} f(\alpha) = n \sum_{\beta=0}^{m-1} f(n\beta) + \cdots,$$

where the terms concealed by $\cdots$ are either expressions involving finite differences or derivatives. In the former case the results are Lubbock type

---

[4] Chebyshev [1874], pp. 19–34 or *Works*, Vol. III, pp. 49–62. It is in this article that Chebyshev defines his functions as $(1/2^{n-1}) \cos (n \arccos x)$. The remainder term in his formula appears in Milne–Thomson [1933], pp. 177–180.

[5] Jordan [1939], pp. 521–523. He notes, moreover, that for $n = 8$ and 10, e.g., the roots are not always real. This, of course, makes the utility of the method for $n > 7$ questionable. One big advantage of the procedure is that all observations $f(x_i)$ receive equal weight. The remainder term is calculated in Milne–Thomson [1933], pp. 177–178.

[6] Bronwin [1849]. What he does here is to find formulas for the coefficients of a finite Fourier series; cf. Gauss's results in Chapter IV, pp. 246ff.

and in the latter Woolhouse.[7] There are also a variety of numerical integration formulas due to men such as Weddle, Shovelton, and Hardy. Each of these has some particular merit but we do not have the space to discuss them. The interested reader may consult one of the standard texts.[8]

5.3. Jacobi on the Euler–Maclaurin Formula

Jacobi wrote on this topic in order to make the validity of the formula more precise.[9] In previous discussions we have generally seen only infinite series given with no suggestion of error terms. Jacobi now made a detailed investigation of the remainder term when the Euler–Maclaurin expansion is terminated. This study is particularly interesting since he introduced into the analysis a sequence of polynomials which are now known as the Bernoulli polynomials. We discussed them previously in Chapter II on pp. 96ff. These polynomials were named in Bernoulli's honor by J. L. Raabe (cf. Raabe [1848], and also J. W. Glaisher [1898].) Jacobi starts with the expressions

$$\psi(x + h) = \psi(x) + \psi'(x)h + \psi''(x)\frac{h^2}{1\cdot 2} + \cdots + \psi^{(n)}(x)\frac{h^n}{\Pi(n)}$$
$$+ \int_0^h \frac{(h - t)^n}{\Pi(n)}\psi^{(n+1)}(x + t)\,dt,$$
$$\psi(x - h) = \psi(x) - \psi'(x)h + \psi''(x)\frac{h^2}{1\cdot 2} - \cdots + (-1)^n\psi^{(n)}(x)\frac{h^n}{\Pi(n)}$$
$$+ (-1)^{n+1}\int_0^h \frac{(h - t)^n}{\Pi(n)}\psi^{(n+1)}(x - t)\,dt,$$

where $\Pi(n) = 1\cdot 2\cdot 3\cdots n$, and sets

$$\psi(x) = \int_a^x f(x)\,dx, \qquad \psi(x) - \psi(x - h) = \varphi(x).$$

[7] Lubbock [1829], and Woolhouse [1888]. The original result was improved by de Morgan [1842].

[8] Steffensen [1850], pp. 166–170; Whittaker, *WR*, pp. 150–151 or Milne–Thomson [1933], pp. 171–173.

[9] Cf. Jacobi VI [1834], pp. 64–75. The paper is entitled "De usu legitimo formulae summatoriae Maclaurinianae." The subject was also treated by Poisson [1823], pp. 571–602. In previous discussions we have seen only infinite series given with no suggestion of error terms. Jacobi now made a detailed investigation of the remainder term when the Euler–Maclaurin expansion is terminated. The interested reader may wish to consult the analysis of this error term in Milne–Thomson [1933], pp. 187–190 or Nörlund [1924], pp. 30–32.

Then

$$\varphi(a + h) + \varphi(a + 2h) + \cdots + \varphi(x) = \psi(x) - \psi(a) = \psi(x).$$

He now defines a sum operator $\sum$ so that $\sum_a^x \varphi(x)$ is the left-hand member of this relation above. Hence

$$\sum\nolimits_a^x \varphi(x) = \psi(x) = \int_a^x f(x)\,dx,$$

and

$$\varphi(x) = \psi'(x)h - \psi''(x)\frac{h^2}{1\cdot 2} + \cdots + (-1)^{n-1}\psi^{(n)}(x)\frac{h^n}{\Pi(n)}$$

$$+ (-1)^n \int_0^h \frac{(h-t)^n}{\Pi(n)} \psi^{(n+1)}(x - t)\,dt.$$

But $\psi'(x) = f(x)$, $\psi^{(m+1)}(x) = f^{(m)}(x)$. Therefore he has

$$\sum\nolimits_a^x \frac{\varphi(x)}{h} = \int_a^x \frac{f(x)}{h}\,dx$$

$$= \sum\nolimits_a^x \left\{ f(x) - f'(x)\frac{h}{2} + f''(x)\frac{h^2}{2\cdot 3} - \cdots + (-1)^{n-1} f^{(n-1)}(x)\frac{h^{n-1}}{\Pi(n)} \right\}$$

$$+ (-1)^n \int_0^h \frac{(h-t)^n}{h\Pi(n)} \sum\nolimits_a^x f^{(n)}(x - t)\,dt. \tag{5.4}$$

Jacobi now resorted to a very adroit "trick". He successively replaced $f(x)$ by $f(x)$, $\frac{1}{2}f'(x)h$, $\alpha_1 f''(x)h^2$, $-\alpha_2 f''''(x)h^4, \ldots, (-1)^{m+1}\alpha_m f^{(2m)}(x)h^{2m}$, and n by $n, n-1, n-2, n-3, \ldots, n-2m$, where the $\alpha_1, \alpha_2, \ldots$ are the coefficients indicated in the expansion

$$\frac{1}{2}\frac{e^{(1/2)h} + e^{-(1/2)h}}{e^{(1/2)h} - e^{-(1/2)h}} = \frac{1}{2} + \frac{1}{e^h - 1} = \frac{1}{h} + \alpha_1 h - \alpha_2 h^3 + \alpha_3 h^5 - \cdots.^{10} \tag{5.5}$$

(Jacobi gave the values for the α_i which he found by a multiplication by the series expansion of $e^h - 1 = h + \cdots$. They are $\alpha_n = (-1)^{n+1}B'_{2n}/(2n)!$.) If the resulting integrals are added together, there results the expression

$$\int_a^x dx \left\{ \frac{f(x)}{h} + \frac{1}{2}f'(x) + \alpha_1 f''(x)h - \alpha_2 f''''(x)h^3 + \cdots \right.$$

$$\left. + (-1)^{m+1}\alpha_m f^{(2m)}(x)h^{2m-1} \right\}$$

$$= \sum\nolimits_a^x f(x) + \int_0^h T_m \sum\nolimits_a^x f^{(2m+2)}(x - t)\,dt, \tag{5.6}$$

where T_m represents the expression

$$T_m = \frac{(h-t)^{2m+2}}{h\Pi(2m+2)} - \frac{1}{2}\frac{(h-t)^{2m+1}}{\Pi(2m+1)} + \alpha_1 \frac{(h-t)^{2m}h}{\Pi(2m)} - \alpha_2 \frac{(h-t)^{2m-2}h^3}{\Pi(2m-2)}$$

---

[10] Jacobi VI [1834], p. 67.

$$+ \alpha_3 \frac{(h-t)^{2m-4}h^5}{\Pi(2m-4)} - \cdots + (-1)^{m+1}\alpha_m \frac{(h-t)^2 h^{2m-1}}{\Pi(2)}. \tag{5.7}$$

To estimate the sizes of α_m Jacobi integrates the relation (5.5) and has

$$\log\left(e^{(1/2)h} - e^{-(1/2)h}\right) = \log h + \frac{1}{2}\alpha_1 h^2 - \frac{1}{4}\alpha_2 h^4 + \frac{1}{6}\alpha_3 h^6 - \cdots$$

$$= \log h + \log\left[1 + \frac{1}{\Pi(3)}\left(\frac{h}{2}\right)^2 + \frac{1}{\Pi(5)}\left(\frac{h}{2}\right)^4 + \cdots\right].$$

He then writes $e^{h/2} - e^{h/2}$ as an infinite product and takes the logarithm, finding thereby

$$\frac{1}{2}\alpha_1 h^2 - \frac{1}{4}\alpha_2 h^4 + \frac{1}{6}\alpha_3 h^6 - \cdots = \sum_1^\infty \log\left(1 + \frac{h^2}{4p^2\pi^2}\right).$$

Hence he has an expression for α_m as

$$\frac{1}{2}\alpha_m = \frac{1}{(2\pi)^{2m}} \sum_1^\infty \frac{1}{p^{2m}} = \frac{1}{(2\pi)^{2m}}\left[1 + \frac{1}{2^{2m}} + \frac{1}{3^{2m}} + \frac{1}{4^{2m}} + \cdots\right],$$

and by a little manipulating he has the inequalities

$$\frac{1}{(2\pi)^{2m}} < \frac{1}{2}\alpha_m < \frac{1}{(2\pi)^{2m}}\left[1 + \frac{1}{2^{2m}}\left(\frac{\pi^2}{6} - 1\right)\right].$$

(Cf. p. 131.) Now he defines polynomials χ_{2m+1} by means of the relations

$$\chi_{2m+1}(x) = \frac{x^{2m+2}}{\Pi(2m+2)} + \frac{1}{2}\frac{x^{2m+1}}{\Pi(2m+1)} + \alpha_1 \frac{x^{2m}}{\Pi(2m)} - \alpha_2 \frac{x^{2m-2}}{\Pi(2m-2)} + \cdots$$

$$+ (-1)^{m+1}\alpha_m \frac{x^2}{\Pi(2)}. \tag{5.8}$$

Then

$$T_m = h^{2m+1}\chi_{2m+1}\left(\frac{t-h}{h}\right). \tag{5.9}$$

Now if we apply the formula (5.6) to $f(x) = x^{2m+1}/\Pi(2m+1)$, and let x be any positive integer, then, clearly, for $a = 0$, $h = 1$,

$$\chi_{2m+1}(x) = \sum_0^x \frac{x^{2m+1}}{\Pi(2m+1)}.$$

Moreover, it is evident from the definition of $\sum$ that

$$\chi_{2m+1}(x+1) = \chi_{2m+1}(x) + \frac{(x+1)^{2m+1}}{\Pi(2m+1)}$$

for integral x. Moreover, by this and (5.9) it follows that

$$h^{2m+1}\chi_{2m+1}\left(\frac{t}{h}\right) = T_m + \frac{t^{2m+1}}{\Pi(2m+1)},$$

so that T_m is given by the series

$$T_m = \frac{t^{2m+2}}{h\Pi(2m+2)} - \frac{1}{2}\frac{t^{2m+1}}{\Pi(2m+1)} + \alpha_1\frac{t^{2m}h}{\Pi(2m)} - \alpha_2\frac{t^{2m-2}h^3}{\Pi(2m-2)} + \cdots$$
$$+ (-1)^{m+1}\alpha_m\frac{t^2h^{2m-1}}{\Pi(2)}.$$

Hence there result the relations

$$T_m = h^{2m+1}\chi_{2m+1}\left(\frac{t-h}{h}\right) = h^{2m+1}\chi_{2m+1}\left(-\frac{t}{h}\right),$$

and so $\chi_{2m+1}(x-1) = \chi_{2m+1}(-x)$, where $x = (t-h)/h$. Jacobi now examines the signs of the polynomials χ_{2m+1} for $0 \leq x \leq 1$. He first notes that for x integral,

$$\frac{1-e^{xz}}{1-e^z} = \sum_0^x e^{z(x-1)}.$$

Therefore

$$\frac{1}{2}\left\{\frac{1-e^{xz}}{1-e^z} - \frac{1-e^{-xz}}{1-e^{-z}}\right\} = z\chi_1(x-1) + z^3\chi_3(x-1) + z^5\chi_5(x-1) + \cdots \quad .$$

Now he sets $x' = 1 - x$ and writes

$$\frac{1-e^{xz}}{1-e^z} - \frac{1-e^{-xz}}{1-e^{-z}} = \frac{1-e^{xz}}{1-e^z} + \frac{e^z - e^{x'z}}{1-e^z}$$
$$= \frac{(1-e^{xz})(1-e^{x'z})}{1-e^z}$$
$$= -\frac{(e^{(1/2)xz} - e^{-(1/2)xz})(e^{(1/2)x'z} - e^{-(1/2)x'z})}{e^{(1/2)z} - e^{-(1/2)z}}.$$

Using the infinite product expansion of $e^u - e^{-u}$, he readily finds that this expression has the form

$$-zxx'\Pi\frac{(1+(x^2z^2/4p^2\pi^2))(1+(x'^2z^2/4p^2\pi^2))}{(1+(z^2/4p^2\pi^2))}$$
$$= 2[z\chi_1(x-1) + z^3\chi_3(x-1) + \cdots]. \quad (5.10)$$

Set $y = -z^2/4p^2\pi^2$; then each factor in the infinite product above is of the form

$$\frac{(1-x^2y)(1-x'^2y)}{(1-y)} = 1 + (1 - x^2 - x'^2)y + \frac{(1-x^2)(1-x'^2)y^2}{1-y}$$
$$= 1 + 2xx'y + xx'(2+xx')\frac{y^2}{1-y}.$$

When this is expressed as a series in y, it is evident that each coefficient is positive if xx' is positive. But $y = -z^2/4p^2\pi^2$ and the infinite product in the left-hand member of (5.10) is multiplied by a factor $-zxx'$. Thus the expansion of the left-hand member is a series in odd powers z^{2m+1} of z whose coefficients are positive expressions multiplied by $(-1)^{m+1}$. Thus $\chi_{2m+1}(x-1)$ (the coefficient of z^{2m+1} in the expansion of (5.10)) is positive for m odd and negative for m even for $0 < x < 1$. Also, if $x = t/h$ then $T_m(t)$ is positive for m odd and negative for m even, provided that $0 < t < h$.

This enables Jacobi to conclude from the relation (5.6) that the expression

$$\int_a^x dx\left\{\frac{f(x)}{h} + \frac{1}{2}f'(x) + \alpha_1 f''(x)h - \alpha_2 f''''(x)h^3 + \cdots \right.$$
$$\left. + (-1)^{m+1}\alpha_m f^{(2m)}(x)h^{2m-1}\right\} - \sum_a^x f(x)$$

has the same sign as $\sum_\alpha^x f^{(2m+2)}(x-t)$ if m is odd, and the opposite sign if m is even. (Note that Jacobi has written $f^{(\alpha)}(x-t) = \partial^\alpha f(x-t)/\partial x^\alpha$) provided $\sum_\alpha^x f^{(2m+2)}(x-t)$ is neither infinite nor changes sign for $0 < t < h$. Hence if the integral in the last expression is called S_m, and if $\sum_\alpha^x f^{(2m)}(x-t)$, $\sum_\alpha^x f^{(2m+2)}(x-t)$ $(0 < t < h)$ are neither infinite nor change sign, then the sum $\sum_a^x f(x)$ lies between S_{m-1} and S_m. Jacobi also has the result that if $\sum_\alpha^x f^{(2m+2)}(x-t)$ for $0 < t < h$ neither changes sign nor is infinite, then $S_m - \sum_\alpha^x f(x)$ has the opposite sign to $(-1)^m\alpha_{m+2}\int_\alpha^x f^{(2m+2)}(x)\,dx$.

In this paper Jacobi does not go any further with an analysis of the χ_{2m+1}. It is however worth saying a word about them. It is not hard to show that

$$1^v + 2^v + \cdots + x^v = \int_0^{x+1} B_v(z)\,dz = \frac{B_{v+1}(x+1) - B'_{v+1}}{v+1},$$

where $B_\mu(z)$ is the μth Bernoulli polynomial and B'_μ is the corresponding Bernoulli number.[11] But $\chi_{2m+1}(x)$ is, by definition,

$$\frac{1}{(2m+1)!}(1^{2m+1} + 2^{2m+1} + \cdots + x^{2m+1}),$$

and thus

$$\chi_{2m+1}(x) = \frac{1}{(2m+2)!}[B_{2m+2}(x+1) - B'_{2m+2}].^{12}$$

5.4. Jacobi on Linear Equations

Jacobi wrote several papers on systems of linear equations as a result of his interest in Gauss's work on least squares and more generally in physical

[11] Nörlund [1924], p. 19 or p. 97 above.

[12] The remainder term in the Euler–Maclaurin series is discussed in many texts such as Steffensen [1950], pp. 131ff. A valuable list of references to papers on the subject appears in Nörlund [1924], p. 30.

systems undergoing small oscillations. He also interested his former student Ludwig Seidel, who taught in Munich, in the topic; out of this arose the Gauss–Seidel method mentioned earlier on p. 224. In this section we intend to discuss Jacobi's efforts in connection with systems of linear equations. The first paper we consider is particularly interesting since it was unknown to von Neumann or me, and we worked on the same problem with Prof. E. J. Murray, Jr. Out of this effort came a rediscovery of Jacobi's method which I presented at a meeting on numerical analysis at the University of California, Los Angeles. At the end of the presentation Prof. Ostrowski of Basel told me this had been done a century earlier by Jacobi.[13]

Jacobi considers a system of linear equations

$$\begin{aligned}(00)x + (01)x_1 + (02)x_2 + \cdots &= (0m),\\ (10)x + (11)x_1 + (12)x_2 + \cdots &= (1m),\\ (20)x + (21x)_1 + (22)x_2 + \cdots &= (2m),\\ \text{etc.} \qquad \text{etc.} \qquad &\text{etc.,}\end{aligned}$$

where all the off-diagonal coefficients (ik) are small compared to the diagonal ones (ii). He then concludes that a first approximation to the solution is given by

$$(00)x = (0m), \qquad (11)x_1 = (1m), \qquad (22)x_2 = (2m), \text{ etc.}$$

Let this solution be called a, a_1, a_2, etc. Then if Δ, Δ_1, Δ_2, etc. designate errors, he has approximately

$$\begin{aligned}(00)\Delta &= -\{(01)a_1 + (02)a_2 + \cdots\},\\ (11)\Delta_1 &= -\{(10)a + (12)a_2 + \cdots\},\\ \text{etc.} \qquad &\text{etc.}\end{aligned}$$

In fact, the solution is given as

$$\begin{aligned}x &= a + \Delta + \Delta^2 + \Delta^3 + \cdots,\\ x_1 &= a_1 + \Delta_1 + \Delta_1^2 + \Delta_1^3 + \cdots,\\ x_2 &= a_2 + \Delta_2 + \Delta_2^2 + \Delta_3^2 + \cdots,\\ \text{etc.} \qquad &\text{etc.}\end{aligned}$$

where the superscripts are used to signify "ever smaller corrections." Thus, e.g.,

$$\begin{aligned}(00)\Delta^{i+1} &= -\{(01)\Delta_1^i + (02)\Delta_2^i + \cdots\},\\ (11)\Delta_1^{i+1} &= -\{(10)\Delta^i + (12)\Delta_2^i + \cdots\},\\ (22)\Delta_2^{i+1} &= -\{(20)\Delta^i + (21)\Delta_1^i + (23)\Delta_3^i + \cdots\},\\ \text{etc.} \qquad &\text{etc.}\end{aligned}$$

Jacobi notes that often in least square problems the off-diagonal elements are relatively small compared to those on the diagonal, and that the so-called

---

[13] Jacobi III [1845], pp. 468–478. Cf. also, Goldstine [1959], pp. 59–96.

normal equations are always symmetric — $(ik) = (ki)$ in his notation. He then sets

$$x = \cos\alpha\cdot\eta + \sin\alpha\cdot\eta_1,$$
$$x_1 = \sin\alpha\cdot\eta - \cos\alpha\cdot\eta_1,$$

and remarks that

$$(00)x + (01)x_1 = \{(00)\cos\alpha + (01)\sin\alpha\}\eta + \{(00)\sin\alpha - (01)\cos\alpha\}\eta_1,$$
$$(10)x + (11)x_1 = \{(10)\cos\alpha + (11)\sin\alpha\}\eta + \{(10)\sin\alpha - (11)\cos\alpha\}\eta.$$

He then determines the angle α so that

$$\frac{1}{2}\tan 2\alpha = \frac{(01)}{(00) - (11)}.$$

This gives him for his first two equations

$$\{(00)\cos^2\alpha + 2(01)\cos\alpha\sin\alpha + (11)\sin^2\alpha\}\eta + \{(02)\cos\alpha + (12)\sin\alpha\}x_2 + \cdots = (0m)\cos\alpha + (1m)\sin\alpha,$$
$$\{(00)\sin^2\alpha - 2(01)\sin\alpha\cos\alpha + (11)\cos^2\alpha\}\eta_1 + \{(02)\sin\alpha - (12)\cos\alpha\}x_2 + \cdots = (0m)\sin\alpha - (1m)\cos\alpha.$$

Furthermore, the other coefficients can easily be calculated. Jacobi then observes that the coefficients of η and η_1 are expressible as

$$\frac{(00) + (11)}{2} + \sqrt{R}, \qquad \frac{(00) + (11)}{2} - \sqrt{R},$$

where

$$R = \left\{\frac{(00) - (11)}{2}\right\}^2 + (01)^2.$$

The effect of this transformation is to change the coefficient (01) into zero. Jacobi remarks that the sum $(00) + (11)$ remains invariant and hence that the sum of the squares of the off-diagonal elements is reduced by $2\cdot(01)^2$. From this he concludes that this sum becomes smaller and smaller. He also sees that the sum of the squares of the diagonal elements and the sum of the squares of all elements are invariants. He observes that it is best to rotate in such a manner that the (numerically) largest off-diagonal coefficient is eliminated.

Jacobi recommends his procedure particularly for the solution of systems of the form

$$\begin{aligned}\{(00) - G\}x + (01)x_1 + (02)x_2 + \cdots &= 0,\\ (10)x + \{(11) - G\}x_1 + (12)x_2 + \cdots &= 0,\\ (20)x + (21)x_1 + \{(22) - G\}x_2 + \cdots &= 0,\\ \text{etc.} \qquad\qquad &\text{etc.}\end{aligned}$$

i.e., for finding the characteristic values and vectors of a matrix. He says that by elimination of the unknowns x, x_1, x_2, etc. there results an equation whose roots give the different values of G and that for these values the x, x_1, x_2, etc. are determinate.

Jacobi remarked that his method obviated the need to find and solve the characteristic equation although he did not call it by any name. He was pleased with this method since it enabled him to analyze the secular perturbations of the seven planets. In this connection Seidel carried out the numerical work for him. In this paper, however, he was content to apply the method to a very simple problem of Gauss:

$$\begin{aligned} 27p + 6q + * - 88 &= 0, \\ 6p + 15q + r - 70 &= 0, \\ * + q + 54r - 107 &= 0.^{14} \end{aligned}$$

He chose his first angle α of rotation to be 23°30′ and had

$$\begin{aligned} p &= 0.92390y + 0.38268y', \\ q &= 0.38268y - 0.92390y'; \end{aligned}$$

and the new system was

$$\begin{aligned} 29.4853y + \qquad\qquad + 0.38268r - 108.0901 &= 0, \\ + 12.5147y' - 0.92390r + 30.9967 &= 0, \\ 0.38268y - 0.92390y' + 54r - 107 &= 0. \end{aligned}$$

Jacobi then took, as an approximation to the roots, $y = 108.0901/29.4853$, $y' = -30.9967/12.5147$, $r = 107/54$, and recorded their logarithms base 10 as 0.56419; 0.39389n: 0.29699. (His notation "n" after the second number means that y' is negative.) He then found a second approximation to the roots, which he recorded as $\log y = 0.56114$, $\log y' = 0.36746$n and $\log r = 0.28174$. Then after two "easy corrections one finds the improved values $\log y = 0.56125$, $\log y' = 0.36836$n, $\log r = 0.28233$, from which follow the values $\log p = 0.39276$, $\log q = 0.55036$." (These give $p = 2.47036$, $q = 3.55108$, $r = 1.91571$.) He concludes that the logarithms of their weights (in the least-square sense) are 1.39092; 1.13565; 4.73239. These correspond to the squares of the precisions.

We should note in passing that Kummer wrote on the characteristic equation for quadric surfaces in three dimensions.[15] Jacobi discussed the problem when the axes were not necessarily at right angles to each other in 1827.[16] The general n-dimensional case was worked out by Carl W. Borchardt in 1845.[17] The most interesting paper on the subject, however, is perhaps one in the same issue as the Borchardt paper where Jacobi discusses the secular perturbations of the seven planets. He applied his rotation of axes method for symmetric matrices to the numerical data of Leverrier in the *Connaissance*

---

[14] Gauss, *Theoria* (Eng. trans.), p. 268. It is in Book II, Section 3, Paragraph 184.

[15] Kummer [1843].

[16] Jacobi III [1884], pp. 46–53.

[17] Borchardt [1846].

des temps 1843, pp. 31ff. As mentioned above, Jacobi was assisted in the numerical calculations by Seidel. The calculation sharpened Leverrier's results very decisively.[18] The basic problem is one of finding the characteristic values and associated vectors of a symmetric matrix of order 7. Jacobi repeats in this paper his algorithm for rotating axes but includes an interesting estimate for the rate of convergence. Let n be the order of the matrix and let $S = \sum_{i \neq j}, a_{ij}^2$. He remarks that the largest $a_{ij}^2 (i \neq j)$ must be $> S/(n-2)(n+1)$; and hence if one removes this by a transformation, the new sum of squares must be less than $S[1 - 2/(n-2)(n+1)]$; in the next transformation it must be less than $S[1 - 2/(n-2)(n+1)]^2$, and in the ith less than

$$S[1 - 2/(n-2)(n+1)]^i.$$

Thus with increasing i the off-diagonal sum of squares converges to zero. Jacobi then remarks that the elements remaining on the diagonal are the characteristic values and that they are real.[19] He performed 10 rotations of the original system and then resorted to yet another scheme to finish off his calculation. He supposes that the system of n equations

$$\begin{array}{lllllll}
\{(a,a)-x\}\alpha + & (a,b)\beta & + & (a,c)\gamma & + \cdots + & (a,p)\varpi & = 0,\\
(b,a)\alpha & + \{(b,b)-x\}\beta + & & (b,c)\gamma & + \cdots + & (b,p)\varpi & = 0,\\
(c,a)\alpha & + & (c,b)\beta & + \{(c,c)-x\}\gamma & + \cdots + & (c,p)\varpi & = 0,\\
\vdots & & & & & & \\
(p,a)\alpha & + & (p,b)\beta & + & (p,c)\gamma & + \cdots + \{(p,p)-x\}\varpi & = 0
\end{array}$$

is such that the off-diagonal elements (a, b) (a, c), etc. are "small numbers of the first order." He now describes a scheme which converges quickly. Let u be a quantity expressible as

$$u = \Delta^0 u + \Delta^1 u + \Delta^2 u + \Delta^3 u + \cdots,$$

where $\Delta^0 u$ is a first approximation to u, and $\Delta^1 u$, $\Delta^2 u$, etc. are successive corrections of different orders. Thus

$$\Delta^0 x + \Delta^1 x = (a, a), \qquad \Delta^0 \frac{\beta}{\alpha} = 0, \qquad \Delta^0 \frac{\gamma}{\alpha} = 0, \ldots$$

$$\Delta^1 \frac{\beta}{\alpha} = \frac{(b, a)}{(a, a) - (b, b)}, \qquad \Delta^1 \frac{\gamma}{\alpha} = \frac{(c, a)}{(a, a) - (c, c)}, \cdots \quad .$$

---

[18] Jacobi VII [1846]. On p. 142 Jacobi compares his residuals to Leverrier's to show how substantial is his improvement of the latter's values.

[19] Goldstine [1959]. There we note that the average sized off-diagonal square is $\sum_{i \neq j} a_{ij}^2 / n(n-1)$. Hence if we rotate to remove an element above average size, then the new $S < S[1 - 2/n(n-1)] < Se^{-2/n(n-1)}$.

Jacobi now finds the second order corrections to β/α, γ/α, etc. from the equations

$$\{(a, a) - (b, b)\}\Delta^2 \frac{\beta}{\alpha} = (b, c)\Delta^1 \frac{\gamma}{\alpha} + (b, d)\Delta^1 \frac{\delta}{\alpha} + \cdots,$$

$$\{(a, a) - (c, c)\}\Delta^2 \frac{\gamma}{\alpha} = (c, b)\Delta^1 \frac{\beta}{\alpha} + (c, d)\Delta^1 \frac{\delta}{\alpha} + \cdots,$$

$$\text{etc.} \qquad\qquad \text{etc.}$$

These enable him to find the second and third order corrections to the eigenvalue x from

$$\Delta^2 x + \Delta^3 x = (a, b)\left\{\Delta^1 \frac{\beta}{\alpha} + \Delta^2 \frac{\beta}{\alpha}\right\} + (a, c)\left\{\Delta^1 \frac{\gamma}{\alpha} + \Delta^2 \frac{\gamma}{\alpha}\right\} + \cdots,$$

and eventually $\Delta^6 x + \Delta^7 x$ from values for $\Delta^i\beta/\alpha$, $\Delta^i\gamma/\alpha$, etc., for $i = 1, 2, \ldots, 6$. These he uses to find both x and his characteristic vectors. He apparently did not realize that the product of his successive unitary transformations gave him the characteristic directions.[20] After his tenth transformation Jacobi had brought the sum of the squares of his off-diagonal elements down to less than 0.03 compared to the characteristic values which lie between -23 and -2.

Jacobi also made some observations on how to deal with certain nonsymmetric matrices.[21] Let the system be again

$$\begin{aligned} u &= (00)x + (01)x_1 + (02)x_2 + \cdots - (0m) = 0, \\ u_1 &= (10)x + (11)x_1 + (12)x_2 + \cdots - (1m) = 0, \\ u_2 &= (20)x + (21)x_1 + (22)x_2 + \cdots - (2m) = 0, \\ &\text{etc.} \qquad\qquad\qquad\qquad \text{etc.} \end{aligned}$$

He now sets

$$\begin{aligned} \cos 2\Delta\cdot x &= \cos(\alpha + \Delta)\cdot\eta + \sin(\alpha - \Delta)\cdot\eta_1, \\ \cos 2\Delta\cdot x_1 &= \sin(\alpha + \Delta)\cdot\eta - \cos(\alpha - \Delta)\cdot\eta_1, \end{aligned}$$

where the angles α and Δ are determined by the equations

$$\begin{aligned} \rho\cos 2\alpha &= (00) - (11), \\ \rho\sin 2\alpha &= (01) + (10), \\ \rho\sin 2\Delta &= (10) - (01). \end{aligned}$$

He next sets

$$\begin{aligned} v &= \cos(\alpha - \Delta)\cdot u + \sin(\alpha - \Delta)\cdot u_1, \\ v_1 &= \sin(\alpha + \Delta)\cdot u - \cos(\alpha + 2)\cdot u_1, \end{aligned}$$

and considers the transformed system $v = 0$, $v_1 = 0$, $u_2 = 0$, $u_3 = 0$, etc.

[20] Goldstine [1959], pp. 580ff.

[21] Jacobi III [1845], pp. 475–477.

By his choice of angles the coefficient of η_1 in the equation $v = 0$ vanishes, and in $v_1 = 0$ the coefficient of η vanishes. Thus

$$\begin{aligned} v &= [00]\eta + \quad * \quad + [02]x_2 + [03]x_3 + \cdots, \\ v_1 &= \quad * \quad + [11]\eta_1 + [12]x_2 + [13]x_3 + \cdots, \\ u_2 &= [20]\eta + [21]\eta_1 + [22]x_2 + [23]x_3 + \cdots, \\ &\quad \text{etc.} \qquad\qquad\qquad\qquad \text{etc.,} \end{aligned}$$

where

$$[00] = \frac{(00) + (11)}{2} + \frac{\rho}{2}\cos 2\Delta,$$

$$[11] = \frac{(00) + (11)}{2} - \frac{\rho}{2}\cos 2\Delta,$$

$$\begin{aligned} [02] &= (02)\cos(\alpha - \Delta) + (12)\sin(\alpha - \Delta), \\ [12] &= (02)\sin(\alpha + \Delta) - (12)\cos(\alpha + \Delta), \\ [20]\cos 2\Delta &= (20)\cos(\alpha + \Delta) + (21)\sin(\alpha + \Delta), \\ [21]\cos 2\Delta &= (20)\sin(\alpha - \Delta) - (21)\cos(\alpha - \Delta). \end{aligned}$$

It follows at once that

$$\begin{aligned} [00] + [11] &= (00) + (11), \\ [00]^2 + [11]^2 &= (00)^2 + (11)^2 + 2(01)(10), \\ [02][20] + [12][21] &= (02)(20) + (12)(21). \end{aligned}$$

Jacobi sums this up by noting that no matter how often we transform our matrix the sums $\sum [ii]$, $\sum \{(ii)(ii) + 2(ik)(ki)\}$ are invariant. Thus if $2\sum (ik)(ki)$ becomes ever smaller, then $\sum (ii)^2$ becomes ever larger. This is in effect his convergence proof. He does not discuss what happens when his definition of $\rho \sin 2\Delta$ leads to the case where $|(10) - (01)|/\rho > 1$, i.e., where $\sin 2\Delta$ cannot be defined; nor does he illustrate his procedure, but he does say that for the success of the method the coefficients (ik) and (ki) should not be too different in size or, at least, should both have the same sign. What Jacobi is saying is that he needs to have

$$[(10) - (01)]^2 \leq [(00) - (11)]^2 + [(01) + (10)]^2,$$

i.e.,

$$-4(01)(10) \leq [(00) - (11)]^2.$$

5.5. Cauchy on Interpolation

A reasonable evaluation of Cauchy's contributions to our field is not an easy task, since he wrote so extensively that we are faced with a veritable embarrassment of riches. Instead of attempting to examine all of his work on our subject I have decided to be critically selective and limit the discussion to a sampling of a few of his noteworthy accomplishments: his work on divided differences; on Newton's method for solving equations numerically; on trigonometric interpolation: and on the solution of difference equations, which includes some of his developments of symbolic methods.

In 1840 he wrote on divided differences and in doing so introduced a set of functions he attributed to Ampère who called them "interpolatory functions."[22] He considers a function f defined on the set of points $a, b, c, \ldots, h, k$, and the functions

$$f(a, b) = \frac{f(a) - f(b)}{a - b}, \qquad f(a, b, c) = \frac{f(a, b) - f(a, c)}{(b - c)}, \ldots .$$

Cauchy proves the usual results, but then considers the case when certain values of the independent variables approach each other. Thus he shows that

$$f(x, x) = f'(x), \qquad f(x, x, x) = \frac{f''(x)}{1 \cdot 2}, \qquad f(x, x, x, x) = \frac{f'''(x)}{1 \cdot 2 \cdot 3}, \ldots .$$

He then proves several results leading up to an estimate of the error caused by truncating a finite difference series. Thus if the values $a, b, c, \ldots, h, k$ are equally spaced and written as $a, a + h, a + 2h, \ldots, a + (n - 1)h, a + nh$, Cauchy shows that

$$f(x) = f(a) + (x - a)\frac{\Delta f(a)}{h} + \frac{(x - a)(x - a - h)}{1 \cdot 2}\frac{\Delta^2 f(a)}{h^2} + \cdots$$
$$+ \frac{(x - a)(x - a - h) \cdots [x - a - (n - 1)h]}{1 \cdot 2 \cdot 3 \cdots n} f^{(n)}(u),$$

where "the letter u represents a mean between the quantities $a, a + nh, x$."[23]

In a short note to his *Cours d'Analyse*, Cauchy took up Lagrange's interpolation formula and generalized it significantly.[24] Cauchy first derives Lagrange's well-known result and then considers a rational function

$$u = \frac{a + bx + cx^2 + \cdots + hx^{n-1}}{\alpha + \beta x + \gamma x^2 + \cdots + \theta x^m},$$

whose coefficients $a, b, c, \ldots, h, \alpha, \beta, \gamma, \ldots, \theta$ or rather

$$\frac{a}{\alpha}, \quad \frac{b}{\alpha}, \quad \frac{c}{\alpha}, \ldots, \frac{h}{\alpha}, \quad \frac{\beta}{\alpha}, \quad \frac{\gamma}{\alpha}, \ldots, \frac{\theta}{\alpha}$$

are to be determined by knowing $n + m$ values $u_0, u_1, u_2, \ldots, u_{n+m-1}$ of the

[22] Cauchy 1, V [1840], pp. 409–424. Ampère's work is said by Cauchy to appear in *Annales* de M. Gergonne (1826) [*Annales de Mathématiques pures et appliquées*].
[23] Cauchy 1, V [1840], p. 423. He points out here that when one sets $h = 0$, one finds Taylor's series with a remainder term.
[24] Cauchy 2, III [1821], pp. 429–433. The volume of his works in question contains his *Cours d'Analyse*.

function u corresponding to the values $x_0, x_1, x_2, \ldots, x_{n+m-1}$ of x. He shows that the solution is unique and exhibits it in the form

$$\frac{u_0u_1\cdots u_m \dfrac{(x-x_{m+1})(x-x_{m+2})\cdots(x-x_{m+n-1})}{(x_0-x_{m+1})\cdots(x_0-x_{m+n-1})\cdots(x_m-x_{m+1})\cdots(x_m-x_{m+n-1})}+\cdots}{u_0u_1\cdots u_{m-1}\dfrac{(x_0-x)(x_1-x)\cdots(x_{m-1}-x)}{(x_0-x_m)\cdots(x_0-x_{m+n-1})\cdots(x_{m-1}-x_m)\cdots(x_{m-1}-x_{m+n-1})}+\cdots}.$$

In the case where $m = 0$ the denominator is to be replaced by unity; and in the case $n = 1$ the numerator by the product $u_0u_1\cdots u_m$. Thus, e.g., for $m = 0$ his formula becomes

$$u = u_0\frac{(x - x_1)(x - x_2)\cdots(x - x_{n-1})}{(x_0 - x_1)(x_0 - x_2)\cdots(x_0 - x_{n-1})} + \cdots.$$ [25]

Cauchy also derived a result on trigonometric interpolation analogous to those of Gauss and Hermite. (Cf. the relations (4.61) and (4.61′) above.) Cauchy's result is

$$f(t) = \frac{\sin((t - t_2)/2)\sin((t - t_3)/2)\cdots\sin((t - t_n)/2)}{\sin((t_1 - t_2)/2)\sin((t_1 - t_3)/2)\cdots\sin((t_1 - t_n)/2)}f(t_1) + \cdots$$
$$+ \frac{\sin((t - t_1)/2)\sin((t - t_2)/2)\cdots\sin((t - t_{n-1})/2)}{\sin((t_n - t_1)/2)\sin((t_n - t_2)/2)\cdots\sin((t_n - t_{n-1})/2)}f(t_n).$$ [26]

5.6. Cauchy on the Newton–Raphson Method

Cauchy was perhaps the first to formulate conditions under which Newton's method for finding the root of a numerical equation converges. He wrote several papers on the topic.[27] To see what he did let us express a function f as a finite Taylor's series in the form

$$f(a + i) = f(a) + \frac{i}{1}f'(a) + \cdots + \frac{i^{n-1}}{1\cdot 2\cdot 3\cdots(n - 1)}f^{(n-1)}(a)$$
$$+ \frac{i^n}{1\cdot 2\cdot 3\cdots n}[f^{(n)}(a) + I],$$

[25] An elegant study of the Cauchy interpolation formula was made by Jacobi III [1846′], pp. 479–511. The paper is followed by G. Rosenhain [1846], pp. 157–165, on this subject. Rosenhain uses Cauchy's result to find conditions that $f = 0$, $\varphi = 0$ have a common root by considering the quotient f/φ when f and φ are polynomials. It is also relevant to mention the work of Padé on rational approximation, particularly since it is of interest today. Padé [1892], pp. 1–93. We need also to mention Thiele's important work on his reciprocal differences. Cf. Thiele [1906] or Nörlund [1924], pp. 415–438.

[26] Cauchy 1, VI [1841], p. 71. This is the same as Gauss's result (4.61) above.

[27] Cauchy 1, V [1840′], pp. 455–473; 2, III [1821′], pp. 378–425; and 2, IV [1829], pp. 573–609. (In Série 1, Vol. V of his *Oeuvres*, pp. 431–493, there are a number of papers by Cauchy on numerical methods. None is very important. They all are from the *Comptes Rendus*, Vol. XI (1840).)

where "the modulus of I is very small." Now if $a + i$ is a root of $f(x) = 0$, and if he neglects I relative to $f^{(n)}(a)$, then Cauchy can write out the equation

$$f(a) + \frac{i}{1} f'(a) + \frac{i^2}{1 \cdot 2} f''(a) + \cdots + \frac{i^{n-1}}{1 \cdot 2 \cdot 3 \cdots (n-1)} f^{(n-1)}(a) + \frac{i^n}{1 \cdot 2 \cdot 3 \cdots n} f^{(n)}(a) = 0;$$

he particularly remarks on the cases $n = 1, 2$, and 3. He then states the following theorem which serves to show the state of analysis in this period. "If f is continuous on the interval from $x = a$ to $x = a + 2i$, and has at those end-points values of opposite sign, and if $f'(x)$ does not change sign on the interval, then the equation $f(x) = 0$ admits one and only one real root in the interval."

Cauchy argues that $f(x)$ must increase or decrease steadily from $x = a$ to $x = a + 2i$, since $f'(x)$ does not change sign. He says moreover that f must take on all values intermediate between $f(a)$ and $f(a + 2i)$ and hence zero. Next he states a theorem involving f''. It is this: choose a number B at least equal to the largest value assumed by $|f''(x)|$ on the interval $[a, a + 2i]$. Now if $f'(a) > 2Bi$ where

$$i = -\frac{f(a)}{f'(a)},$$

then $f(x) = 0$ has one and only one real root in the interval. His proof again uses a Taylor expansion. He sets $x = a + i + z$ and has

$$\begin{aligned} f(x) &= f(a) + (i + z)f'(a) + \frac{(i + z)^2}{2} f''[a + \theta(i + z)] \\ &= zf'(a) + \frac{(i + z)^2}{2} f''[a + \theta(i + z)], \end{aligned}$$

$$f'(x) = f'(a) + (i + z)f''[a + \Theta(i + z)],$$

where "θ, Θ designate numbers less than 1." He then argues that $f'(x)$ has the same sign as $f'(a)$ for all z between $z = -i$, $z = +i$. Moreover, at $x = a$, $a + 2i$, f has the values $-if'(a)$ and $i[f'(a) + 2if''(a + 2\theta i)]$, which are of opposite sign. Hence the first theorem applies.

Cauchy then proceeds to discuss a result of Fourier [1818] on the rate of convergence of Newton's method. He remarks that the number of decimals at least doubles with each iteration. He next takes up the case where

$$f(a) + if'(a) + \frac{i^2}{2} f''(a) = 0$$

is solved for i and works through the results. He then analyses the case where

$$f(a) + if'(a) + \frac{i^2}{2} f''(a) + \frac{i^3}{6} f'''(a) = 0.$$

It is not without interest to state one of his lemmas as follows: if $\alpha, \beta, \gamma, \delta$ are four real numbers, then the sum $\alpha\gamma + \beta\delta$ always lies between the limits

$$-(\alpha^2 + \beta^2)^{1/2}(\gamma^2 + \delta^2)^{1/2}, \qquad (\alpha^2 + \beta^2)^{1/2}(\gamma^2 + \delta^2)^{1/2}.$$

This is, of course, the Cauchy–Schwarz inequality. As a scholium to the lemma Cauchy states the general result that the expression

$$\alpha\alpha_1 + \beta\beta_1 + \gamma\gamma_1 + \cdots$$

is in absolute value less than or equal to

$$(\alpha^2 + \beta^2 + \gamma^2 + \cdots)^{1/2}({\alpha_1}^2 + {\beta_1}^2 + {\gamma_1}^2 + \cdots)^{1/2}$$

for all real $\alpha, \beta, \gamma, \ldots, \alpha_1, \beta_1, \gamma_1, \ldots$.[28]

With the help of his inequality he next takes up the case where f is a complex-valued function and carries through the analysis of Newton's method analogously to the real-valued case. He closes the paper with a numerical example: $x^5 + 10x - 1 = 0$, which has no negative roots, one positive root and four imaginary ones; he proceeds to find them numerically.

5.7. Cauchy on Operational Methods

Cauchy made very impressive and skillful use of operational methods for a variety of purposes, including the solutions of total and partial differential and difference equations with constant coefficients. In particular, he considers polynomial functions of the operators $Df = df/dx$ and $\Delta f = f(x + h) - f(x)$.[29] Thus he has

$$\begin{aligned} F(D)e^{rx} &= e^{rx}F(r), \\ F(\Delta)e^{rx} &= e^{rx}F(e^{rh} - 1), \\ F(D, \Delta)e^{rx} &= e^{rx}F(r, e^{rh} - 1), \end{aligned}$$

where the functions F are polynomials and r is real or complex. On the left-hand side, of course, they are to be viewed as operators acting on the exponential whereas on the right-hand side they are numerical-valued. Then it is not hard to see that

$$\begin{aligned} F(D)[e^{rx}f(x)] &= e^{rx}F(r + D)f(x), \\ F(\Delta)[e^{rx}f(x)] &= e^{rx}F[e^{rh}(1 + \Delta) - 1]f(x), \\ F(D, \Delta)[e^{rx}f(x)] &= e^{rx}F[r + D, e^{rh}(1 + \Delta) - 1]f(x). \end{aligned}$$

Moreover, out of the result on Fourier integrals

$$f(x) = \frac{1}{2\pi}\int_{-\infty}^{\infty}\int_{-\infty}^{\infty} e^{\alpha(x-\lambda)\sqrt{-1}}f(\lambda)\,d\alpha\,d\lambda, \tag{5.11}$$

---

[28] Cauchy 2, IV [1829], p. 589.

[29] Cauchy 2, VII [1827], pp. 198–235 and 2, VII [1827′], pp. 236–254.

Cauchy is able to deduce the following integral relations:

$$F(D)f(x) = \frac{1}{2\pi}\int_{-\infty}^{\infty}\int_{-\infty}^{\infty} e^{\alpha(x-\lambda)\sqrt{-1}}F(\alpha\sqrt{-1})f(\lambda)\,d\alpha\,d\lambda,$$

$$F(\Delta)f(x) = \frac{1}{2\pi}\int_{-\infty}^{\infty}\int_{-\infty}^{\infty} e^{\alpha(x-\lambda)\sqrt{-1}}F(e^{h\alpha\sqrt{-1}} - 1)f(\lambda)\,d\alpha\,d\lambda,$$

$$F(D,\Delta)f(x) = \frac{1}{2\pi}\int_{-\infty}^{\infty}\int_{-\infty}^{\infty} e^{\alpha(x-\lambda)\sqrt{-1}}F(\alpha\sqrt{-1}, e^{h\alpha\sqrt{-1}} - 1)f(\lambda)\,d\alpha\,d\lambda.$$

In the case where φ is an "arbitrary" function he now defines a new function of x using a somewhat peculiar operator notation,

$$\varphi(\alpha)f(\bar{x}) = \frac{1}{2\pi}\int_{-\infty}^{\infty}\int_{-\infty}^{\infty} e^{\alpha(x-\lambda)\sqrt{-1}}\varphi(\alpha)f(\lambda)\,d\alpha\,d\lambda,$$

and then finds

$$\varphi(\alpha)e^{a\bar{x}\sqrt{-1}} = \frac{1}{2\pi}\int_{-\infty}^{\infty}\int_{-\infty}^{\infty} e^{\alpha(x-\lambda)\sqrt{-1}}e^{a\lambda\sqrt{-1}}\varphi(\alpha)\,d\alpha\,d\lambda$$

with the help of the relation (5.11) above. (Note that $\bar{x}$ does not mean x conjugate.) Thus he has shown that

$$\varphi(\alpha)e^{a\bar{x}\sqrt{-1}} = e^{ax\sqrt{-1}}\varphi(\alpha).$$

He now assumes that $a, b, c, \ldots$ are real constants; $\varphi(\alpha)$, $\psi(\alpha)$ are "arbitrary" functions of α; and $A, B, C, \ldots$ are real or imaginary coefficients. Then he has

$$\begin{aligned}&\varphi(\alpha)[Ae^{a\bar{x}\sqrt{-1}} + Be^{b\bar{x}\sqrt{-1}} + Ce^{c\bar{x}\sqrt{-1}} + \cdots]\\ &= Ae^{ax\sqrt{-1}}\varphi(a) + Be^{bx\sqrt{-1}}\varphi(b) + Ce^{cx\sqrt{-1}}\varphi(c) + \cdots,\end{aligned}$$

and consequently

$$\varphi(\alpha)\sum e^{a\bar{x}\sqrt{-1}}\psi(a) = \sum e^{ax\sqrt{-1}}\varphi(a)\cdot\psi(a).$$

He chooses two distinct real numbers r_0, R and concludes from this that

$$\varphi(\alpha)\int_{r_0}^{R} e^{r\bar{x}\sqrt{-1}}\psi(r)\,dr = \int_{r_0}^{R} e^{rx\sqrt{-1}}\varphi(r)\psi(r)\,dr.$$

Next he takes $r_0 = -\infty$, $R = \infty$. After a little manipulation he concludes that

$$\begin{aligned}F(D)f(x) &= F(\alpha\sqrt{-1})f(\bar{x}),\\ F(\Delta)f(x) &= F(e^{h\alpha\sqrt{-1}} - 1)f(\bar{x}),\\ F(D,\Delta)f(x) &= F(\alpha\sqrt{-1}, e^{h\alpha\sqrt{-1}} - 1)f(\bar{x}).\end{aligned}$$

Cauchy next applies these results to the integration of differential, difference, and mixed equations. Thus, e.g., he integrates the first order linear equation with constant coefficients $(\Delta - r)y = f(x)$ and finds that

$$y = (1 + r)^{(x/h)-1}[\textstyle\sum (1 + r)^{-(x/h)}f(x) + \omega(x)],$$

where $h = \Delta x$ and $\omega(x)$ is periodic (period h) but otherwise arbitrary. In general for F polynomial he writes the equation $F(\Delta)y = f(x)$, as

$$(\Delta - r_1)(\Delta - r_2)\cdots(\Delta - r_n)y = \frac{f(x)}{a_0},$$

where a_0 is the leading coefficient in F. To solve this he proceeds inductively:

$$\begin{aligned}(\Delta - r_1)y_{n-1} &= \frac{f(x)}{a_0},\\ (\Delta - r_2)y_{n-2} &= y_{n-1},\\ &\vdots\\ (\Delta - r_{n-1})y_1 &= y_2,\\ (\Delta - r_n)y &= y_1.\end{aligned}$$

Thus, e.g., the equation $\Delta^2 y - 2\Delta y + y = f(x)$ can be written as $\Delta z - z = f(x)$, $\Delta y - y = z$, He has then

$$z = 2^{(x/h)-1}\sum 2^{-(x/h)}f(x), \qquad y = 2^{(x/h)-1}\sum 2^{-(x/h)}z.$$

To simplify this result Cauchy notes that

$$\sum u\Delta v = uv - \sum (v + \Delta v)\,\Delta u,$$

and accordingly

$$\sum\left[\sum 2^{-(x/h)}f(x)\right] = \frac{x}{h}\sum 2^{-(x/h)}f(x) - \sum\frac{x+h}{h}2^{-(x/h)}f(x);$$

he then writes his solution as

$$y = 2^{(x/h)-2}\left\{\frac{x}{h}\left[\varpi_1(x) + \sum 2^{-(x/h)}f(x)\right] - \left[\varpi_2(x) + \sum\frac{x+h}{h}2^{-(x/h)}f(x)\right]\right\}$$

In the *Addition* to this paper Cauchy proceeds rather more boldly.[30] Thus, e.g., he first notes that the differential equation

$$Dy - ry = \frac{dy}{dx} - ry = f(x)$$

has $l(y) = rx +$ const. when $f(x) \equiv 0$ and hence

$$y = \mathfrak{S}e^{rx},$$

where $\mathfrak{S}$ is any arbitrary constant and l means log. He now remarks that this last relation still is the solution even when $f(x) \not\equiv 0$ provided we take $\mathfrak{S}$ as a function of x. He writes $y = e^{rx}z$ and finds $Dz = e^{rx}f(x)$ so that symbolically

$$z = \frac{e^{-rx}f(x)}{D} = \int e^{-rx}f(x)\,dx.$$

[30] Cauchy 2, VII [1827′], pp. 236–254.

In the same way he first finds the solution of $\Delta y - ry = 0$ to be

$$y = (1 + r)^{x/h}\mathfrak{S};$$

and then in case $f(x) \not\equiv 0$, he writes $y = (1 + r)^{x/h}z$, where z is given symbolically by

$$z = \frac{(1 + r)^{-(x/h)-1}f(x)}{\Delta} = \sum (1 + r)^{-(x/h)-1}f(x)$$
$$= (1 + r)^{-1} \sum (1 + r)^{-(x/h)}f(x),$$

and the sum is understood to contain an arbitrary periodic function. Cauchy found this result as follows: let

$$y = (1 + r)^{x/h}z;$$

and note that

$$(\Delta - r)[(1 + r)^{x/h}z] = f(x)$$

implies that

$$(\Delta - r)[(1 + r)^{x/h}z] = (1 + r)^{(x/h)+1}\Delta z.$$

Hence he has

$$(1 + r)^{(x/h)+1}\,\Delta z = f(x), \qquad \Delta z = (1 + r)^{-(x/h)-1}f(x).$$

From this it is immediately clear that his result obtains.

At the end of his *Addition* he remarks that when x is not a multiple of h and $1 + r$ is negative, then the functions

$$(1 + r)^{(x/h)}, \qquad (1 + r)^{-(x/h)}$$

are not well-defined in his sense.[31] He replaces them by

$$(-1 - r)^{x/h}e^{\pi(x/h)\sqrt{-1}}, \qquad (-1 - r)^{-(x/h)}e^{-\pi(x/h)\sqrt{-1}}$$

whenever $(1 + r)$ is less than zero. Thus if in the difference equation above r is -2, he replaces

$$(1 + r)^{(x/h)-1}, \qquad (1 + r)^{-(x/h)}$$

by

$$e^{\pi((x/h)-1)\sqrt{-1}}, \qquad e^{-\pi(x/h)\sqrt{-1}}$$

and the general integral becomes

$$y = e^{\pi((x/h)-1)\sqrt{-1}} \sum e^{-\pi(x/h)\sqrt{-1}}f(x).$$

---

[31] Cauchy 2, VI [1826], p. 13 and 2, VII [1827′], p. 253. (The functions in question are now complex-valued and may not be single valued.)

For $f(x) = e^{ax}$, the difference equation becomes $\Delta y + 2y = e^{ax}$, and its solution is

$$y = e^{\pi((x/h)-1)\sqrt{-1}} \sum e^{(a-(\pi/h)\sqrt{-1})x}$$
$$= -e^{(\pi x/h)\sqrt{-1}} \sum e^{(a-(\pi/h)\sqrt{-1})x} = -e^{(\pi x/h)\sqrt{-1}} \left[\frac{e^{(a-(\pi/h)\sqrt{-1})x}}{e^{ah-\pi\sqrt{-1}} - 1} + \varpi(x) \right].$$

Cauchy sets

$$-e^{(\pi x/h)\sqrt{-1}}\varpi(x) = \psi(x),$$

and has

$$y = \frac{e^{ax}}{e^{ah} + 1} + \psi(x).$$

There are many other papers where Cauchy analyses both difference and differential equations with considerable finesse and it is unfortunate that they cannot be dealt with here.

5.8. Other Nineteenth Century Results

There were a considerable number of interesting and some important papers done by others besides those mentioned so far. In fact the list of these men reads like a Who's Who of mathematics; it includes Bendixson, Cayley, Hermite, Méray, Poincaré, Schwarz, Teixeira and many others.[32] However, it is not possible in this book to include these works no matter how worthy. I shall mention only the so-called Ruffini–Horner and the Dandelin, Lobachevsky, and Graeffe methods for solving algebraic equations.[33] The works of Ruffini and Horner are clearly independent, that of the latter showing no evidence of an acquaintance with the former, even though Paolo Ruffini (1765–1822) won a gold medal for his accomplishment. In any event the method is now taught as a standard way of finding roots.[34] (Ruffini also worked in an anticipatory way on Abel's famous problem in the theory of equations on the insolvability of equations of degree five or more by radicals.) The root-squaring method of Dandelin, Lobachevsky, and Graeffe also has some utility.[35] The three authors independently discovered a simple but useful process which replaces a given polynomial equation by one whose roots are high powers of the given ones. This separates the roots and makes

[32] The interested reader may profitably examine the bibliography in Nörlund [1924].
[33] Ruffini [1804], [1813] and Horner [1819].
[34] L. E. Dickson [1922], pp. 86–90 and Cajori [1911].
[35] Dandelin [1826], pp. 48ff., Lobachevsky [1834], pp. 349–359 and Graeffe [1837]. Cf. also, Encke [1888] and [1841]. In the paper by the Belgian Dandelin there is also a discussion of conditions under which Newton's method is valid (cf. p. 278 above).

it easy to solve the equation. The algorithm depends upon systematically finding an equation whose roots are the squares of the given ones.[36] In this cursory look at numerical aspects of the theory of equations it would be wrong not to mention both Jacques C. F. Sturm (1803–1855), a Swiss who taught at the Sorbonne, and Cauchy, both of whom made major contributions to the field.

5.9. Integration of Differential Equations

Let us turn now to the subject of numerical integration and particularly to the integration of total and partial differential equations. Runge and Willers in their encyclopedia article (Runge [1915], pp. 47–176) cite some preliminary work by Newton, Leibniz, Kästner, and Euler, but it is the material we discussed in Chapter III (pp. 141ff.) by Euler which is basically responsible for the present-day methods. This work of Euler was made into a viable mathematical procedure for showing the existence of a solution of a differential equation by Cauchy in lectures at the École Polytechnique.[37] Although the fact seems to have disappeared from the literature, Weierstrass independently worked out more or less the same procedure.[38]

The fundamental idea is this: let the equation be

$$\frac{dy}{dx} = f(x, y), \qquad y(x_0) = y_0;$$

let f and $\partial f/\partial y$ be real and continuous functions on a rectangle $|x - x_0| \leq a$, $|y - y_0| \leq b$, and let $x_i = x_0 + ih$. Then Cauchy defines a sequence $y_0, y_1, \ldots, y_n$ by the relations $y_{i+1} = y_i + hf(x_i, y_i)$ $(i = 0, 1, \ldots, n - 1)$. He showed that under these hypotheses the polygonal arc whose end-points are $(x_0, y_0), (x_1, y_1), \ldots$ converges to the solution of the differential equation, and he extended the proof to a system of n equations.

This argument of Cauchy was improved by Rudolf Lipschitz, who replaced Cauchy's conditions on f and $\partial f/\partial y$ by one on f: namely, there is a bound k such that $|f(x, y) - f(x, y')| < k|y - y'|$ for x, y and x, y' in the rectangle.[39]

---

[36] For a discussion of the process see Whittaker, *WR*, pp. 106–126 or Willers [1948], pp. 258–265. In some sense E. Waring [1770], pp. 39 and 53, in his *Meditationes Algebraicae*, anticipated Graeffe's method. He knew how to separate the roots of an equation, and in the latter place he shows how to find an equation whose roots are the squares of the roots of a given equation.

[37] These lectures were written up and published by F. N. M. Moigno [1840/61], Vol. 2, Leçons 26–28, 33, Paris, 1844. Cf. also Cauchy *Oeuvres*, Série 1, Vol. IV–VII and X, where he showed that the formal solution to an analytic differential equation converges in some region. This proof of Cauchy was improved by Briot and Bouquet [1855], pp. 787ff.

[38] A good discussion of the literature and proofs appears in Moulton, *DE*.

[39] Moulton, *DE*, pp. 225–231. The original paper on the Lipschitz condition appears in Lipschitz [1877], Vol. II, pp. 500ff.

This method of constructing a polygonal approximation to the solution curve for the differential equation not only provides a proof of the existence of the solution, but also gives a reasonable method for finding numerically a solution. Its main flaw, of course, is that the interval size h needs, in general, to be made quite small.

The developments in the theory of heat by Fourier and in celestial mechanics by Adams, Bessel, Cauchy, Gauss, Lagrange, Laplace, Legendre, Leverrier, Poincaré and others reached the point where it became very important to have reasonable schemes for solving differential equations numerically. Also the theory of exterior ballistics led to a considerable interest in and a need for exactly the same sort of procedures.

We may divide the methods of numerically solving differential equations into two general classes: firstly, those which are similar to the Euler–Cauchy–Lipschitz method, where the values of $y_1, y_2, \ldots$ at the points $x_1, x_2, \ldots$ are calculated *seriatim* without any iteration or correction taking place. These have sometimes been called "marching procedures" since they move forward one step at a time: secondly, a set of schemes in which successive approximations $y_i^{(j)}$ $(j = 0, 1, 2, \ldots)$ to y_i are formed at a given value x_i, the procedure being iterated at that point until some exactness criterion is satisfied.

The modern successors to the Euler–Cauchy–Lipschitz procedure are the Heun and Runge–Kutta methods.[40] Let us first examine Heun's paper.[41] His aim was to generalize the Gaussian quadrature scheme (Chapter IV, pp. 224ff. above) from the differential equation

$$\frac{dy}{dx} = f(x)$$

to the general first-order one

$$\frac{dy}{dx} = f(x, y).$$

As he remarks, the Gaussian scheme consists of determining parameters $\alpha_\nu, \varepsilon_\nu$ so that

$$y(x + \Delta x) - y(x) = \Delta y = \sum_{\nu=1}^{\nu=n} \alpha_\nu f(x + \varepsilon_\nu \Delta x)\cdot\Delta x$$

is as close as possible in some sense to the "true value" in the interval Δx. His idea is to determine parameters $\alpha_1, \alpha_2, \ldots, \alpha_n; \varepsilon_1, \varepsilon_2, \ldots, \varepsilon_n; \varepsilon_1', \varepsilon_2', \ldots, \varepsilon_n'; \varepsilon_1'', \varepsilon_2'', \ldots, \varepsilon_n''$, etc. so that the approximation

$$\Delta y = \sum_{\nu=1}^{\nu=n} \{\alpha_\nu f(x + \varepsilon_\nu \Delta x, y + \Delta' y)\}\cdot\Delta x$$

[40] Collatz [1951] or Runge, *RK*.
[41] Heun [1900], pp. 23–88.

is as good as possible, where

$$\begin{aligned}\Delta'_\nu y &= \varepsilon_\nu f(+\varepsilon'_\nu\cdot\Delta x, y + \Delta''_\nu y)\cdot\Delta x,\\ \Delta''_\nu y &= \varepsilon''_\nu f(x + \varepsilon''_\nu\cdot\Delta x, y + \Delta'''_\nu y)\cdot\Delta x,\\ &\vdots\\ \Delta^{(m)}_\nu y &= \sum_\nu^{(m-1)} f(x, y)\cdot\Delta x.\end{aligned}$$ [42]

In the case where $m = 1$ Heun has

$$\begin{aligned}\Delta y &= \sum_{\nu=1}^{\nu=n}\{\alpha_\nu f(x + \varepsilon_\nu\cdot\Delta x, y + \Delta'_\nu y)\}\,\Delta x,\\ \Delta'_\nu y &= \varepsilon_\nu f(x, y)\cdot\Delta x.\end{aligned}$$

He then expands Δy with the help of Taylor's theorem and finds that $\sum\alpha = 1$, $\sum\alpha\varepsilon = 1/2, \ldots$. These equations contain $2n$ unknowns. He then examines the cases $n = 1$ and 2 and finds for $n = 1$ that

$$\Delta y = f\left(x + \frac{1}{2}\Delta x, y + \frac{1}{2}f\cdot\Delta x\right)\cdot\Delta x.$$

For $n = 2$ he arbitrarily sets $\alpha_1 = \alpha_2$, and considers the cases $\varepsilon_1 = 0$, $\varepsilon_2 = 1$ and $\varepsilon_1 = \frac{1}{3}$, $\varepsilon_2 = \frac{2}{3}$. They yield the results

$$\Delta y = \frac{1}{2}\{f(x, y) + f(x + \Delta x, y + f\cdot\Delta x)\}\cdot\Delta x,$$

$$\Delta y = \frac{1}{2}\left\{f\left(x + \frac{1}{3}\cdot\Delta x, y + \frac{1}{3}f\cdot\Delta x\right) + f\left(x + \frac{2}{3}\Delta x, y + \frac{2}{3}f\cdot\Delta x\right)\right\}\cdot\Delta x.$$

He then proceeds to the cases when $m = 2$ and 3, as well as to the integration of simultaneous systems. His method is not so widely-known or used as is the Runge–Kutta procedure.[43]

Runge worked out the algorithm for the case of a single equation $y' = f(x, y)$, and Kutta proceeded to generalize the result to a general system of first-order equations. The method is a generalization of Simpson's Rule, as we shall see.[44] Let us use the notations and methods of Runge's 1894 paper in the *Annalen*.[45] Runge starts out by remarking that the scheme $\Delta y = f(x_0 y_0)\,\Delta x$, etc., is less accurate than the one given by

$$\Delta y = f\left(x_0 + \frac{1}{2}\Delta x, y_0 + \frac{1}{2}f(x_0 y_0)\,\Delta x\right)\Delta x, \text{ etc.}$$

[42] Heun [1900], p. 27.

[43] Runge [1895], pp. 167–178 and [1905], pp. 252–257. Cf. also, Kutta [1901], pp. 435–453. The first system of total differential equations solved by the ENIAC was integrated using Heun's method.

[44] Collatz [1951], pp. 437–439, Tables I–III. Here are collected many of the standard integration formulas in handy form.

[45] Runge [1895], pp. 167–178. This is his most primitive paper on the subject. It should be contrasted with his 1924 work in Runge–König, *RK*, pp. 286–300.

Still another scheme is the trapezoidal one

$$\Delta y = \frac{f(x_0 y_0) + f(x_0 + \Delta x, y_0 + f(x_0 y_0)\,\Delta x)}{2}\,\Delta x, \text{ etc.;}$$

and he goes on to consider the problem systematically. He notes that the "true" value of Δy is given by the relation

$$\Delta y = f\,\Delta x + (f_1 + f_2 f)\frac{\Delta x^2}{2!} + (f_{11} + 2f_{12}f + f_{22}f^2 + f_2(f_1 + f_2 f))\frac{\Delta x^3}{3!} + \cdots,$$

where f_1, f_2 are the partial derivatives f_x, f_y of the first order, and f_{11}, f_{12}, f_{22} are the corresponding ones of the second order, etc. The three special cases above clearly correspond to the approximations for Δy

$$\begin{aligned} &f\,\Delta x, \\ &f\,\Delta x + (f_1 + f_2 f)\frac{\Delta x^2}{2} + (f_{11} + 2f_{12}f + f_{22}f^2)\frac{\Delta x^3}{8} + \cdots, \\ &f\,\Delta x + (f_1 + f_2 f)\frac{\Delta x^2}{2} + (f_{11} + 2f_{12}f + f_{22}f^2)\frac{\Delta x^3}{4} + \cdots \quad . \end{aligned} \tag{5.12}$$

Runge then proceeds by analogy to Simpson's Rule, combining his last two results, and writes

$$f\,\Delta x + (f_1 + f_2 f)\frac{\Delta x^2}{2} + (f_{11} + 2f_{12}f + f_{22}f^2)\frac{\Delta x^3}{6} + \cdots \quad .$$

However he intends to attack the problem in a slightly different way. He forms

$$\frac{\Delta' y + \Delta''' y}{2},$$

where $\Delta' y = f(x_0 y_0)\,\Delta x$, and $\Delta'' y$, $\Delta''' y$ are given by the relations

$$\Delta'' y = f(x_0 + \Delta x, y_0 + \Delta' y)\,\Delta x, \qquad \Delta''' y = f(x_0 + \Delta x, y_0 + \Delta'' y)\,\Delta x.$$

This gives as the value of Δy

$$f\Delta x + (f_1 + f_2 f)\frac{\Delta x^2}{2} + (f_{11} + 2f_{12}f + f_{22}f^2 + 2f_2(f_1 + f_2 f))\frac{\Delta x^3}{4} + \cdots \quad .$$

The difference between this and the second of the relations (5.12) is, evidently,

$$\left[\frac{1}{8}(f_1 + 2f_{12}f + f_{22}f^2) + \frac{1}{2}f_2(f_1 + f_2 f)\right]\Delta x^3 + \cdots \quad .$$

Runge then proposes to add a third of this difference to the second of the relations (5.12). This gives him

$$f\Delta x + (f_1 + f_2 f)\frac{\Delta x^2}{2} + (f_{11} + 2f_{12}f + f_{22}f^2 + f_2(f_1 + f_2 f))\frac{\Delta x^3}{6} + \cdots,$$

which he says agrees at least through terms of order three with the true value.

Kutta in his 1901 paper in the *Zeitschrift* picked up Runge's and Heun's ideas and systematically merged them.[46] He forms the expressions

$$\begin{aligned}
\Delta' &= f(x, y)\,\Delta x,\\
\Delta'' &= f(x + \kappa\,\Delta x, y + \kappa\,\Delta')\,\Delta x,\\
\Delta''' &= f(x + \lambda\,\Delta x; y + \rho\,\Delta'' + (\lambda - \rho)\,\Delta')\,\Delta x,\\
\Delta'''' &= f(x + \mu\,\Delta x; y + \rho\,\Delta''' + \tau\,\Delta'' + (\mu - \sigma - \tau)\,\Delta')\,\Delta x,\\
\Delta^{\mathrm{V}} &= f(x + \nu\,\Delta x, y + \varphi\,\Delta'''' + \chi\,\Delta''' + \psi\,\Delta'' + (\nu - \varphi - \chi - \psi)\,\Delta')\,\Delta x,\\
&\vdots \qquad\qquad \vdots
\end{aligned}$$

and then sets

$$\Delta y = a\Delta' + b\Delta'' + c\Delta''' + d\Delta'''' + e\Delta^{\mathrm{V}} + \cdots,$$

where he views $\kappa, \lambda, \mu, \nu, \ldots; \rho, \tau, \varphi, \chi, \psi, \ldots; a, b, c, d, e, \ldots$ as parameters to be determined so that the Taylor's expansion of y will agree up to any desired order. Thus for $a = 1$ Kutta has $\Delta y = f(x, y)\,\Delta x$, the Euler–Cauchy–Lipschitz case which gives first-order agreement. For $a + b = 1$, $b\kappa = 1/2$, he has one of Heun's cases which gives second-order agreement. For

$$a + b + c = 1, \qquad b\kappa + c\lambda = \frac{1}{2}, \qquad b\kappa^2 + c\lambda^2 = \frac{1}{3}, \qquad c\rho\kappa = \frac{1}{6},$$

with

$$\rho = \frac{\lambda(\lambda - \kappa)}{\kappa(2 - 3\kappa)}, \qquad a = \frac{6\kappa\lambda - 3(\kappa + \lambda) + 2}{6\kappa\lambda},$$

$$b = \frac{2 - 3\lambda}{6\kappa(\kappa - \lambda)}, \qquad c = \frac{2 - 3\kappa}{6\lambda(\lambda - \kappa)},$$

he has third-order agreement. In this case Kutta remarks that there is a two-parameter family of solutions — evidently κ, λ are the free parameters. When he sets $\rho = \lambda$ he has

$$\text{(I)} \qquad \begin{aligned} \lambda &= 3\kappa(1 - \kappa), \qquad & a &= \frac{2 - 12\kappa + 27\kappa^2 - 18\kappa^3}{18\kappa^2(1 - \kappa)},\\ b &= \frac{3\kappa - 1}{6\kappa^2}, & c &= \frac{1}{18\kappa^2(1 - \kappa)}, \end{aligned}$$

which is another one of Heun's cases.

---

[46] Kutta [1901], p. 437.

He then takes up some special cases, namely:

$$\text{(II)} \qquad \kappa = \frac{2}{3}, \qquad \lambda = 0, \qquad b = \frac{3}{4}, \qquad c = \frac{1}{4\rho}, \qquad a = \frac{1}{4} - \frac{1}{4\rho};$$

$$\text{(III)} \qquad \kappa = \frac{2}{3}, \qquad \lambda = \frac{2}{3}, \qquad a = \frac{1}{4}, \qquad c = \frac{1}{4\rho}, \qquad b = \frac{3}{4} - \frac{1}{4\rho};$$

$$\text{(IV)} \qquad \lambda = \frac{2}{3}, \qquad a = \frac{1}{4}, \qquad b = 0, \qquad c = \frac{3}{4}, \qquad \rho = \frac{2}{9\kappa};$$

$$\text{(V)} \qquad \lambda = \frac{3\kappa - 2}{3(2\kappa - 1)}, \qquad a = 0.$$

In case (III) Kutta sets $\rho = \frac{2}{3}$ and has $a = \frac{1}{4}$, $b = \frac{3}{8}$, $c = \frac{3}{8}$ and

$$\begin{aligned}
\Delta y &= \frac{2\Delta' + 3\Delta'' + 3\Delta'''}{8}, \\
\Delta' &= f(x, y)\,\Delta x, \\
\Delta'' &= f\left(x + \frac{2}{3}\Delta x,\, y + \frac{2}{3}\Delta'\right)\Delta x, \\
\Delta''' &= f\left(x + \frac{2}{3}\Delta x,\, y + \frac{2}{3}\Delta''\right)\Delta x.
\end{aligned}$$

When he sets $\kappa = \frac{1}{3}$, $\lambda = \frac{2}{3}$, $\rho = \frac{2}{3}$ in cases (I) or (IV) he has Heun's relations

$$\begin{aligned}
\Delta y &= \frac{\Delta' + 3\Delta'''}{4}, \\
\Delta' &= f(x, y)\,\Delta x, \\
\Delta'' &= f\left(x + \frac{1}{3}\Delta x,\, y + \frac{1}{3}\Delta'\right)\Delta x, \\
\Delta''' &= f\left(x + \frac{2}{3}\Delta x,\, y + \frac{2}{3}\Delta''\right)\Delta x.
\end{aligned}$$

He also works out the cases $\kappa = \frac{2}{3}$, $\lambda = \frac{2}{3}$, $\rho = \frac{1}{3}$ and $\kappa = \frac{1}{2}$, $\lambda = 1$. Now, to obtain an approximation good through the fourth order Kutta finds the relations

$$\begin{aligned}
a + b + c + d &= 1, \\
b\kappa + c\lambda + d\mu &= \frac{1}{2}, \\
b\kappa^2 + c\lambda^2 + d\mu^2 &= \frac{1}{3}, \\
c\rho\kappa + d(\sigma\lambda + \tau\kappa) &= \frac{1}{6}, \\
b\kappa^3 + c\lambda^3 + d\mu^3 &= \frac{1}{4},
\end{aligned}$$

$$c\rho\kappa\lambda + d(\sigma\lambda + \tau\kappa)\mu = \frac{1}{8},$$

$$c\rho\kappa^2 + d(\sigma\lambda^2 + \tau\kappa^2) = \frac{1}{12},$$

$$d\rho\sigma\kappa = \frac{1}{24}.$$

In general, these equations may be solved in a number of different ways for the parameters in terms of κ, λ. He finds, e.g., that a solution is

$$a = 1 - b - c - d, \qquad b = \frac{1 - 2\lambda}{12\kappa(\kappa - \lambda)(1 - \kappa)},$$

$$c = \frac{1 - 2\kappa}{12\lambda(\lambda - \kappa)(1 - \lambda)}, \qquad d = \frac{6\kappa\lambda - 4(\kappa + \lambda) + 3}{12(1 - \lambda)(1 - \kappa)};$$

$$\mu = 1, \qquad \rho = \frac{\lambda(\lambda - \kappa)}{2\kappa(1 - 2\kappa)}, \qquad \rho = \frac{1}{24\kappa\rho d}, \qquad \tau = \frac{1}{6\kappa d} - \frac{\lambda\sigma}{\kappa} - \frac{c\rho}{d}.$$

Kutta, following Runge, asks, however, for a symmetrical solution, i.e., one for which $a = d$, $b = c$. This yields the relations

$$\lambda = 1 - \kappa, \qquad \mu = 1, \qquad \rho = \frac{1 - \kappa}{2\kappa}, \qquad \sigma = \frac{\kappa}{6\kappa(1 - \kappa) - 1};$$

$$\tau = \frac{(1 - \kappa)(2\kappa - 1)}{2\kappa[6\kappa[1 - \kappa) - 1]}, \qquad a = d = \frac{6\kappa(1 - \kappa) - 1}{12\kappa(1 - \kappa)}, \qquad b = c = \frac{1}{12\kappa(1 - \kappa)}.$$

If $\kappa = \frac{1}{3}$, $\lambda = \frac{2}{3}$, then the integration scheme becomes

$$\Delta y = \frac{\Delta' + 3\Delta'' + 3\Delta''' + \Delta''''}{8},$$

$$\Delta' = f(x, y)\,\Delta x,$$

$$\Delta'' = f\left(x + \frac{\Delta x}{3}, y + \frac{1}{3}\Delta'\right)\Delta x,$$

$$\Delta''' = f\left(x + \frac{2}{3}\Delta x, y + \frac{3\Delta'' - \Delta'}{3}\right)\Delta x,$$

$$\Delta'''' = f(x + \Delta x, y + \Delta''' - \Delta'' + \Delta')\,\Delta x.$$

Consider now with Kutta the generalization of Simpson's Rule. He takes up four possible generalizations of which one is this: $\lambda = \frac{1}{2}$. Then still another solution is $\mu = 1$; $a = \frac{1}{6}$, $b = 0$, $c = \frac{2}{3}$, $d = \frac{1}{6}$; $\sigma = 2$, $\rho = \frac{1}{8}\kappa$,

$\tau = -\frac{1}{2}\kappa$. Hence for $\kappa = \frac{1}{4}$,

$$\Delta y = \frac{\Delta' + 4\Delta''' + \Delta''''}{6},$$

$$\Delta' = f(x, y)\,\Delta x,$$

$$\Delta'' = f\left(x + \frac{\Delta x}{4}, y + \frac{\Delta'}{4}\right)\Delta x,$$

$$\Delta''' = f\left(x + \frac{\Delta x}{2}, y + \frac{\Delta''}{2}\right)\Delta x,$$

$$\Delta'''' = f(x + \Delta x, y + 2\Delta''' - 2\Delta'' + \Delta')\,\Delta x.$$

He next considers three other special cases of the Simpson's Rule which we do not have space for here.[47]

In Section IV of his paper Kutta sets up the apparatus for fifth-order approximation and works some cases through in detail. These do not seem to have survived in the current literature. In Section V he again takes up a generalized Simpson's rule which he shows is nearly good through the fifth order. It is this:

$$\Delta y = \frac{\Delta' + 2\Delta'' + 2\Delta''' + \Delta''''}{6},$$

$$\Delta' = f(x, y)\,\Delta x,$$

$$\Delta'' = f\left(x + \frac{\Delta x}{2}, y + \frac{\Delta'}{2}\right)\Delta x,$$

$$\Delta''' = f\left(x + \frac{\Delta x}{2}, y + \frac{\Delta''}{2}\right)\Delta x,$$

$$\Delta'''' = f(x + \Delta x, y + \Delta''')\,\Delta x.$$

The paper concludes with a summary by Kutta which is worth repeating. The first he calls Euler's method:

$$\Delta' = f(x, y)\cdot\frac{\Delta x}{4};$$

$$\Delta'' = f\left(x + \frac{\Delta x}{4}, y + \Delta'\right)\cdot\frac{\Delta x}{4};$$

$$\Delta''' = f\left(x + \frac{\Delta x}{2}, y + \Delta'' + \Delta'\right)\cdot\frac{\Delta x}{4};$$

$$\Delta'''' = f\left(x + \frac{3\Delta x}{4}, y + \Delta''' + \Delta'' + \Delta'\right)\cdot\frac{\Delta x}{4};$$

$$\Delta y = \Delta' + \Delta'' + \Delta''' + \Delta''''.$$

[47] Kutta [1901], pp. 442–443.

The second, Runge's method:

$$\begin{aligned}
\Delta' &= f(x, y)\,\Delta x;\\
\Delta'' &= f\left(x + \frac{\Delta x}{2}, y + \frac{\Delta'}{2}\right)\Delta x;\\
\Delta''' &= f(x + \Delta x, y + \Delta')\,\Delta x;\\
\Delta'''' &= f(x + \Delta x, y + \Delta''')\,\Delta x;\\
\Delta y &= \frac{\Delta' + 4\Delta'' + \Delta''''}{6}.
\end{aligned}$$

The third, Heun's method:

$$\begin{aligned}
\Delta' &= f(x, y)\,\Delta x;\\
\Delta'' &= f\left(x + \frac{\Delta x}{3}, y + \frac{\Delta'}{3}\right)\Delta x;\\
\Delta''' &= f\left(x + \frac{2\Delta x}{3}, y + \frac{2\Delta''}{3}\right)\Delta x;\\
\Delta y &= \frac{\Delta' + 3\Delta'''}{4}.
\end{aligned}$$

The fourth, the Kutta method:

$$\begin{aligned}
\Delta' &= f(x, y)\,\Delta x;\\
\Delta'' &= f\left(x + \frac{\Delta x}{3}, y + \frac{\Delta'}{3}\right)\Delta x;\\
\Delta''' &= f\left(x + \frac{2\Delta x}{3}, y + \Delta'' - \frac{\Delta'}{3}\right)\Delta x;\\
\Delta'''' &= f(x + \Delta x, y + \Delta''' - \Delta'' + \Delta)\,\Delta x;\\
\Delta y &= \frac{\Delta' + 3\Delta'' + 3\Delta''' + \Delta''''}{8}.
\end{aligned}$$

(Evidently what Kutta refers to as his rule is a generalization of the so-called Three-eighth rule.) Kutta compares the accuracy of these various methods on the differential equation

$$\frac{dy}{dx} = \frac{y - x}{y + x}, \qquad y(0) = 1,$$

whose solution is known to be (in his notation)

$$\text{lg nat}\,(x^2 + y^2) - 2\,\text{arc tg}\,\frac{x}{y} = 0.$$[48]

[48] Collatz [1951], pp. 26–34 or Runge, *RK*, pp. 286–300. In this latter place a number of schemes are worked through in some detail, but the one below appears on p. 294 and Runge remarks that it is "the generalization of Simpson's rule."

Kutta concludes from a numerical calculation that his result is the most accurate of the four given. Today the Runge–Kutta method is usually given as

$$
\begin{aligned}
\Delta' &= f(x, y)\,\Delta x,\\
\Delta'' &= f\left(x + \frac{\Delta x}{2}, y + \frac{\Delta'}{2}\right)\Delta x,\\
\Delta''' &= f\left(x + \frac{\Delta x}{2}, y + \frac{\Delta''}{2}\right)\Delta x,\\
\Delta'''' &= f(x + \Delta x, y + \Delta''')\,\Delta x,\\
\Delta y &= \frac{\Delta' + 2\Delta'' + 2\Delta''' + \Delta''''}{6}.
\end{aligned}
$$

5.10. Successive Approximation Methods

The solution of differential equations by successive approximations is also a nineteenth-century development. Cauchy considered the second-order linear equations

$$\frac{d^2y}{dx^2} = X(x)y,$$

$$y(x_0) = y_0, \qquad y'(x_0) = y'_0,$$

and used an iterative procedure to solve it.[49] To this end he considered the system

$$\frac{d^2y_{n+1}}{dx^2} = X(x)y_n,$$

$$y_n(x_0) = y_0, \qquad y'_n(x_0) = y'_0.$$

This method was picked up by Picard who carried through an interesting analysis both for systems of total differential equations and also for Poisson-type partial differential equations.[50] Picard's method was restricted to showing the existence of a solution very near to $x = x_0$. Bendixson and Lindelöf showed the true extent of the interval of convergence.[51] Out of these methods of successive approximations grew practical numerical schemes for carrying out the integration of total differential equations. One of the earliest and best

[49] Moigno [1840/61], Vol. 2, (1844), pp. 702–707.

[50] Picard [1890], pp. 145–210, [1890′], p. 231; and [1891], Vol. II, pp. 340–346. Cf. also, Moulton, *DE*, pp. 182–197.

[51] Bendixson [1893], pp. 599–612 and Lindelöf [1894], pp. 117–128.

known is called Adams' method. It was devised by John Couch Adams, the codiscoverer of Neptune, in an investigation of capillary action.[52]

Adams was not only a remarkable astronomer: he was a computer *par excellence*. Thus, e.g., he calculated Euler's constant to 515 places. He also determined the first 62 Bernoulli numbers to 110 places each. These appear in Adams I [1878′] and I [1877].

Before we describe this method, consider with Adams the differential equation

$$\frac{dy}{dt} = q = f(y, t), \qquad y(0) = y_0.$$

In general we can express y as a power series about $x = x_0$ with the help of the given equation, since

$$y'(0) = q_0 = f(y_0, 0),$$
$$y''(0) = \left(\frac{dq}{dt}\right)_0 = \left(\frac{\partial f}{\partial t}\right)_0 + \left(\frac{\partial f}{\partial y}\right)_0 q_0 + \cdots \quad .$$

But usually the series will only converge in a small neighborhood of x_0, or one must evaluate prohibitively many terms in the series. This method, which has been used since early times, is therefore of limited utility. A more practical procedure is a form of analytic continuation: one finds the solution in a small neighborhood of x_0, then reevaluates the solution and its derivatives or differences at a point near the boundary of this neighborhood. This gives a new expansion of the solution that, one may hope, carries the solution outside the original neighborhood. Successive application of the procedure carries the solution through to its natural boundary. This scheme in various forms is what is often used. Let us see a few examples.

The algorithm which is sometimes styled the Adams–Bashforth method (incorrectly, since it seems to be due to Adams alone) consists of several parts: firstly a unique procedure to get started and secondly a way to proceed *seriatim* from the initial values obtained by the first procedure. Let us consider with Adams the differential equation

$$\frac{dy}{dt} = q = f(y, t),$$

and let us assume that we have found the values $\ldots y_{-4}, y_{-3}, y_{-2}, y_{-1}, y_0$ corresponding to the quantities $\ldots t_{-4}, t_{-3}, t_{-2}, t_{-1}, t_0$. Moreover let $\ldots q_{-4}$, $q_{-3}, q_{-2}, q_{-1}, q_0$, be the corresponding values of q. Next choose ω "so small that the successive differences of q soon become small enough to be neglected,"

[52] Bashforth [1883], pp. 13ff. (As we see from the title of the book, Adams devised his scheme to integrate Bashforth's equation.)

and let $t = t_0 + n\omega$. Adams now forms the differences of the q_n and with their help expresses q in the form

$$q = q_0 + \Delta q_0 \frac{n}{1} + \Delta^2 q_0 \frac{n(n+1)}{1 \cdot 2} + \Delta^3 q_0 \frac{n(n+1)(n+2)}{1 \cdot 2 \cdot 3}$$
$$+ \Delta^4 q_0 \frac{n(n+1)(n+2)(n+3)}{1 \cdot 2 \cdot 3 \cdot 4} + \&\text{c.},$$

"provided that n be taken between limits for which this series remains convergent." He then infers that y is expressible as

$$y = \int q\, dt = \omega \int q\, dn$$

and hence as

$$y = y_0 + \omega \left\{ q_0 n + \Delta q_0 \frac{n^2}{2} + \Delta^2 q_0 \int \frac{n(n+1)}{1 \cdot 2}\, dn \right.$$
$$\left. + \Delta^3 q_0 \int \frac{n(n+1)(n+2)}{1 \cdot 2 \cdot 3}\, dn + \&\text{c.} \right\},$$

"where all the integrals are supposed to vanish when $n = 0$." From this result for $n = -1$ and $n = +1$ he finds

$$y_0 - y_{-1} = \omega \left\{ q_0 - \frac{1}{2} \Delta q_0 - \frac{1}{12} \Delta^2 q_0 - \frac{1}{24} \Delta^3 q_0 - \frac{19}{720} \Delta^4 q_0 \right.$$
$$- \frac{3}{160} \Delta^5 q_0 - \frac{863}{60480} \Delta^6 q_0 - \frac{275}{24192} \Delta^7 q_0$$
$$\left. - \frac{33953}{3628800} \Delta^8 q_0 - \frac{8183}{1036800} \Delta^9 q_0 - \&\text{c.} \right\};$$

and

$$y_1 - y_0 = \omega \left\{ q_0 + \frac{1}{2} \Delta q_0 + \frac{5}{12} \Delta^3 q_0 + \frac{3}{8} \Delta^3 q_0 + \frac{251}{720} \Delta^4 q_0 \right.$$
$$+ \frac{95}{288} \Delta^5 q_0 + \frac{19087}{60480} \Delta^6 q_0 + \frac{5257}{17280} \Delta^7 q_0$$
$$\left. + \frac{1070017}{3628800} \Delta^8 q_0 + \frac{2082753}{7257600} \Delta^9 q_0 + \&\text{c.} \right\}.$$

(He mentions that in practice it is "expedient to choose ω so small as to render it unnecessary to proceed beyond the fourth order of differences.")[53]

Adams now notes that the relation for $y_0 - y_{-1}$ may be viewed as a means of improving an estimate for y_0. Thus he supposes that (y_0) is such an estimate

---

[53] Notice that Adams' second formula is the Laplace integration formula. (These results appear on p. 18 of Bashforth [1883].) The first formula is often called a "corrector" and the second a "predictor." Adams notation $\Delta^i q_0$ differs from ours; e.g., his $\Delta q_0 = q_0 - q - 1$, a backward difference.

for y_0 and that $y_0 = (y_0) + \eta$, "where η is so small that its square may be neglected." He also forms $(q_0) = f[(y_0), t_0]$ and assumes that $q_0 = (q_0) + k\eta$, where k is the value of $\partial q/\partial y$ at $y = (y_0)$, $t = t_0$. He then writes

$$\begin{aligned} \Delta q_0 &= \Delta(q_0) + k\eta, \\ \Delta^2 q_0 &= \Delta^2(q_0) + k\eta, \\ \Delta^3 q_0 &= \Delta^3(q_0) + k\eta, \\ \&\text{c.} &= \&\text{c.}, \end{aligned}$$

and notes that this gives

$$\begin{aligned} (y_0) - y_{-1} + \eta = \omega\Big\{(q_0) - \frac{1}{2}\Delta(q_0) - \frac{1}{12}\Delta^2(q_0) - \frac{1}{24}\Delta^3(q_0) \\ - \frac{19}{720}\Delta^4(q_0) - \&\text{c.}\Big\} \\ + \omega k\eta\Big\{1 - \frac{1}{2} - \frac{1}{12} - \frac{1}{24} - \frac{19}{720} - \&\text{c.}\Big\} \end{aligned}$$

Let ε be "the excess of the quantity

$$\omega\Big\{(q_0) - \frac{1}{2}\Delta(q_0) - \frac{1}{12}\Delta^2(q_0) - \frac{1}{24}\Delta^3(q_0) - \frac{19}{720}\Delta^4(q_0) - \&\text{c.}\Big\}$$

over the quantity $(y_0) - y_{-1}$. Thus approximately

$$\eta = \frac{\varepsilon}{1 - \frac{251}{720}\omega k}, \qquad k\eta = \frac{k\varepsilon}{1 - \frac{251}{720}\omega k}."$$

This gave Adams a method for correcting an approximate value and also one for advancing to the next time step.

As far as I can find it was apparently the impetus given to computing during World War I that resulted in improvements upon this scheme. The first of these was due to F. R. Moulton, and is worth describing even though it carries us beyond 1900. It should not be thought, however, that there were no other means for numerical integration beyond those we have cited. The astronomers certainly made use of a procedure due to Encke [1852].[54] It is in part peculiarly adapted to calculating orbits but could be used more generally. However, let us look briefly at Moulton's method as typical of a class of formulas developed in the early part of this century.[55] The procedure is a combination of an extrapolational method to get a first guess for q_{n+1} in Adams's equation above and a sequence of iterated interpolations to improve that value. In this respect it is essentially Adams's scheme. The main differences lie in the fact that Moulton considers a variety of formulas, and

---

[54] Brouwer [1961], pp. 167–186.

[55] Moulton, *NM*, pp. 60ff. and *DE*, pp. 199ff.

in his book on differential equations shows the convergence of the method. There are other methods due, e.g., to Milne and others.[56]

Real insight into the numerical integration of partial differential equations particularly of hyperbolic and parabolic types did not come until 1929, when Courant, Friedrichs, and Lewy published a revolutionary paper on the subject.[57] Moreover, it was not much discussed until World War II, when von Neumann revived these ideas in a highly useful form.

5.11. Hermite

Charles Hermite did a number of very elegant things in numerical analysis that we should take cognizance of. First let us examine his studies of the Bernoulli numbers.[58] There is an elegant theorem due to Clausen and von Staudt which states that

$$(-1)^n B_n = A_n + \frac{1}{2} + \frac{1}{\alpha} + \frac{1}{\beta} + \cdots + \frac{1}{\lambda}, \tag{5.13}$$

where A_n is an integer, B_n is a Bernoulli number and the denominators $\alpha, \beta, \ldots, \lambda$ are all the primes for which $(\alpha - 1)/2, (\beta - 1)/2, \ldots, (\lambda - 1)/2$ are divisors of n.[59] Hermite remarks that the defining relations for the Bernoulli numbers suffice to determine the integers A_n. He writes

$$(2n + 1)_2 B_1 - (2n + 1)_4 B_2 + (2n + 1)_6 B_3 - \cdots$$
$$+ (-1)^{n-1}(2n + 1)_{2n} B_n = n - \frac{1}{2},$$

where $(2n + 1)_2, (2n + 1)_4, \ldots$ are the coefficients of $x^2, x^4, \ldots$ in the series development of $(1 + x)^{2n+1}$. Into this relation Hermite substitutes the series (5.13) and considers the consequences. Thus, e.g., the terms containing the factor 1/2 in the resulting relations are

$$\frac{1}{2}[(2n + 1)_2 + (2n + 1)_4 + \cdots + (2n + 1)_{2n} - 1];$$

but he has by definition

$$(2n + 1)_2 + (2n + 1)_4 + \cdots + (2n + 1)_{2n} = 2^{2n} - 1.$$

---

[56] Collatz [1951], pp. 433ff. In particular the Tables on pp. 437–439, in which many methods are listed.

[57] Courant [1928], pp. 32–74. Since then a very extensive literature has developed on the so-called Courant condition and related topics of numerical stability and convergence.

[58] Hermite III [1876], pp. 211–214. In Vol. IV of his *Oeuvres* are several other interesting papers on this topic. Cf. IV [1895], pp. 405–411, IV [1896], pp. 427–447, where Hermite takes up in part the paper of Jacobi VI [1834] and gives a form for the remainder term for the Euler–Maclaurin formula. Cf. also, Hermite IV [1894], pp. 393–396.

[59] von Staudt [1845]. Cf. also, Glaisher [1871], [1898]. Notice that Hermite's B_n is what today is usually called $|B'_{2n}|$. (Cf. Jordan [1939], p. 234, or by contrast Whittaker, *WW*, p. 125.)

Thus the expression (5.14) reduces to the integer $2^{2n-1} - 1$. He then looks at the terms containing the general factor $1/p$. They yield the Bernoulli numbers whose indices are multiples of $(p - 1)/2$. Thus he has

$$S_p = \frac{1}{p}[(2n + 1)_{p-1} + (2n + 1)_{2p-2} + (2n + 1)_{3p-3} + \cdots]$$

an integer. To see this let ω be any root of the equation $x^{p-1} = 1$; then the sum $\sum (1 + \omega)^{2n+1}$ over the roots ω has the value

$$(p - 1)[1 + (2n + 1)_{p-1} + (2n + 1)_{2p-2} + (2n + 1)_{3p-3} + \cdots].$$

After some manipulation Hermite concludes that this sum over the roots is congruent to -1 modulo p and hence that S_p is integral. From this he reaches the relation

$$\begin{aligned}(2n + 1)_2 A_1 + (2n + 1)_4 A_2 + \cdots + (2n + 1)_{2n} A_n \\ = 1 - n - 2^{2n-1} - S_3 - S_5 - \cdots - S_p,\end{aligned}$$

where the $S_3, S_5, \ldots, S_p$ are calculated for all the odd primes up to $2n + 1$. This enables him to find the A_i.

He now takes the case $n = 4$ as an example. The primes in question are of course 3, 5, 7 and

$$S_3 = \frac{1}{3}(36 + 126 + 84 + 9) = 85,$$

$$S_5 = \frac{1}{5}(126 + 9) = 27,$$

$$S_7 = \frac{1}{7}84 = 12.$$

Hence $12A_1 + 42A_2 + 28A_3 + 3A_4 = -85$. To solve this he notes that

$$\begin{aligned} A_1 &= -1, \\ 2A_1 + A_2 &= -3, \\ 3A_1 + 5A_2 + A_3 &= -9. \end{aligned}$$

Thus $A_1 = A_2 = A_3 = A_4 = -1$, and he then calculates the Bernoulli numbers $B_1 = 1/6$, $B_2 = 1/30$, $B_3 = 1/42$, $B_4 = 1/30$. (Note that the nineteenth-century definition of these numbers differs from ours on p. 127 above. In fact, we see that $B_\alpha = |B'_{2\alpha}|$ for $\alpha = 1, 2, \ldots$.)

It should be remarked that Adams used von Staudt's result to calculate the first 62 Bernoulli numbers. He said in his paper of 1877:

> A remarkable theorem, due to Staudt, gives at once the fractional part of any one of Bernoulli's numbers, and thus greatly facilitates the finding of those numbers, by reducing all the requisite calculations to operations with integers only.

The theorem may be thus stated:

If 1, 2, a, a'...$2n$ be all the divisors of $2n$, and if unity be added to each of these divisors so as to form the series 2, 3, $a+1$, $a'+1$...$2n+1$, and if from this series only the prime numbers 2, 3, p, p'... be selected, then the fractional part of the nth number of Bernoulli will be

$$(-1)^n\left(\frac{1}{2}+\frac{1}{3}+\frac{1}{p}+\frac{1}{p'}+\cdots\right).$$[60]

Hermite also wrote on the Bernoulli polynomials in 1875.[61] In yet another letter to Borchardt, Hermite took up the Euler–Maclaurin formula.[62] The basic idea in Hermite's mind was to establish certain properties of the Bernoulli polynomials by a different method than Malmsten had used.[63] This is a key paper since it contains a superior method of handling the Bernoulli polynomials. Hermite considers the fundamental relation

$$\frac{e^{\lambda x}-1}{e^{\lambda}-1}=S(x)_0+\frac{\lambda}{1}S(x)_1+\frac{\lambda^2}{1\cdot 2}S(x)_2+\cdots, \tag{5.14}$$

where $S(x)_n = 1^n + 2^n + 3^n + \cdots + (x-1)^n$ for integer x. He writes

$$\begin{aligned}\frac{e^{i\lambda x}-1}{e^{i\lambda}-1}&=\frac{e^{(1/2)i\lambda x}(e^{(1/2)i\lambda x}-e^{-(1/2)i\lambda x})}{e^{(1/2)i\lambda}(e^{(1/2)i\lambda}-e^{-(1/2)i\lambda})}=\frac{e^{(1/2)i\lambda(x-1)}\sin\frac{1}{2}\lambda x}{\sin\frac{1}{2}\lambda}\\&=\frac{\sin\frac{1}{2}\lambda x\cos\frac{1}{2}\lambda(x-1)}{\sin\frac{1}{2}\lambda}+i\,\frac{\sin\frac{1}{2}\lambda x\sin\frac{1}{2}\lambda(x-1)}{\sin\frac{1}{2}\lambda}.\end{aligned} \tag{5.15}$$

From the real and imaginary parts of this he concludes that

$$\frac{\sin\frac{1}{2}\lambda x\sin\frac{1}{2}\lambda(x-1)}{\sin\frac{1}{2}\lambda}=\lambda S(x)_1-\frac{\lambda^3}{(3)}S(x)_3+\frac{\lambda^5}{(5)}S(x)_5-\cdots,$$

$$\frac{\sin\frac{1}{2}\lambda x\cos\frac{1}{2}\lambda(x-1)}{\sin\frac{1}{2}\lambda}=S(x)_0-\frac{\lambda^2}{(2)}S(x)_2+\frac{\lambda^4}{(4)}S(x)_4-\cdots.$$

(Note Hermite's notation for factorials.) He is seeking the form of the functions $S(x)_n$ appearing above, and to find it he first notes that

$$\begin{aligned}\log\frac{\sin\frac{1}{2}\lambda x\sin\frac{1}{2}\lambda(x-1)}{\sin\frac{1}{2}\lambda}&=\log\frac{1}{2}\lambda x(x-1)+[1-x^2-(1-x)^2]\frac{B_1}{(2)}\frac{\lambda^2}{2}\\&\quad+[1-x^4-(1-x)^4]\frac{B_2}{(4)}\frac{\lambda^4}{4}\\&\quad+\cdots\\&\quad+[1-x^{2n}-(1-x)^{2n}]\frac{B_n}{(2n)}\frac{\lambda^{2n}}{2n}\\&\quad+\cdots,\end{aligned}$$

[60] Adams I [1866], pp. 426–427, and I [1877], pp. 454–458.

[61] Hermite III [1875], pp. 215–221. The paper is extracted from a letter to Borchardt and improves upon a paper by Raabe [1851].

[62] Hermite III [1876'], pp. 425–431.

[63] Malmsten [1884].

since

$$\log \sin \frac{1}{2} x = \log \frac{1}{2} x - \frac{B_1}{(2)} \frac{x^2}{2} - \frac{B_2}{(4)} \frac{x^4}{4} - \cdots - \frac{B_n}{(2n)} \frac{x^{2n}}{2n} - \cdots .$$

(This follows by integrating the usual defining relation in terms of cotan $\frac{1}{2}x$:

$$\frac{1}{2} \operatorname{cotan} \frac{1}{2} x = \frac{1}{x} - \frac{B_1 x}{(2)} - \frac{B_2 x^3}{(4)} - \frac{B_3 x^5}{(6)} - \cdots .$$

Note that these quantities $B_1, B_2, \ldots$ are Hermite's form of the Bernoulli numbers.) Setting $X_n = 1 - x^{2n} - (1 - x)^{2n}$, Hermite finds the result

$$\frac{\sin \frac{1}{2}\lambda x \sin \frac{1}{2}\lambda(x-1)}{\sin \frac{1}{2}\lambda} = -\frac{\lambda}{4} X_1 e^{(B_1 X_1/(2))(\lambda^2/2) + (B_2 X_2/(4))(\lambda^4/4) + \cdots},$$

where, of course, the B_n are the Bernoulli numbers. From these relations Hermite infers that

$$S(x)_1 = -\frac{1}{4} X_1,$$

$$S(x)_3 = \frac{1}{16} X_1^2,$$

$$S(x)_5 = -\frac{1}{192} (2X_1X_2 + 5X_1^3),$$

$$S(x)_7 = \frac{1}{2304} (16X_1X_3 + 42X_1^2X_2 + 35X_1^4),$$

$$\ldots .$$

Now the functions X_n vanish at $x = 0$ and $x = 1$ and have maxima at $x = 1/2$. From these facts Hermite concludes that the polynomials $(-1)^{n-1}S(x)_{2n+1}$ are positive on the interval from 0 to 1, and assume their maxima at $x = 1/2$.

He next takes up the real part of (5.15) and writes it as

$$\frac{1}{2} + \frac{\sin \frac{1}{2}\lambda(2x-1)}{2 \sin \frac{1}{2}\lambda}.$$

He finds that

$$\log \frac{\sin \frac{1}{2}\lambda(2x-1)}{\sin \frac{1}{2}\lambda} = \log (2x-1) + \frac{B_1 X_1^0}{(2)} \frac{\lambda^2}{2} + \frac{B_2 X_2^0}{(4)} \frac{\lambda^4}{4} + \cdots,$$

where $X_n^0 = 1 - (2x-1)^{2n}$. He can then determine all the coefficients in the definition (5.14). The functions

$$\frac{(-1)^n S(x)_{2n}}{2x-1} \quad (n = 1, 2, 3, \cdots)$$

are positive on (0, 1) with a single maximum at $x = \frac{1}{2}$.

Hermite then remarks that the polynomials $S(x)_n$ are very useful in deriving the Euler–Maclaurin formula. (Before proceeding, however, let us recall that these functions are closely connected with the Bernoulli polynomials. In fact as we saw on p. 97 above

$$S(x)_n = \frac{1}{n+1}[B_{n+1}(x) - B_{n+1}(0)].)$$

Hermite notes that the relation

$$\int U^{2n}V\,dx = U^{2n-1}V - U^{2n-2}V' + \cdots - UV^{2n-1} + \int UV^{2n}\,dx$$

holds when the U, V are any two functions whatever and U^k, V^k mean kth derivatives. He sets

$$\Phi(x) = U^{2n-1}V + U^{2n-3}V'' + \cdots + U'V^{2n-2},$$
$$\Psi(x) = U^{2n-2}V' + U^{2n-4}V''' + \cdots + UV^{2n-1},$$

and observes that

$$\int U^{2n}V\,dx = \Phi(x) - \Psi(x) + \int UV^{2n}\,dx.$$

He then chooses

$$U = \frac{\sin\frac{1}{2}\lambda x \sin\frac{1}{2}\lambda(x-1)}{\sin\frac{1}{2}\lambda} = \lambda S(x)_1 - \frac{\lambda^3}{1\cdot 2\cdot 3}S(x)_3 + \cdots,$$

but lets V remain arbitrary and fixes his limits of integration to be $x = 0$ and $x = 1$. After considerable manipulation he has

$$\int_0^1 \lambda^{2n}\frac{\cos\frac{1}{2}\lambda(2x-1)}{2\sin\frac{1}{2}\lambda}V\,dx = \frac{1}{2}\varphi(\lambda) + \frac{\operatorname{cotan}\frac{1}{2}\lambda}{2}\psi(\lambda) + (-1)^n\int_0^1 UV^{2n}\,dx,$$

where φ and ψ are two functions defined by the relations

$$\Phi(1) - \Phi(0) = \frac{(-1)^n}{2}\varphi(\lambda),$$

$$\Psi(1) - \Psi(0) = \frac{(-1)^{n-1}\operatorname{cotan}\frac{1}{2}\lambda}{2}\psi(\lambda).$$

To obtain the Euler–Maclaurin result he now sets $V = f(x_0 + hx)$, and has

$$V_1^k = h^k f^k(x_0 + h), \qquad V_k^0 = h^k f^k(x_0),$$

and

$$\frac{1}{2}\operatorname{cotan}\frac{1}{2}\lambda = \frac{1}{\lambda} - \frac{B_1\lambda^1}{(2)} - \frac{B_2\lambda^3}{(4)} - \frac{B_3\lambda^5}{(6)}\cdots\quad.$$

It follows that

$$\int_0^1 f(x_0 + hx)\,dx = \frac{1}{2}[f(x_0 + h) + f(x_0)] - \frac{B_1 h}{(2)}[f'(x_0 + h) - f'(x_0)]$$
$$+ \frac{B_2 h^3}{(4)}[f'''(x_0 + h) - f'''(x_0)] + \cdots$$
$$+ (-1)^{n-1}\frac{B_{n-1}h^{2n-3}}{(2n-2)}[f^{2n-3}(x_0 + h) - f^{2n-3}(x_0)]$$
$$- \frac{h^{2n}}{(2n-1)}\int_0^1 f^{2n}(x_0 + hx)S(x)_{2n-1}\,dx.$$

He notes in closing that Malmsten's result permits one to write the remainder term by means of the relation

$$\int_0^1 f^{2n}(x_0 + hx)S(x)_{2n-1}\,dx = f^{2n}(x_0 + \theta h)\int_0^1 S(x)_{2n-1}\,dx$$
$$= -\frac{1}{2n}f^{2n}(x_0 + \theta h)B'_{2n}.$$[64]

In another extract from a letter to Borchardt, Hermite considers a generalized interpolation result.[65] He wishes to find a polynomial F of degree $n - 1$ satisfying the relations

$$\begin{array}{llll} F(a) = f(a), & F'(a) = f'(a), & \ldots, & F^{\alpha-1}(a) = f^{\alpha-1}(a), \\ F(b) = f(b), & F'(b) = f'(b), & \ldots, & F^{\beta-1}(b) = f^{\beta-1}(b), \\ \vdots & \vdots & \vdots & \vdots \\ F(l) = f(l), & F'(l) = f'(l), & \ldots, & F^{\lambda-1}(l) = f^{\lambda-1}(l), \end{array}$$

where $f(x)$ is a given function and $\alpha + \beta + \cdots + \lambda = n$. He considers a region s containing the points $a, b, \ldots, l$ and x and supposes that in the interior of the region the function f is analytic without poles. Then

$$F(x) - f(x) = \frac{1}{2i\pi}\int_s \frac{f(z)(x-a)^\alpha(x-b)^\beta\cdots(x-l)^\lambda}{(x-z)(z-a)^\alpha(z-b)^\beta\cdots(z-l)^\lambda}\,dz.$$

He sets $\Phi(x) = (x-a)^\alpha(x-b)^\beta\cdots(x-l)^\lambda$ and

$$\varphi(z) = \frac{f(z)\Phi(x)}{(x-z)\Phi(z)}.$$

Then the integral above will be the sum of the residues of $\varphi(z)$ for the values $z = a, b, \ldots, l$ and $z = x$. Moreover, the last of these residues is $-f(x)$. Hermite first evaluates the other residues. To do this he seeks the coefficient

64 Nörlund [1924], pp. 29–32. His treatment of the Euler–Maclaurin summation formula is very similar to that of Hermite.
65 Hermite III [1878], pp. 432–443.

of $1/h$ in the Laurent development of $\varphi(a + h)$. Let us look for a moment at the residue at $z = a$. Clearly we have

$$f(z) = f(a) + (z - a)f'(a) + \cdots,$$

$$\frac{1}{x - z} = \frac{1}{x - a + a - z} = \frac{1}{x - a}\left[1 + \frac{z - a}{x - a} + \left(\frac{z - a}{x - a}\right)^2 + \cdots\right],$$

and hence the coefficient of $(z - a)^{\alpha - 1}$ in $f(z)/(x - z)$ is

$$\frac{f(a)}{(x - a)^\alpha} + \frac{f'(a)}{(x - a)^{\alpha - 1}} + \frac{f''(a)}{2!\,(x - a)^{\alpha - 2}} + \cdots + \frac{f^{(\alpha - 1)}(a)}{(\alpha - 1)!\,(x - a)^1}.$$

We see then that the residue at $z = a$ is this expression times $\Phi(x)/(a - b)^\beta (a - c)^\alpha \cdots (a - l)^\lambda$, i.e., it is

$$\frac{(x - b)^\beta (x - c)^\gamma \cdots (x - l)^\lambda}{(a - b)^\beta (a - c)^\gamma \cdots (a - l)^\lambda}\left[f(a) + f'(a)(x - a) + \frac{1}{2!}f''(a)(x - a)^2 + \cdots + \frac{1}{(\alpha - 1)!}f^{(\alpha - 1)}(a)(x - a)^{\alpha - 1}\right].$$

This is clearly a polynomial of degree $\alpha - 1 + \beta + \cdots + \lambda = n - 1$, and the residues at $x = b, c, \ldots, l$ are evidently similar. Hence their sum $F(x)$ is also such a polynomial and

$$F(x) - f(x) = \frac{1}{2i\pi}\int_s \frac{f(z)\Phi(x)}{(x - z)\Phi(z)}\,dz.$$

It is not difficult to see that F is the function Hermite sought, since $\Phi(x)$ vanishes along with its derivatives up to the order $\alpha - 1$ at $x = a$, etc.

He next considers several special cases. Firstly, he takes $\Phi(x)$ to be $(x - a)^\alpha$, and remarks that in this case the remainder term in a finite Taylor's expansion of $f(x)$ is, in his notation,

$$R = \frac{1}{2i\pi}\int_s \frac{f(z)(x - a)^\alpha}{(x - z)(z - a)^\alpha}\,dz = \int_x^a \frac{(x - a)^{\alpha - 1} f^{(\alpha)}(a)\,da}{1 \cdot 2 \cdots \alpha - 1}.$$

Thus the choice of $\Phi(x) = (x - a)^\alpha$ gives for $F(x)$ the Taylor's expansion:

$$f(a) + (x - a)f'(a) + \frac{1}{2!}(x - a)^2 f''(a) + \cdots + \frac{1}{(\alpha - 1)!}(x - a)^{\alpha - 1}(a).$$

Secondly, he takes $\Phi(x)$ to be $(x - a)(x - b) \cdots (x - l)$, and observes that in this case $F(x)$ is Lagrange's interpolation function. This is not difficult to see. Thirdly, he takes $\Phi(x)$ to be $(x - a)^\alpha (x - b)^\beta$ and works out in detail the form of $\int_a^b f(x)\,dx$. He finds for $\alpha + \beta = m$ that

$$\int_a^b f(x)\,dx = \frac{\theta(b) - \theta(a)}{1 \cdot 2 \cdots m} + \frac{(-1)^m}{1 \cdot 2 \cdots m}\int_a^b f^m(x)(x - a)^\beta (x - b)^\alpha\,dx,$$

where

$$\frac{\theta(a)}{1\cdot 2\cdots m} = -\frac{\alpha}{m}(b-a)f(a) - \frac{\alpha(\alpha-1)}{m(m-1)}\frac{(b-a)^2 f'(a)}{1\cdot 2} - \frac{\alpha(\alpha-1)(\alpha-2)}{m(m-1)(m-2)}\frac{(b-a)^3 f''(a)}{1\cdot 2\cdot 3} - \cdots,$$

$$\frac{\theta(b)}{1\cdot 2\cdots m} = -\frac{\beta}{m}(a-b)f(b) - \frac{\beta(\beta-1)}{m(m-1)}\frac{(a-b)^2 f'(b)}{1\cdot 2} - \frac{\beta(\beta-1)(\beta-2)}{m(m-1)(m-2)}\frac{(a-b)^3 f''(b)}{1\cdot 2\cdot 3} - \cdots .$$

The last term he writes as

$$\int_a^b f^m(x)(x-a)^\beta (x-b)^\alpha\,dx = f^m(\xi)\int_a^b (x-a)^\beta (x-b)^\alpha\,dx,$$

where ξ lies between a and b, by the Law of the mean. The integral in this remainder term is evidently expressible with the help of the relations

$$\int_a^b (x-a)^{p-1}(b-x)^{q-1}\,dx = (b-a)^{p+q-1}\frac{\Gamma(p)\Gamma(q)}{\Gamma(p+q)} = (b-a)^{p+q-1}B(p,q).$$

(If we view $\alpha = p-1, \beta = q-1$ as parameters subject to the condition that $\alpha+\beta = m$, then, this integral has its minimum when $\alpha = \beta$.) Hermite works through the two cases $\alpha = \beta = 1$ and $\alpha = \beta = 2$, since they lead to interesting numerical integration formulas. In the former he obtains the quadrature formula

$$\int_a^b f(x)\,dx = \frac{1}{2}(b-a)[f(a)+f(b)] - \frac{1}{12}(b-a)^3 f''(\xi),$$

and in the latter

$$\int_a^b f(x)\,dx = \frac{1}{2}(b-a)[f(a)+f(b)] + \frac{1}{12}(b-a)^2[f'(a)-f'(b)] + \frac{1}{720}(b-a)^5 f^{\mathrm{IV}}(\xi).$$

He closes his paper with a *postscriptum* in which he works out "the most general interpolation formula with the following remainder term":

$$f(x) - F(x) = \frac{\Phi(x)}{\Gamma(\alpha)\Gamma(\beta)\cdots\Gamma(\lambda)}\int_0^1 dt_n \int_0^{t_n} dt_{n-1}\cdots\int_0^{t_2} f^{\alpha+\beta+\cdots+\lambda}(u)\theta\,dt_1,$$

where

$$\Phi(x) = (x-a_1)^\alpha (x-a_2)^\beta\cdots(x-a_n)^\lambda,$$
$$\theta = (t_2-t_1)^{\alpha-1}(t_3-t_2)^{\beta-1}\cdots(1-t_n)^{\lambda-1},$$
$$u = (x-a_1)t_1 + (a_1-a_2)t_2 + \cdots + (a_{n-1}-a_n)t_n + a_n.$$

(Note the shift from contour to definite integrals.) To do this Hermite shows that if $\Pi(z) = (z - x)(z - a_1)\cdots(z - a_n)$ then, provided f is continuous in the interior of s, the remainder term is expressible in real form, since

$$\frac{1}{2i\pi}\int_s \frac{f(z)}{\Pi(z)}\,dz = \int_0^1 dt_n \int_0^{t_n} dt_{n-1} \int_0^{t_{n-1}} dt_{n-2}\cdots \int_0^{t_2} f^n(u)\,dt_1.$$

We mentioned earlier (cf. relation (4.61′) above) a trigonometric interpolation formula sometimes attributed to Hermite but actually first given by Gauss.[66] His starting point is a result of the well-known numerical analyst J. W. L. Glaisher [1880], who discussed the relation

$$\begin{aligned}\frac{\sin(a-f)\sin(a-g)\sin(a-h)}{\sin(a-b)\sin(a-c)} &+ \frac{\sin(b-f)\sin(b-g)\sin(b-h)}{\sin(b-a)\sin(b-c)}\\ &+ \frac{\sin(c-f)\sin(c-g)\sin(c-h)}{\sin(c-a)\sin(c-b)}\\ &+ \frac{\sin(f-a)\sin(f-b)\sin(f-c)}{\sin(f-g)\sin(f-h)}\\ &+ \frac{\sin(g-a)\sin(g-b)\sin(g-c)}{\sin(g-f)\sin(g-h)}\\ &+ \frac{\sin(h-a)\sin(h-b)\sin(h-c)}{\sin(h-f)\sin(h-g)} = 0,\end{aligned}$$

where a, b, c, f, g, h are arbitrary parameters.[67] Hermite considers the function

$$f(x) = \frac{\sin(x-f)\sin(x-g)\cdots\sin(x-s)}{\sin(x-a)\sin(x-b)\cdots\sin(x-l)},$$

in which the number of factors in numerator and denominator is the same. He lets $A, B, \ldots, L$ be the residues corresponding to the poles $a, b, \ldots, l$. Then

$$A = \frac{(\sin(a-f)\sin(a-g)\cdots\sin(a-s)}{\sin(a-b)\sin(a-c)\cdots\sin(a-l)},$$

$$B = \frac{\sin(b-f)\sin(b-g)\cdots\sin(b-s)}{\sin(b-a)\sin(b-c)\cdots\sin(b-l)},$$

$$\cdots .$$

Hermite then makes use of the relation

$$A + B + \cdots + L = \frac{H - G}{2i},$$

---

[66] Hermite IV [1885], pp. 206–208. (Hermite did not seem to notice its utility as an interpolation formula.)

[67] Glaisher [1880].

where G and H are the values of $f(x)$ when $z = e^{ix}$ with $z = 0$ and ∞.[68] To see this, Hermite notes that

$$\frac{\sin(x - f)}{\sin(x - a)} = \frac{z^2e^{-if} - e^{if}}{z^2e^{-ia} - e^{ia}};$$

and therefore for $z = 0$ and ∞, he has $e^{i(f-a)}$ and $e^{-i(f-a)}$. If

$$u = a + b + \cdots + l,$$
$$v = f + g + \cdots + s,$$

then

$$G = e^{-i(u-v)}, \qquad H = e^{i(u-v)},$$
$$A + B + \cdots + L = \sin(u - v).$$

By permuting a and f, b and $g, \ldots, l$ and s. Hermite obtains Glaisher's result, which is in effect

$$A + B + \cdots + L + F + G + \cdots + S = 0.$$

In a paper on Stirling's formula Hermite shows how to use Raabe's integral expression

$$\int_a^{a+1} \log \Gamma(x)\, dx = a \log a - a + \log \sqrt{2\pi}$$

to find the results of Stirling and Gauss, that

$$\log \Gamma(a) = \left(a - \frac{1}{2}\right) \log a - a + \log \sqrt{2\pi} + \sum \frac{(-1)^{n-1}B_n}{2n(2n-1)a^{2n-1}},$$

$$\log \Gamma\left(a + \frac{1}{2}\right) = a \log a - a + \log \sqrt{2\pi} + \sum \frac{(-1)^n(2^{2n-1} - 1)B_n}{2n(2n-1)(2a)^{2n-1}},$$

where $B_1, B_2, \ldots$ designate the Bernoulli numbers $1/6, 1/30, \ldots$.[69]

He also wrote another interesting and important work on interpolation entitled "Sur l'interpolation."[70] The paper was another demonstration of a result of Chebyshev: given a function $F(x)$ and $n + 1$ values of that function at $x = x_0, x_1, x_2, \ldots, x_n$, to find a polynomial of degree $m \leq n$ such that the sum of the squares of the differences between the polynomial and F at $x = x_0, x_1, x_2, \ldots, x_n$, each multiplied by given numbers, shall be a minimum.[71] Hermite connects his method directly with Lagrange's interpolation formula, in the following way. Let $f(x) = (x - x_0)(x - x_1)\cdots(x - x_n)$. Then the function of Lagrange,

$$\frac{f(x)}{x - x_0}\frac{u_0}{f'(x_0)} + \frac{f(x)}{x - x_1}\frac{u_1}{f'(x_1)} + \cdots + \frac{f(x)}{x - x_n}\frac{u_n}{f'(x_n)},$$

---

[68] He proves this in his *Cours d'Analyse*; cf. Hermite [1873], p. 328.

[69] Hermite IV [1893], pp. 378–388.

[70] Hermite II [1859], pp. 87–92.

[71] Chebyshev's method, presented to the Academy of St. Petersburg in 1855, depends on the use of continued fractions. See Chebyshev [1855].

is a polynomial of degree n which takes on the values $u_0, u_1, \ldots, u_n$ at the points $x = x_0, x_1, \ldots, x_n$. Hermite now chooses an arbitrary polynomial $\theta(x)$, and writes

$$f_i(x) = \frac{f(x)\theta(x)}{(x - x_i)f'(x_i)\theta(x_i)},$$

and

$$\Pi(x) = f_0(x)u_0 + f_1(x)u_1 + \cdots + f_n(x)u_n.$$

He then introduces some new quantities. Let

$$\begin{aligned} u_0 &= a_0 v_0 + b_0 v_1 + \cdots + l_0 v_n, \\ u_1 &= a_1 v_0 + b_1 v_1 + \cdots + l_1 v_n, \\ &\vdots \\ u_n &= a_n v_0 + b_n v_1 + \cdots + l_n v_n, \end{aligned}$$

and suppose that the coefficients are so chosen that

$$u_0^2 + u_1^2 + \cdots + u_n^2 = v_0^2 + v_1^2 + \cdots + v_n^2.$$

Then he sets

$$\begin{aligned} \Phi_0(x) &= a_0 f_0(x) + a_1 f_1(x) + \cdots + a_n f_n(x), \\ \Phi_1(x) &= b_0 f_0(x) + b_1 f_1(x) + \cdots + b_n f_n(x), \\ &\vdots \\ \Phi_n(x) &= l_0 f_0(x) + l_1 f_1(x) + \cdots + l_n f_n(x), \end{aligned}$$

and notices that $\Pi(x) = \Phi_0(x)v_0 + \Phi_1(x)v_1 + \cdots + \Phi_n(x)v_n$, and hence that

$$\Pi^2(x_0) + \Pi^2(x_1) + \cdots + \Pi^2(x_n) = v_0^2 + v_1^2 + \cdots + v_n^2.$$

Hermite next observes that the functions $\Phi_m(x)$, which arise out of the Lagrange Interpolation formula, have important orthonormality properties. Namely, for $m \neq m'$,

$$\sum_{i=0}^{i=n} \Phi_m(x_i)\Phi_{m'}(x_i) = 0, \qquad \sum_{i=0}^{i=n} \Phi_m^2(x_i) = 1.$$

He goes on to note that all the $\Phi_m(x)$ have a common factor $\theta(x)$, so that he writes $\Phi_m(x) = \varphi_m(x)\theta(x)$, and considers the system of $n + 1$ polynomials of degree n, $\varphi_0(x), \varphi_1(x), \ldots, \varphi_n(x)$. He now takes $m + 1$ of these polynomials, e.g., $\varphi_0(x), \varphi_1(x), \ldots, \varphi_m(x)$, and considers minimizing the sum of the squares of the differences

$$F(x) - A\varphi_0(x) - B\varphi_1(x) - \cdots - H\varphi_m(x), \qquad x = x_0, x_1, \ldots, x_n,$$

multiplied by the weights $\theta(x_0), \theta(x_1), \ldots, \theta(x_n)$. That is, he seeks to minimize the expression

$$\begin{aligned} &\sum_{i=0}^{i=n} [F(x_i) - A\varphi_0(x_i) - B\varphi_1(x_i) - \cdots - H\varphi_m(x_i)]^2\theta^2(x_i) \\ &\quad = \sum_{i=0}^{i=n} [F(x_i)\theta(x_i) - A\Phi_0(x_i) - B\Phi_1(x_i) - \cdots - H\Phi_m(x_i)]^2, \end{aligned}$$

by proper choice of the parameters $A, B, \ldots, H$. He differentiates with respect to these parameters and finds

$$A = \sum_{i=0}^{i=n} \Phi_0(x_i)F(x_i)\theta(x_i),$$

$$B = \sum_{i=0}^{i=n} \Phi_1(x_i)F(x_i)\theta(x_i),$$

$$\vdots$$

$$H = \sum_{i=0}^{i=n} \Phi_m(x_i)F(x_i)\theta(x_i).$$

He then sets

$$\pi(x) = A\varphi_0(x) + B\varphi_1(x) + \cdots + H\varphi_m(x),$$

and concludes that

$$\sum_{i=0}^{i=n} \pi^2(x_i)\theta^2(x_i) = \sum_{i=0}^{i=n} F^2(x_i)\theta^2(x_i).$$

Hermite closes with a discussion of the form of the φ in terms of invariants of quadratic forms.

In the same volume of Hermite's works is a paper in which he defines the now well-known Hermite polynomials by means of the recurrence relation

$$U_{n+1} + 2xU_n + 2nU_{n-1} = 0,$$

and shows that U_n satisfies the differential equation

$$\frac{d^2U_n}{dx^2} - 2x\frac{dU_n}{dx} + 2nU_n = 0.\text{[72]}$$

He goes on to show that

$$\int_{-\infty}^{+\infty} e^{-x^2}U_nU_{n'}\,dx = \begin{cases} 0 & n \neq n' \\ 2\cdot 4\cdot 6\cdots 2n\cdot\sqrt{\pi} & n = n', \end{cases}$$

and then indicates how to expand various functions such as polynomials, $\sin 2\omega x$, or $\cos 2\omega x$ in terms of his polynomials. We will not pursue this further since it lies outside of the topic under discussion.

5.12. Sums

There is a very extensive literature on sums of a function $f(x)$. This is basically the problem of solving the difference equation $\varphi(x + 1) - \varphi(x) = f(x)$. We have seen a number of examples of how this was done earlier. In the

[72] Hermite II [1802], pp. 293–308.

nineteenth century the problem received renewed attention from a number of authors including Plana, Abel, and Cauchy. They were then followed by Guichard, Malmsten, Lindelöf and many others, but principally by Nörlund, who established the modern theory.[73] His work, however, is grounded in earlier fundamental discoveries of Cauchy and Lindelöf. In fact, it was the application by Cauchy of the theory of functions of a complex variable to the summation problem that made Nörlund's work possible. It is not possible to discuss the subject in depth without recourse to Cauchy's Residue theorem, which makes it possible to relate a sum (of residues) with an integral. There is hardly room here for even a brief glimpse of that theory since it is so comprehensive as well as extensive.

Abel wrote a number of very nice papers on the subject which we can profitably review. His aim was to show the utility of definite integrals. Let us consider with him, how to express the Bernoulli numbers as definite integrals.[74] He first writes on the Euler expansion, where he uses $A_1, A_2, A_3, \ldots$ for the quantities Euler called the Bernoulli numbers:

$$1 - \frac{u}{2}\cot\frac{u}{2} = A_1\frac{u^2}{2} + A_2\frac{u^4}{2\cdot 3\cdot 4} + \cdots + A_n\frac{u^{2n}}{2\cdot 3\cdot 4\cdots 2n} + \cdots .$$

He makes use of another expansion of Euler (p. 130 above), which says that

$$1 - \frac{u}{2}\cot\frac{u}{2} = 2u^2\left(\frac{1}{4\pi^2 - u^2} + \frac{1}{4\cdot 4\pi^2 - u^2} + \frac{1}{9\cdot 4\pi^2 - u^2} + \cdots\right).^{75}$$

He now expands each expression in the right-hand member as a series in u^2 and collects terms. This gives him Euler's formula

$$\frac{A_n}{1\cdot 2\cdot 3\cdots 2n} = \frac{1}{2^{2n-1}\pi^{2n}}\left(1 + \frac{1}{2^{2n}} + \frac{1}{3^{2n}} + \cdots\right).$$

Next he expresses $1/(e^t - 1)$ in the form

$$\frac{1}{e^t - 1} = e^{-t} + e^{-2t} + e^{-3t} + \cdots$$

and then finds

$$\int \frac{t^{2n-1}\,dt}{e^t - 1} = \int e^{-t}t^{2n-1}\,dt + \int e^{-2t}t^{2n-1}\,dt + \cdots + \int e^{-kt}t^{2n-1}\,dt + \cdots .$$

He notes that

$$\int_0^{1/0} e^{-kt}t^{2n-1}\,dt = \frac{\Gamma(2n)}{k^{2n}}.$$

[73] Nörlund [1924]. In his bibliography there are listed papers by all these authors. The Nörlund theory is also fully discussed in Milne–Thomson [1933]. There is an excellent discussion of the subject and of the authors in E. Lindelöf [1905], pp. 68–69. His Chap. II, pp. 52–86, is devoted to a discussion of summation formulas.

[74] Abel I [1823], pp. 11–21.

[75] Abell I [1823], p. 21.

(Notice how Abel writes infinity; he finds the relation by changing variables in Legendre's definition of

$$\Gamma a = \int_0^1 dx \left(\log \frac{1}{x}\right)^{a-1}.)$$

This gives Abel the relation

$$\int_0^{1/0} \frac{t^{2n-1}\,dt}{e^t - 1} = \Gamma(2n)\left(1 + \frac{1}{2^{2n}} + \frac{1}{3^{2n}} + \cdots\right)$$

$$= \frac{\Gamma(2n)}{\Gamma(2n+1)}\, 2^{2n-1}\pi^{2n} A_n = \frac{2^{2n-1}\pi^{2n}}{2n} A_n.$$

Therefore he has his formula

$$A_n = \frac{2n}{2^{2n-1}\pi^{2n}} \int_0^{1/0} \frac{t^{2n-1}\,dt}{e^t - 1} = \frac{2n}{2^{2n-1}} \int_0^{1/0} \frac{t^{2n-1}\,dt}{e^{\pi t} - 1}.$$

(This latter relation follows by replacing t by $t\pi$.)

Abel turns now to the Euler–Maclaurin formula

$$\sum \varphi x = \int \varphi x \cdot dx - \frac{1}{2}\varphi x + A_1 \frac{\varphi' x}{1 \cdot 2} - A_2 \frac{\varphi''' x}{1 \cdot 2 \cdot 3 \cdot 4} + \cdots$$

and substitutes into it the integrals he just found. This gives him the relation

$$\sum \varphi x = \int \varphi x \cdot dx - \frac{1}{2}\varphi x + \int_0^{1/0} \frac{dt}{e^{\pi t} - 1}\left(\varphi' x \frac{t}{2} - \frac{\varphi''' x}{1 \cdot 2 \cdot 3} \frac{t^3}{2^3} + \cdots\right).$$

He notes that the expression in the parentheses in this relation is expressible as

$$\frac{1}{2\sqrt{-1}}\left[\varphi\left(x + \frac{t}{2}\sqrt{-1}\right) - \varphi\left(x - \frac{t}{2}\sqrt{-1}\right)\right],$$

and thus that

$$\sum \varphi x = \int \varphi x \cdot dx - \frac{1}{2}\varphi x$$

$$+ \int_0^{1/0} \frac{\varphi(x + (t/2)\sqrt{-1}) - \varphi(x - (t/2)\sqrt{-1})}{2\sqrt{-1}} \frac{dt}{e^{\pi t} - 1}.$$

Abel uses this relation to evaluate the definite integral appearing in his result for the cases where φx is e^x, e^{mx}, $1/x$ and $\sin ax$. He then turns to the task of summing the series

$$S = \varphi(x+1) - \varphi(x+2) + \varphi(x+3) - \varphi(x+4) + \cdots$$

by means of definite integrals. He says: "One can easily see that S is expressible as follows,

$$S = \frac{1}{2}\varphi x + A_1 \varphi' x + A_2 \varphi'' x + A_3 \varphi''' x + \cdots \quad ."$$

He now seeks a new form for the Bernoulli numbers and finds directly that

$$A_1 = 2\int_0^{1/0} \frac{t\,dt}{e^{\pi t} - e^{-\pi t}},$$

$$A_3 = -\frac{2}{2\cdot 3}\int_0^{1/0} \frac{t^3\,dt}{e^{\pi t} - e^{-\pi t}},$$

$$A_5 = \frac{2}{2\cdot 3\cdot 4\cdot 5}\int_0^{1/0} \frac{t^5\,dt}{e^{\pi t} - e^{-\pi t}},$$

etc.;

this gives him the partial result

$$S = \frac{1}{2}\varphi x + 2\int_0^{1/0} \frac{dt}{e^{\pi t} - e^{-\pi t}}\left\{t\varphi' x - \frac{t^3}{2\cdot 3}\varphi''' x + \frac{t^5}{2\cdot 3\cdot 4\cdot 5}\varphi^{(V)} x - \cdots\right\},$$

which he simplifies as before to find his relation

$$\varphi(x+1) - \varphi(x+2) + \varphi(x+3) - \varphi(x+4) + \cdots$$
$$= \frac{1}{2}\varphi x + 2\int_0^{1/0} \frac{dt}{e^{\pi t} - e^{-\pi t}} \frac{\varphi(x + t\sqrt{-1}) - \varphi(x - t\sqrt{-1})}{2\sqrt{-1}},$$

which holds, e.g., for $x = 0$.

In a later paper, Abel I [1825], pp. 34–39, generalizes his previous results to n-fold summations with the help of some generalized Bernoulli numbers. In the case $n = 1$ Abel notes that his formula becomes

$$\sum \varphi x = C + \int \varphi x \cdot dx - \frac{1}{2}\varphi x$$
$$+ 2\int_0^{1/0} \frac{dt}{e^{2\pi t} - 1} \frac{\varphi(x + t\sqrt{-1}) - \varphi(x - t\sqrt{-1})}{2\sqrt{-1}},$$

and if $\sum \varphi x$, $\int \varphi x\,dx$ vanish for $x = a$, then

$$C = \frac{1}{2}\varphi a - 2\int_0^{1/0} \frac{dt}{e^{2\pi t} - 1} \frac{\varphi(a + t\sqrt{-1}) - \varphi(a - t\sqrt{-1})}{2\sqrt{-1}}.$$

Moreover, if they also vanish for $x = \infty$, then Abel writes

$$\varphi a + \varphi(a+1) + \varphi(a+2) + \varphi(a+3) + \cdots \text{ in inf.}$$
$$= \int_a^{1/0} \varphi x \cdot dx + \frac{1}{2}\varphi a - 2\int_0^{1/0} \frac{dt}{e^{2\pi t} - 1} \frac{\varphi(a + t\sqrt{-1}) - \varphi(a - t\sqrt{-1})}{2\sqrt{-1}}.$$ [76]

In a short paper Abel gives integral formulas to express the sums $\sum 1/a^2$, $\sum 1/a^3, \ldots, \sum 1/a^n$. These are simple and elegant.[77] In yet another note he

[76] Abel I [1825], pp. 34–39.
[77] Abel II [1881], pp. 1–6.

considered series of the form $\varphi(0) + \varphi(1)x + \varphi(2)x^2 + \varphi(3)x^3 + \cdots + \varphi(n)x^n$, with n finite or infinite, and φ a rational function of n.[78] He was not aware of similar work by Plana which preceded his by a few years.[79]

Lindelöf says that Cauchy worked in ignorance of the results of Plana and Abel when he established his own fundamental discoveries. In what follows, we outline Cauchy's basic results as given by Lindelöf. Consider f to be an analytic function of a complex variable z in some domain and take in that region a simply closed curve C containing the points $m, m+1, \ldots, n$, but no other integer points. Assume further that in the domain f has only a finite number of singularities distinct from $m, m+1, \ldots, n$. Then by the theory of residues Cauchy finds

$$\frac{1}{2\pi i}\int_C \pi \operatorname{cotan} \pi z f(z)\, dz = \sum_m^n f(\nu) + \mathcal{E}_C\, \pi \operatorname{cotan} \pi z\{f(z)\},$$

where he uses the peculiar symbol in the second part of the right-hand member to mean the sum of the residues of π cotan $[\pi z f(z)]$ relative to the singularities of f.

Similarly he has the companion result

$$\sum_m^n (-1)^\nu f(\nu) = \frac{1}{2\pi i}\int_C \frac{\pi}{\sin \pi z} f(z)\, dz - \mathcal{E}_C\, \frac{\pi}{\sin \pi z}\{f(z)\}.$$

Thus if $f(z) = 1/(x+z)^2$, a little calculation shows that Cauchy's formula gives

$$\sum_{-\infty}^{+\infty} \frac{1}{(x+\nu)^2} = \frac{\pi^2}{\sin^2 \pi x}.$$

It is possible under suitable hypotheses to reduce these formulas of Cauchy and Lindelöf to those of Plana and Abel.[80] Thus Lindelöf has the relations

$$\sum_m^n f(\nu) = \int_\alpha^\beta f(\tau)\, d\tau - 2\int_0^\infty [Q(\alpha, t) - Q(\beta, t)]\, dt,$$

$$\sum_m^n (-1)^\nu f(\nu) = 2\int_0^\infty [Q_1(\beta, t) - Q_1(\alpha, t)]\, dt,$$

$$\sum_m^n f(\nu) = \frac{1}{2}[f(m) + f(n)] + \int_m^n f(\tau)\, d\tau - 2\int_0^\infty \frac{q(m, t) - q(n, t)}{e^{2\pi t} - 1}\, dt,$$

[78] Abel II [1881′], pp. 14–18.
[79] Plana [1820], pp. 403–418.
[80] Lindelöf [1905], pp. 55–61. Cf. Cauchy 1, II [1827], pp. 12–19.

where q, Q, and Q_1 are defined with the help of the relations

$$\Psi(\tau, t) = \frac{\cos 2\pi\tau\,(1 - e^{-2\pi t})}{e^{2\pi t} - 2\cos 2\pi\tau + e^{-2\pi t}},$$

$$\Psi_1(\tau, t) = \frac{\cos \pi\tau(e^{\pi t} - e^{-\pi t})}{e^{2\pi t} - 2\cos 2\pi\tau + e^{-2\pi t}},$$

$$X(\tau, t) = \frac{\sin 2\pi\tau}{e^{2\pi t} - 2\cos 2\pi\tau + e^{-2\pi t}},$$

$$X_1(\tau, t) = \frac{\sin \pi\tau(e^{\pi t} + e^{-\pi t})}{e^{2\pi t} - 2\cos 2\pi\tau + e^{-2\pi t}},$$

$$p\,(\tau, t) = \frac{1}{2}\,[f(\tau + it) + f(\tau - it)],$$

$$q(\tau, t) = \frac{1}{2i}\,[f(\tau + it) - f(\tau - it)],$$

$$Q(\tau, t) = p(\tau, t)X(\tau, t) + q(\tau, t)\Psi(\tau, t),$$

$$Q_1(\tau, t) = p(\tau, t)X_1(\tau, t) + q(\tau, t)\Psi_1(\tau, t).$$

After some analysis Lindelöf shows that out of the three summation formulas above he can deduce the Euler–Maclaurin, the Boole, the Sonin–Hermite and the Hermite relations, respectively, in the forms

$$\sum_m^n f(\nu) = \frac{1}{2}\,[f(m) + f(n)] + \int_m^n f(\tau)\,d\tau + \sum_{\nu=1}^k (-1)^{\nu-1}\frac{B_\nu}{2\nu}\,\frac{f^{(2\nu-1)}(n) - f^{(2\nu-1)}(m)}{(2\nu-1)!} + R;$$

$$\sum_m^n (-1)^\nu f(\nu) = \frac{1}{2}\,[(-1)^m f(m) + (-1)^n f(n)] + \sum_1^k (-1)^{\nu-1}\frac{(2^{2\nu}-1)B_\nu}{2\nu}\,\frac{(-1)^n f^{(2\nu-1)}(n) - (-1)^m f^{(2\nu-1)}(m)}{(2\nu-1)!} + R;$$

$$\sum_m^n f(\nu) = \int_\alpha^\beta f(\tau)\,d\tau + \sum_1^{2k}\frac{(-1)^\nu}{\nu}\,\frac{\bar{\varphi}_\nu(\beta)f^{(\nu-1)}(\beta) - \bar{\varphi}_\nu(\alpha)f^{(\nu-1)}(\alpha)}{(\nu-1)!} + R;$$

$$\sum_m^n (-1)^\nu f(\nu) = \sum_0^{2k-1} (-1)^\nu\,\frac{\bar{\chi}_\nu(\beta)f^{(\nu)}(\beta) - \bar{\chi}_\nu(\alpha)f^{(\nu)}(\alpha)}{\nu!} + R,$$

where

$$\varphi_{2k}(x) = (-1)^{k+1}4k\int_0^\infty t^{2k-1}\Psi(x, t)dt \qquad (k = 1, 2, \cdots),$$

$$\varphi_{2k+1}(x) = (-1)^{k+1}(4k + 2)\int_0^\infty t^{2k}X(x, t)dt \qquad (k = 0, 1, 2, \cdots),$$

$$\chi_{2k}(x) = (-1)^{k}2\int_0^\infty t^{2k}X_1(x, t)dt$$

$$\qquad\qquad\qquad\qquad (k = 0, 1, 2, \cdots).$$

$$\chi_{2k+1}(x) = (-1)^{k+1}2\int_0^\infty t^{2k+1}\Psi_1(x, t)dt$$

Furthermore, he has the exact forms for the remainders and nice bounds on their sizes. Lindelöf assumes that f satisfies the conditions: 1. the function $f(\tau + it)$ is holomorphic for $\tau \geq \alpha$ and t arbitrary; 2. f is such that

$$\lim_{t=\pm\infty} e^{-2\pi|t|}f(\tau + it) = 0$$

uniformly for $\alpha \leq \tau \leq \beta$, with β arbitrarily large; 3. f is also such that

$$\lim_{t=\infty}\int_{-\infty}^{+\infty} e^{2\pi|t|}|f(\tau + it)|\, dt = 0.$$

Notice that if $F(x) = \sum_{\alpha=1}^{x-1} f(\alpha)$, then $\Delta F = F(x + 1) - F(x) = f(x)$. Thus the problem of summing a function and that of solving a very simple difference equation are equivalent. We have shown earlier how Laplace and others solved more complicated linear difference equations; and there is an extensive theory of difference equations which parallels the corresponding one of differential equations. The interested reader may wish to consult Nörlund [1924], or Milne–Thomson [1933]. Since space does not permit us to discuss the theory in any detail, let us close with an elegant asymptotic result of Poincaré on linear difference equations.[81] His work was the foundation for much modern research in the field. He considers the difference equation

$$P_kU_{n+k} + P_{k-1}u_{n+k-1} + \cdots + P_1u_{n+1} + P_0u_n = 0,$$

where the coefficients P_i are polynomials of order p. He also writes this equation in the form

$$R_k\Delta^k u_n + R_{k-1}\Delta^{k-1}u_n + \cdots + R_1\Delta u_n + R_0u_n = 0.$$

He calls A_i the coefficient of n^p in the polynomial P_i and then considers the algebraic equation

$$A_kz^k + A_{k-1}z^{k-1} + \cdots + A_1z + A_0 = 0.$$

If the roots of this indicial equation are distinct and no two roots have equal moduli, then Poincaré asserts his well-known result that the ratio u_{n+1}/u_n tends to a root of that equation as n approaches infinity and in general to the root of largest modulus. Perron generalized this result and showed in effect that each root can be approximated by the ratio $u_{n+1}(k)/u_n(k)$ for a suitable solution $u_n(k)$.[82]

---

[81] Poincaré [1885], pp. 213–217.

[82] Perron [1909], pp. 17–37.

Bibliography

Aaboe [1954]:

Asger Aaboe, "Al-Kashi's Iteration Method for the Determination of sin 1°," *Scripta Mathematica*, Vol. 20, 1954, pp. 24–29.

Abel

Oeuvres: Oeuvres complètes de Niels Henrik Abel, new ed. by M. Sylow and S. Lie, Christiania, 1881.

I [1823]: Niels Henrik Abel, "Solution de quelques problèmes à l'aide d'intégrales," *Magazin f. Naturvidenskaberne*, Aar. II, Bd. I, Christiania, 1823, pp. 55–63 = *Oeuvres*, Vol. I, pp. 11–21.

I [1825]: ———, "L'intégrale finie $\sum^n \varphi x$ exprimeé par une intégrale définie simple," *Magazin f. Naturvidenskaberne*, Aar. III, Bd. II., 1825, pp. 182–189 = *Oeuvres* Vol. I, pp. 34–39.

II [1881]: ———, "Les fonctions transcendantes $\sum 1/a^2$, $\sum 1/a^3$, $\sum 1/a^4, \ldots$, $\sum 1/a^n$ exprimées par des intégrales définies," *Oeuvres*, Vol. II, pp. 1–6.

II [1881′]: ———, "Sommation de la série $y = \varphi(0) + \varphi(1)x + \varphi(2)x^2 + \varphi(3)x^3 + \cdots + \varphi(n)x^n$, n étant un nombre entier positif fini ou infini, et $\varphi(n)$ une fonction algébrique rationelle de n," *Oeuvres*, Vol. II, pp. 14–18.

Adams

SP: The Scientific papers of John Couch Adams, ed. by W. G. Adams, 2 Vols., Cambridge, 1896–1900.

I [1866]: John Couch Adams, "On the calculation of the Bernoullian numbers from B_{32} to B_{62}," *App. to Camb. Observ.*, Vol. XXII, 1866–1869, App. I, pp. i–xxii = *SP*, Vol. I, pp. 426–453.

I [1877]: ———, "On the calculation of Bernoulli's numbers up to B_{62} by means of Staudt's theorem," *Brit. Assoc. Report*, 1877, pp. 8–14 and "Table of the values of the first sixty-two numbers of Bernoulli," *Jour. für reine u. angew. Math.*, Vol. 85, 1878, pp. 269–272 = *SP*, Vol. I, pp. 454–458.

[1877′]: ———, "On the calculation of the sum of the reciprocals of the first thousand integers, and on the value of Euler's constant to 260 places of decimals," *Brit. Assoc. Report*, 1877, pp. 14–15.

I [1878]: ———, "Note on the value of Euler's constant . . . ," *Proc. Roy. Soc.*, Vol. XXVII, 1878, pp. 88–94 = *SP*, Vol. I, pp. 459–466.

I [1882]: ———, "On Newton's solution to Kepler's problem," *Mon. Not. Roy. Astr. Soc.*, Vol. XLIII, 1882, pp. 43–49 = *SP*, Vol. I, pp. 289–296.

Bashforth [1883]:

Francis Bashforth, *An Attempt to test the Theories of Capillary Action by comparing the theoretical and measured forms of drops of fluid. With an explanation of the method of integration employed in constructing the tables which give the theoretical forms of such drops*, by J. C. Adams, Cambridge, 1883.

Bayes [1763]:

Thomas Bayes, "An essay towards solving a problem in the doctrine of chances," *Phil. Trans.*, Vol. 53, London, 1763, pp. 370–403. There is also an Appendix on pp. 404–418.

Bendixson [1893]:

Ivar Bendixson, "Sur le calcul des intégrales d'un système d'équations différentielles par des approximations successives," *Stock. Akad. Forh.*, Vol. 51, 1893, pp. 599–612.

Bernoulli, D. [1728]:

Daniel Bernoulli, "Observationes de seriebus recurrentibus," *Comment. acad. sc. Petrop.*, Vol. III, 1728, 1732, pp. 85–100.

Bernoulli, Ja., *AC*:

James Bernoulli, *Ars Conjectandi Opus Posthumum*, Basel, 1713 = *Die Werke von Jakob Bernoulli*, Vol. 3, Basel, 1975, pp. 107–286 = *Wahrscheinlichkeitsrechnung*, German, transl. by R. Haubner, Leipzig, 1899.

Bernoulli, Jo. [1695]:

Johann Bernoulli, *Letter from Joh. Bernoulli to Leibniz*, XIV, Basel, July 17, 1695. It appears in *Leibnizens mathematische Schriften*, ed. by C. J. Gerhardt, Series I, Vol. III, Halle, 1855, pp. 197–205.

Boole [1800]:

George Boole, *A Treatise on the Calculus of Finite Differences*, Cambridge, 1860; 2nd ed. London, 1872, 3rd ed. by J. F. Moulton, London, 1880 (reprinted New York, 1947.)

Borchardt [1846]:

Carl W. Borchardt, "Neue Eigenschaft der Gleichung, mit deren Hülfe man die seculären Störungen der Planeten bestimmt," *Jour. für reine u. angew. Math.*, Vol. XXX, 1846, pp. 38–45.

Bourbaki [1960]:

Nicolas Bourbaki, *Élements d'histoire des mathématiques*, Paris, 1960.

Briggs

LOG: Henry Briggs, *Logarithmorum Chilias Prima. Quam autor typis excudendam curauit, non eo concilio, vt publici iuris fieret; sed partim, vt quorundam suorum necessariorum desiderio priuatim satisfaceret; partim, vt eius adiumento, non solum Chiliadas aliquot insequentes; sed etiam integrum Logarithmorum Canonem, omnium Triangulorum calculo inseruientem commodius absolueret. Habet enim Canonem Sinuum, à seipso, ante Decennium, per aequationes Algebraicas, & differentias, ipsis Sinubus proportionales, pro singulis Gradibus & graduū centesimis, à primis fundamentis accurate extructū: quem vna cum Logarithmis adiunctis, volente Deo, in lucem se daturum sperat, quam primum commode licuerit. Quod autem hi Logarithmi, diuersi sint ab iis, quos Clarissimus inuentor, memoriae semper colendae, in suo edidit Canone mirifico; sperandum, eius librū posthumum, abunde nobis propediem satisfacturum. Qui autori (cum eum domi suae, Edinburgi, bis inuiseret, & apud eum humanissime exceptus, per aliquot septimanas libentissime mansisset; eique horum partem praecipuam quam tum absoluerat ostendisset) suadere non destitit, vt hunc in se laborem susciperet. Cui ille non inuitus morem gessit. In tenui; sed non tenuis, fructusve laborve.* (1617. Small 8°.)

ARITH: ———, *Arithmetica Logarithmica sive Logarithmorum Chiliades Triginta, Pro numeris naturali serie crescentibus ab unitate ad 20,000: et a 90,000 ad 100,000. Quorum ope multa perficiuntur Arithmetica problemata et Geometrica. Hos Numeros Primus Invenit Clarissimus Vir Iohannes Neperus Baro Merchistonii; eos autem ex eiusdem sententia mutavit, eorumque ortum et usum illustravit Henricus Briggius in celeberrima Academia Oxoniensi Geometriae professor Savilianus. Deus Nobis Usuram Vitae Dedit Et Ingenii, Tanquam Pecuniae, Nulla Praestituta Die.* Londoni, Excudebat Gulielmus Iones. (1624. Folio.)

TRIG: ———, *Trigonometria Britannica, sive de Doctrinâ Triangulorum libri duo, quorum prior continet constructionem canonis sinuum, tangentium et secantium, unâ cum logarithmis sinuum et tangentium ad gradus et graduum centesimas et ad minuta et secunda centesimis respondentia.*

A classissimo, doctissimo integerrimoque viro Domino Henrico Briggio, Geometriae in celeberrimâ Academiâ Oxoniensi professore Saviliano dignissimo, paulo antè ipsius è terris emigrationem compositus.

Posterior vero usum sive applicationem canonis in resolutione triangulorum tam planorum quam sphaericorum e geometricis fundamentis petitâ, calculo facillimo, eximiisque compendiis exhibet; ab Henrico Gellibrand Astronomiae in collegio Greshamensi apud Londinenses professore, constructus. Goudae, 1633.

See also Vlacq.

Bronwin [1849]:

B. Bronwin, "On the Determination of the Coefficients in any Series of Sines and Cosines of Multiples of a variable Angle from particular Values of that Series," *Phil. Mag.*, Vol. 54, 1849, pp. 260–268.

Brouncker [1668]:

William Brouncker, "The squaring of the Hyperbola, by an infinite series of Rational Numbers, together with its Demonstration by that Eminent Mathematician, The Right Honourable the Lord Viscount Brouncker," *Phil. Trans. Roy. Soc.*, Vol. III–IV, 1668–1669, pp. 645–649. (Reprinted Amsterdam, 1963–1964.)

Brouwer [1961]:

D. Brouwer and G. M. Clemence, *Methods of Celestial Mechanics*, New York, 1961.

Brown [1896]:

Ernest W. Brown, *An Introductory Treatise on the Lunar Theory*, Cambridge, 1896 (reprinted New York, 1960.)

Bürgi *ANT:*

Joost Bürgi, *Arithmetische und geometrische Progress Tabulen, sambt gründlichem unterricht wie solche nützlich in allerley Rechnungen zugerbrauchen und verstanden werden sol.* Gedruckt in der alten Stadt Prag bei Paul Sessen, der Löblichen Universitet Buchdruckern im Jahr 1620.

See also Gieswald.

Cajori

[1911]: Florian Cajori, "Horner's method of approximation anticipated by Ruffini," *Bull. Am. Math. Soc.*, Vol. 17, 1911, pp. 409–414.

[1919]: ———, *A History of Mathematics*, 2nd ed., New York, 1919.

Cauchy

Oeuvres: Oeuvres complètes d'Augustin Cauchy, 27 Vols., Paris, 1882–1938.

2, III [1821]: Augustin Louis Cauchy, "Sur la formule de Lagrange relative à l'interpolation," *Cours d'Analyse de l'École Royale Polytechnique* (Analyse algébrique): Note V, Paris, 1821 = *Oeuvres*, Sér. 2, Vol. III, 1897, pp. 429–433.

2, III [1821′]: ———, "Sur la resolution numérique des équations," *Cours d'Analyse*: Note III = *Oeuvres*, Sér. 2, Vol. III, 1897, pp. 378–425.

2, VI [1826]: ———, "*Sur l'analyse des sections angulaires,*" *Exercises de Mathématiques* (Anciens Exercises), 1826 = *Oeuvres*, Sér. II, Vol. VI, 1887, pp. 11–22.

2, VII [1827]: ———, "Sur l'analogie des puissances et des différences," *Exercises de Mathématiques* (Anciens Exercises), Seconde Année, Paris, 1827 = *Oeuvres*, Sér. 2, Vol. VII, 1889, pp. 198–235.

2, VII [1827′]: ———, "Addition a l'article précédent," *Exercises de Mathématiques* (Anciens Exercises), Seconde Année, Paris, 1827 = *Oeuvres*, Sér. 2, Vol. VII, 1889, pp. 236–254.

1, II [1827″]: ———, "Mémoire sur le développements des fonctions en séries périodiques," *Mém. Ac. Sc. Paris*, Vol. VI, 1827, pp. 603–612 = *Oeuvres*, Sér. 1, Vol. II, 1908, pp. 12–19.

2, IV [1829]: ———, "Sur la détermination approximative des racines d'une équation algébrique ou transcendante," *Leçons sur le calcul différentiel: Note*, Paris, 1829 = *Oeuvres*, Sér. 1, Vol. IV, 1899, pp. 573–609.

1, V [1840]: ———, "Sur les fonctions interpolaires," *C.R. Ac. Sc. Paris*, Vol. XI, 1840, pp. 775–789 = *Oeuvres*, Sér. 1, Vol. V, 1885, pp. 409–424.

1, V [1840′]: ———, "Sur la résolution numérique des équations algébriques et transcendantes," *C.R. Acad. Sc. Paris*, Vol. XI, 1840, pp. 829–859 = *Oeuvres*, Sér. 1, Vol. V, 1885, pp. 455–473.

1, VI [1841]: ———, "Mémoire sur diverses formules d'Analyse," *C.R. Ac. Sc. Paris*, Vol. XII, 1841, pp. 283–298 = *Oeuvres*, Sér. 1, Vol. VI, 1888, pp. 63–78.

Chebyshev

Works: Collected Works, 5 Vols. Moscow–Leningrad, 1946–1951.

[1855]: Pafnutiy L. Chebyshev, "Sur les fractions continues," *Ac. Sc. St. Peter*, Vol. XII, 1855, pp. 287–288; *Liouville Jour. Math.*, Vol. III, 1859, pp. 289–323.

[1874]: ———, "Sur les quadratures," *Jour. de Math. pures et appl.*, Sér. 2, Vol. XIX, 1874, pp. 19–34 = *Works*, Vol. III, 1948, pp. 49–62.

Clairaut [1759]:

Alexis Claude Clairaut, "Sur l'orbite apparente du Soleil autour de la terre, en ayant égard aux perturbations produites par les actions de la lune & des planètes principales," *Mém. (Hist.) d. Acad. des Sc. Paris*, 1754 (1759), pp. 521–564. See esp. Art. Quatrième: "De la manière de convertir une fonction quelconque T de t en une série telle que $A + B\cos. t + C\cos. 2t + D\cos. 3t + \&c.$," pp. 544–564.

Clemence: *See* Brouwer

Collatz [1951]:

Lothar Collatz, *Numerische Behandlung von Differentialgleichungen*, Berlin–Göttingen–Heidelberg, 1951.

Collins [1712]:

John Collins, *Commercium epistolicum D. Johannis Collins, et aliorum de analysi promota:* London, 1712.

Cooley *CTA:*

James W. Cooley and John W. Tukey, "An algorithm for the machine calculation of complex Fourier series," *Math. of Comput.*, Vol. 19, 1965, pp. 297–301.

Cotes

CAN: Roger Cotes, *Canontechnia sive Constructio Tabularum per Differentias* = *Opera Miscellanea* (appended to R. Smith ed. of Cotes, *Harmonia*).

HAR: ———, *Harmonia Mensurarum, sive Analysis et Synthesis Per Rationum & Angulorum. Mensuras Promotae: Accedunt alia Opuscula Mathematica . . . .* Edidit et Auxit Robt. Smith, Cambridge, 1722. This contains, *inter alia*, "De Methode Differentiali Newtoniana."

Courant

[1928]: Richard Courant, Kurt Friedrichs, and Hans Lewy, "Über die partiellen Differenzengleichungen der Mathematischen Physik," *Math. Ann.*, Vol. 100, 1928, pp. 32–74.

Methoden: Courant u. Hilbert, *Methoden der Mathematischen Physik*, 2 Vols., Berlin, 1st ed. 1924, 2nd ed. 1930/38. English transl. by Courant, New York, 1953.

Cramer [1750]:

Gabriel Cramer, *Introduction à l'analyse des lignes courbes algébraiques*, Geneva, 1750.

Dandelin
Germinal Pierre Dandelin, "Recherches sur la résolution des équations numériques," *Mém. de l'Acad. Bruxelles*, Vol. 3, 1826, pp. 7–71, 153–159.

Delambre *MOD*:
J. B. J. Delambre, *Histoire de l'Astronomie Moderne*, 2 Vols., Paris, 1821, 1827 (reprinted New York, 1969, in 2 Vols.)

De Morgan [1842]:
Augustus De Morgan, *The Differential and Integral Calculus*, London, 1842.

Descartes [1637]:
René Descartes, *La Géometrie*, Appendix to *Discours de la Méthode pour bien conduire sa Raison et chercher la Vérité dans les Sciences*, Leyden, 1637; Engl. transl. by D. E. Smith and M. L. Latham, New York, 1954.

Dickson [1922]:
Leonard E. Dickson, *First Course in the Theory of Equations*, New York, 1922.

Dreyer [1890]:
J. L. E. Dreyer, *Tycho Brahe, A. Picture of Scientific Life and Work in the Sixteenth Century*, Edinburgh, 1890 (New York, reprint 1963.)

Encke
[1841]: Johann F. Encke, "Allgemeine Auflösungen der numerischen Gleichungen," *Berlin Astron. Jahrb.*, 1841, pp. 231–338; *Jour. für reine u. angew. Math.*, Vol. XXII, 1841, pp. 193–248; *Nouv. Ann. Math.*, Vol. XIII, 1854, pp. 81–91.
[1857]: ———, "Über die allgemeinen Störungen der Planeten," *Berlin Astron. Jahrb.*, 1857, pp. 319–397.
[1888]: ———, *Gesammelte mathematische und astronomische Abhandlungen* I, Berlin, 1888, pp. 125–187.

Ephemeris [1952/59]:
Improved Lunar Ephemeris 1952/59. Supplement to American and British Nautical Almanacs, Washington, 1954.

Euler
Opera: Leonhardi Euleri Opera Omnia, 72 Vols., Leipzig and Berlin, 1911–1975.
2, I [1736/42]: Leonhard Euler, *Mechanica sive motus scientia analytice exposita*, 2 Vols., 1736/42 = *Opera*, Ser. 2, Vol. I, 1912, ed. by P. Stäckel.
1, XXII [1738]: ———, "Methodus generalis summandi progressiones," *Comm. Acad. Imp. Petrop.*, Vol. VI, 1738, pp. 68–97.
1, VIII [1748] and 1, IX [1748]: ———, *Introductio in analysin infinitorum*, 2 Vols., 1st ed. Lausanne, 1748, 2nd ed. Lausanne, 1797 (French transls. by Pezzi, Strasbourg, 1786; by J. B. Labèy, Paris, 1796 and 1835; German transls. by J. A. C. Michelsen, 1st ed. Berlin, 1788, 2nd ed. Berlin 1833–1836; by H. Maser, Berlin, 1885) = *Opera*, Ser. 1, Vol. VIII, 1922, ed. by A. Krazer and F. Rudio, 1922 and *Opera*, Ser. 1, Vol. IX, 1945, ed. by A. Speiser.
1, XVII [1749]: ———, "De la controverse entre Mrs. Leibniz et Bernoulli sur les logarithmes des nombres négatifs et imaginaires," *Mém. de l'Acad. Sc. Berlin*, Vol. V, 1749, 1751, pp. 139–179 = *Opera*, Ser. 1, Vol. XVII, 1915, ed. by A. Gutzmer, pp. 195–232.
XXIII [1753]: ———, *Theoria Motus Lunae exhibens omnes eius Inaequalitates*, St. Pétersbourg, 1753 = *Opera*, Ser. 2, Vol. XXIII, 1969, ed. by L. Courvoisier, J. O. Fleckenstein, pp. 64–336.
1, X [1755]: ———, *Institutiones calculi differentialis cum ejus usu in analysi finitorum ac doctrina serierum*, St. Pétersbourg, 1755; Pavia, 1787 (German transl. by J. A. C. Michelsen, Berlin, 1790–1793) = *Opera*, Ser. 1, Vol. X, 1913, ed. by G. Kowalewski.
1, XVII [1756/57]: ———, "De expressione integralium per factores," *Comm. acad. sc. Petrop.*, Vol. VI, 1756/57, 1761, pp. 115–154 = *Opera*, Ser. 1, Vol. XVII, 1936, ed. by H. Dulac, pp. 233–267.

1, XVII [1762/65]: ———, "Observationes circa integralia formularum $\int x^{p-1}\,dx(1 - x^n)^{(q/n)-1}$ posito post integrationem $x = 1$," *Mél. de Phil. et de Math. Soc. Roy. Turin*, Vol. 3_2, 1762/65, 1766, pp. 156–177 = *Opera*, Ser. 1, Vol. XVII, pp. 268–315.

1, XI and 1, XII [1768/69]: ———, *Institutiones calculi integralis*, 2 Vols., St. Pétersbourg, 1st ed. 1768–1769, 2nd ed. 1792, 3rd ed. 1824–1827 (German transl. by J. Salomon, Vienna, 1828–1829) = *Opera*, Ser. 1, Vol. XI–XII, 1913 and 1914, ed. by F. Engel and L. Schlesinger.

1, XV [1783]: ———, "De eximio usu methodi interpolationum in serierum doctrina," *Opuscula analytica*, Vol. 1, St. Pétersbourg, 1783, pp. 345–352 = *Opera*, Ser. 1, Vol. XV, 1927, pp. 435–497.

Everett

[1899]: J. D. Everett, "On the deduction of increase-rates from physical and other tables," *Nature*, Vol. 60, 1899, p. 271.

[1899′]: ———, "On the notation of the calculus of differences," *Brit. Ass. Rep.*, 1899, pp. 645–646.

[1900]: ———, "On a central-difference interpolation formula," *Brit. Ass. Rep.*, 1900, pp. 648–650.

[1900′]: ———, "On Newton's contributions to central-difference interpolation," *Brit. Ass. Rep.*, 1900, p. 650.

[1900″]: ———, "On the algebra of difference-tables," *Quart. J. Pure Appl. Math.*, Vol. 31, 1900, pp. 357–376.

[1901]: ———, "On interpolation formulae," *Quart. J. Pure Appl. Math.*, Vol. 32, 1901, pp. 306–313.

[1901′]: ———, "On a new interpolation formula," *J. Inst. Act.*, Vol. 35, 1901, pp. 452–458.

Fejér [1933]:

L. Fejér, "Mechanische Quadraturen mit positiven Cotesschen Zahlen," *Math. Zeit.*, Vol. 37, 1933, pp. 287–309.

Feller *PROB*:

William Feller, *An Introduction to Probability Theory and its Applications*, 2 Vols., New York, 1st ed. 1950, 2nd ed. 1957, 3rd ed. 1966.

Fourier [1818]:

Joseph B. J. Fourier, "Note relative aux vibrations des surfaces élastiques et au movement des ondes," *Bull. de la Soc. Philomath.*, Paris, 1818, pp. 129–136.

Fraser

[1919]: D. C. Fraser, "Newton's Interpolation formulas," *Jour. Inst. Act.*, Vol. 51, 1918, pp. 77–106, and pp. 211–232.

See also Newton.

Fuss

P.-H. Fuss, *Correspondance Mathématique et Physique de Quelques Célèbres Géometres du XVII ème siècle*, St. Pétersbourg, 1843, Tome 1, *Correspondance entre Léonard Euler et Chr. Goldbach 1729–1763.*

Gauss

Werke: Carl Friedrich Gauss, *Werke*, ed. by E. Schering, F. Klein, M. Brendel, and L. Schlesinger, 12 Vols., Göttingen, 1870–1933.

Theoria: ———, *Theoria Motus corporum coelestium in sectionibus conicis solem ambientium*, Hamburg, 1809. Eng. transl. by C. H. Davis, *Theory of the motion of the Heavenly Bodies moving about the Sun in Conic Sections*, Boston, 1857. (Reprinted New York, 1963); German transl. by C. Haase, Hannover, 1865, French transl. by E. Dubois, 1864. The German transl. is entitled *Theorie der Bewegung der Himmels-Körper welche in Kegelschnitten die Sonne umlaufen.*

KQ: ———, *Abhandlungen zur Methode der Kleinsten Quadrate*, German transl. by A. Börsch and P. Simm. Berlin, 1887; (reprinted Würzburg, 1964.) This is a collection of all Gauss's work on least squares in German translation. A French transl. by J. L. F. Bertrand appeared in Paris, 1855, cf. an unpublished English transl. of this French text by H. Trotter, Princeton. The French edition is entitled *Méthode des moindres carrés. Mémoires sur la combinaison des observations.*

IV *KQ*1: ———, "Theoria combinationis observationum erroribus minimis obnoxiae, *Comm. Soc. Sc. Gött. Math.*, Vol. V, 1819–1822, Part I, pp. 33–62, Part II, pp. 63–90 and "Supplementum theoriae combinationis observationum erroribus minime obnoxiae," *Comm. Soc. Sc. Gött. Math.*, Vol. VI, 1823–1827, pp. 57–98, Vol. VII, pp. 89–148 = *Werke*, Vol. IV, 1880, pp. 1–26, 29–53 and pp. 57–108.

IV *KQ*2: ———, "Bestimmung der Genauigkeit der Beobachtungen," *Zeit. für Ast.*, Vol. I, 1816, pp. 185–197 = *Werke*, Vol. IV, 1880, pp. 109–117.

VI Pallas: ———, "Disquisitio de elementis ellipticis Palladis ex oppositionibus annorum 1803, 1804, 1805, 1807, 1808, 1809," *Comm. Soc. Sc. Gött.*, Vol. I, 1808–1811 = *KQ*, pp. 118–128 = *Werke*, Vol. VI, 1874, pp. 1–24.

III *NI*: ———, "Methodus nova integralium valores per approximationem inveniendi," *Comm. Soc. Sc. Gött. Math.*, Vol. III, 1816, pp. 39–76 = *Werke*, Vol. III, 1876, pp. 163–196, pp. 202–206.

III *HGF*: ———, "Disquisitiones generales circa seriem infinitem

$$1 + \frac{\alpha\beta}{1\cdot\gamma}x + \frac{\alpha(\alpha+1)\beta(\beta+1)}{1\cdot2\cdot\gamma(\gamma+1)}xx + \frac{\alpha(\alpha+1)(\alpha+2)\beta(\beta+1)(\beta+2)}{1\cdot2\cdot3\cdot\gamma(\gamma+1)(\gamma+2)}x^3 + \text{etc.,"}$$

Comm. Soc. Sc. Gött. Math., Vol. II, 1813, pp. 1–43 = *Werke*, Vol. III, 1876, pp. 123–162.

III *TI*: ———, "Theoria interpolationis methodo nova tractata," *Nachlass*, *Werke*, Vol. III, pp. 265–327.

Gerling [1843]:

Christian Ludwig Gerling, *Die Ausgleichungs-Rechnungen der praktischen Geometrie oder die Methode der kleinsten Quadrate mit iherer Anwendungen für geodätische Aufgaben*, Hamburg und Gotha, 1843.

Girard [1629]:

Albert Girard, *Invention nouvelle en l'algèbre tant pour la solution des équations, que pour recognoistre le nombre des solutions quelles reçoivent, avec plusiers choses qui sont necessaires à la perfection de ceste divine science*, Amsterdam, 1629.

Gieswald [1856]:

Gieswald, "Zur Geschichte und Literatur der Logarithmen," *Archiv der Math. u. Phys.*, Vol. 26, 1856, pp. 316–334.

Glaisher

[1871]: J. W. L. Glaisher, "On the constants that occur in certain summations by Bernoulli's series," *Proc. London Math. Soc.* (1) Vol. 4, 1871–1873, pp. 48–56.

[1880]: ———, "Une identité trigonométrique," *Assoc. franc. Congrès de Reims*, Vol. 9, 1880, pp. 222–223.

[1898]: ———, "The Bernoullian functions," *Quart. J. Pure Appl. Math.*, Vol. 29, 1898, pp. 1–168; Vol. 42, 1911, pp. 86–157.

[1898′]: ———, "General summation-formulae in finite differences," *Quart. J. Pure Appl. Math.*, Vol. 29, 1898, pp. 303–328.

[1914]: ———, "On Eulerian numbers (formulae, residues, end-figures) with the values of the first twenty-seven," *Quart. J. Pure Appl. Math.*, Vol. 45, 1914, pp. 1–51.

Goldstine

[1947]: Herman H. Goldstine and John von Neumann, "Numerical Inverting of Matrices of high order," *Bull. Amer. Math. Soc.*, Vol. 53, 1947, pp. 1021–1099.

[1951]: Goldstine and von Neumann, "Numerical Inverting of Matrices of high order II," *Proc. Amer. Math. Soc.*, Vol. 2, 1951, pp. 188–202.

[1959]: Goldstine, F. J., Murray, and von Neumann, "The Jacobi Method for real symmetric Matrices," *J. Assoc. Comp. Mach.*, Vol. 6, 1959, pp. 59–96.

[1972]: Goldstine, *The Computer from Pascal to von Neumann*, Princeton, 1972.

Graeffe [1837]:

Carl Heinrich Graeffe, *Die Auflösung der höheren numerischen Gleichungen*, Zürich, 1837.

Gregory

EG: James Gregory, *Exercitationes geometricae. Appendicula ad veram circuli et hyperbolae quadraturam. N. Mercatoris quadratura hyperbolae geometrice demonstrata, analogia inter lineam meridianam planisphaerii nautici et tangentes artificiales geometrica demonstrata, seu quod secantium naturalium additio efficiat tangentes artificiales. Item, quod tangentium naturalium additio efficiat secantes artificiales. Quadratura conchoidis, quadratura cissoidis, methodus facilis et accurata componendi secantes et tangentes artificiales*, London, 1668.

GU: ———, *Geometriae pars universalis, inserviens quantitatum curvarum transmutationi et mensurae*, Padua, 1668.

GTV: ———, *James Gregory Tercentenary Memorial Volume, containing his correspondence with John Collins and his hitherto unpublished mathematical manuscripts* . . . ed. H. W. Turnbull, London, 1939.

Halley [1695]:

Edmund Halley, "A most compendious and facile Method for Constructing the Logarithms, exemplified and demonstrated from the Nature of Numbers, without any regard to the Hyperbola, with a speedy Method for finding the Number from the Logarithm given," *Phil. Trans.*, Vol. XIX, 1695–1697, pp. 58–67.

Harriot

Thomas Harriot, *De Numeris Triangularibus et inde de Progressionibus Arithmeticis Magisteria magna*, 1611?

See also Lohne.

Heath *ARC*:

T. L. Heath, *The Works of Archimedes*, Cambridge, 1897, plus Supplement, 1912 (reprinted, New York.)

Henderson [1926]:

James Henderson, *Bibliotheca Tabularum Mathematicarum, Being a descriptive catalogue of mathematical tables*, Part I, Logarithmic Tables (A. Logarithms of numbers). Tracts for Computers, No. XIII, Cambridge, 1926. This is an excellent bibliography of logarithmic tables.

Hermite

Oeuvres: Oeuvres de Charles Hermite, ed. by E. Picard, 4 Vols., Paris, 1905–1917.

II [1859]: Charles Hermite, "Sur l'interpolation," *C. R. Ac. Sc. Paris*, Vol. XLVIII, 1859, pp. 62–66 = *Oeuvres*, Vol. II, 1908, pp. 87–92.

II [1862]: ———, "Sur un nouveau développement en série des fonctions," *C. R. Ac. Sc. Paris*, Vol. LVIII, 1864, pp. 93–100 and 266 = *Oeuvres*, Vol. II, 1908, pp. 293–308.

[1873]: ———, *Cours d'Analyse de l'École Polytechnique*, Paris, 1873.

[1875]: ———, "Sur la fonction de Jacob Bernoulli," *Jour. f. reine u. angew. Math.*, Vol. LXXIX, 1875, pp. 339–344 = *Oeuvres*, Vol. III, 1912, pp. 215–221.

III [1876]: ———, "Extrait d'une lettre de M. Ch. Hermite à M. Borchardt, Sur les nombres de Bernoulli," *Jour. f. reine u. angew. Math.*, Vol. LXXXI, 1876, pp. 93–95 = *Oeuvres*, Vol. III, pp. 211–214.

III [1876′]: ———, "Extrait d'une lettre de M. Ch. Hermite à M. Borchardt, Sur la formule de Maclaurin," *Jour. f. reine u. angew. Math.*, Vol. LXXXIV, 1877–1878, pp. 64–69 = *Oeuvres*, Vol. III, pp. 425–431.

III [1878] ———, "Sur la formule d'interpolation de Lagrange," *Jour. f. reine u. angew. Math.*, Vol. LXXXIV, 1878, pp. 70–79 = *Oeuvres*, Vol. III, pp. 432–443.

IV [1885]: ———, "Sur une identité trigonométrique," *Nouv. Ann. de Math.*, Sér. 3, Vol. IV, 1885, pp. 57–59 = *Oeuvres*, Vol. IV, 1917, pp. 206–208.

IV [1893]: ———, "Sur une extension de la formule de Stirling," *Math. Ann.*, Vol. XLI, 1893, pp. 581–590 = *Oeuvres*, Vol. IV, pp. 378–388.

[1894]: ———, "Remarque sur les nombres de Bernoulli et les nombres d'Euler," *Bull. de la Soc. Sc. de Bohême*, 2nd cl., 1894 = *Oeuvres*, Vol. IV, pp. 393–396.

IV [1895]: ———, "Sur les nombres de Bernoulli," *Mathesis*, 2nd Sér., Vol. V., Suppl. 11, 1895, pp. 1–7 = *Oeuvres*, Vol. IV, 1917, pp. 405–411.

IV [1896]: ———, "Sur les polynomes de Bernoulli," *Jour. f. reine u. angew. Math.*, Vol. CXVI, 1896, pp. 133–156 = *Oeuvres*, Vol. IV, pp. 437–447. (This is an extract of a letter to Sonin in St. Petersburg in response to a memoir by Sonin on the subject.)

Heun [1900]:

K. Heun, "Neue Methode zur approximativen Integration der Differentialgleichungen einer unabhängigen Veränderlichen," *Zeit. Math. Phy.*, Vol. 45, 1900, pp. 23–38.

Hilbert: *See* Courant.

Hofmann

St. Vincent: Joseph E. Hofmann, "Das Opus Geometricum des Gregorius a S. Vincentio und seine Einwirkung auf Leibniz," *Abh. der Preuss. Akad. der Wiss.*, 1941, Nr. 13, Berlin, 1942, pp. 1–80.

Mercator: ———, "Nicolaus Mercators Logarithmotechnica (1668)" and "Weiterbildung der logarithmischen Reihe Mercators in England," *Deutsche Math.*, Vol. 3 (1938), pp. 446–466 and 598–605.

Leibniz: ———, *Die Entwicklungsgeschichte der Leibniz-schen Mathematik während des Aufenhalts in Paris* (*1672–1676*), Munich, 1949. Hofmann then produced an English transl., *Leibniz in Paris* 1672–1676, *His Growth to mathematical Maturity*, Cambridge, 1974. References in the text are to the English translation.

Horner [1819]:

W. G. Horner, "A new method of solving numerical equations of all orders, by continuous approximation," *Phil. Trans.*, Vol. 109, 1819, pp. 308–385.

Hutton [1801]:

C. Hutton, *Mathematical Tables; containing the common, hyperbolic, and logistic logarithms. Also sines, tangents, secants and versed sines, both natural and logarithmic. Together with several other tables useful in mathematical calculations*... London, 1st ed. 1785; 2nd ed. 1794; 3rd ed. 1801;... . References are to the 3rd ed. and in particular to the 180-page Introduction.

Jacob [1600]:

Simon Jacob, *Ein New und Wolgegründt Rechenbuch auf den Linien und Ziffern*..., Frankfurt a. M., 1600.

Jacobi

Werke: C.G.J. Jacobi's gesammelte Werke, ed. by K. Weierstrass, 8 Vols., Berlin, 1881–1891.

VI [1826]: Carl Gustav Jacobi, "Über Gauss' neue Methode die Werthe der Integrale näherungsweise zu finden," *Jour. f. reine u. angew. Math.*, Vol. I, 1826, pp. 301–308 = *Werke*, Vol. VI, 1891, pp. 1–11.

VI [1834]: ———, "De usu legitimo formulae summatoriae Maclaurinianae," *Jour. f. reine u. angew. Math.*, Vol. XII, 1834, pp. 263–272 = *Werke*, Vol. VI, 1891, pp. 64–75.

III [1845]: ———, "Über eine neu Auflösungsart der bei der Methode der kleinsten Quadrate vorkommenden linearen Gleichungen," *Astronom. Nachr.*, Vol. XXXII, 1845, pp. 297–306 = *Werke*, Vol. III, 1884, pp. 468–478.

VII [1846]: ———, "Über eine leichtes Verfahren die in der Theorie der Säcularstörungen vorkommenden Gleichungen numerisch aufzulösen," *Jour. f. reine u. Math.*, Vol. XXX, 1846, pp. 51–94 = *Werke*, Vol. VII, 1891, pp. 97–144.

III [1846′]: ———, "Über die Darstellung einer Reihe gegebener Werthe durch eine gebrochene rationale Function," *Jour. f. reine u. angew. Math.*, Vol. XXX, 1846, pp. 127–156 = *Werke*, Vol. III, 1884, pp. 479–511.

III [1884]: ———, "Über die Hauptaxen der Flächen der zweiten Ordnung," *Jour. f. reine u. angew. Math.*, Vol. II, 1827, pp. 227–233 = *Werke*, Vol. III, 1884, pp. 46–53.

Jones [1706]:

William Jones, *Synopsis palmiorum matheseos or new introduction to mathematics*, London, 1706.

See also Newton.

Jordan [1939]:

Charles Jordan, *Calculus of Finite Differences*, Budapest, 1939 (reprinted New York, 1947.)

Kepler

Werke: Johannes Kepler, *Gesammelte Werke*, 19 Vols., ed. by W. von Dyck and M. Caspar, Munich, 1937–1975.

IX *CHIL: Joannis Kepleri Imp. Caes. Ferdinandi II. Mathematici Chilias Logarithmorum Ad Totidem Numeros Rotundos, Praemissâ Demonstratione Legitima Ortus Logarithmorum eorumq usus Quibus Nova Traditur Arithmetica, Seu Compendium, Quo Post Numerorum Notitiam nullum nec admirabilius, nec utilius solvendi pleraq: Problemata Calculatoria, praesertim in Doctrina Triangulorum, citra Multiplicationis, Divisionis, Radicumq; extractionis, in Numeris prolixis, labores molestissimos.* Ad Illustriss. Principem & Dominum Dn. Philippum Landgravium Hassiae Etc. Cum Privilegio Authoris Caesareo. Marpurgi, Excusa Typis Casparis Chemlini MDCXXIV (1624) (4°.), and *Joannis Kepleri, Imp. Caes. Ferdinandi II. Mathematici, Supplementum Chiliadis Logarithmorum, Continens Praecepta De Eorum Usu,* Ad Illustriss. Principem et Dominum, Dn. Philippum Landgravium Hassiae, etc. Marpurgi Ex officina Typographica Casparis Chemlini, MDCXXV, pp. 111–216 (4°.) = *Werke*, Vol. IX, 1960, pp. 275–426.

Maseres, F. *Scriptores Logarithmici, or a Collection of Several Curious Tracts on the Nature and Construction of Logarithms*, Vol. I, London, 1791. (This contains Kepler's work plus Maseres's table.)

X *RT;* ———, *Tabulae Rudolphinae, quibus astronomicae scientiae, temporum longinquitate collapsae Restauratio continetur; A Phoenice illo Astronomorum Tychone, Ex Illustri & Generosa Braheorum in Regno Daniae familia oriundo Equite,...* Opus hoc ad usus praesentium & posteritatis, typis, numericis propriis, caeteris, & praelo Jonae Saurii, Reip. Ulmanae Typographi, in publicum extulit, & Typographicus operis Ulmae curator officit. Ulm, 1627 (Folio.) = *Werke*, Vol. X, 1969.

VII *Epit:* ———, *Epitome astronomiae Copernicanae*, Linz, 1618 = *Werke*, Vol. VII, 1953. (Eng. transl. by C. G. Wallis of Books IV and V in *Great Books of the Western World*, Vol. 16, Chicago, 1952, pp. 845–1004.)

VI *HM:* ———, *Harmonice Mundi...*, Linz, 1619 = *Werke*, Vol. VI, Munich, 1940. (Engl. transl. by C. G. Wallis of Book V in *Great Books of the Western World*, Vol. 16, Chicago, 1952, pp. 1009–1085.)
III AN: ———, *Astronomia Nova αἰτιολογητός seu Physica coelestis, tradita commentariis de Motibus Stellae Martis, Ex Observationibus G. V. Tychonis Brahe*, Prague, 1609. (German transl. by M. Caspar *Neue Astronomie*, Munich–Berlin, 1929.)

König: *See* Runge

Kummer [1843]:
Ernest Eduard Kummer, "Bemerkungen über die cubische Gleichung, durch welche die Haupt-Axen der Flächen zweiten Grades bestimmt werden," *Jour. f. reine u. angew. Math.*, Vol. XXVI, 1843, pp. 268–272.

Kutta [1901]:
W. Kutta, "Beitrag zur näherungsweisen Integration totaler Differentialgleichungen," *Zeit. Math. Phy.*, Vol. 46, 1901, pp. 435–453.

de Lagny [1719]:
Thomas F. de Lagny, "Mémoire sur la quadrature du cercle, et sur la mesure de tout arc, tout secteur et tout segment donné," *Mém. Acad. Sci. Paris* (1719), 1721, pp. 135–145.

Lagrange
Oeuvres: Oeuvres de Lagrange, ed. by J.–A. Serret and G. Darboux, 14 Vols., Paris, 1867–1892.
I [1759]: Joseph Louis Lagrange, "Sur l'intégration d'une équation différentielle à différences finies, qui contient la théorie des suites récurrentes," *Miscell. Taurin.*, Vol. I, 1759, pp. 33–42 = *Oeuvres*, I, 1867, pp. 23–36.
I [1759′]: ———, "Recherches sur la nature et la propagation du son," *Miscell. Taurin.*, Vol. I, 1759 = *Oeuvres*, Vol. I, 1867, pp. 37–148.
I [1760]: ———, "Nouvelles recherches sur la nature...," *Miscell. Taurin.*, Vol. II, 1760–1761, pp. 11–172 = *Oeuvres*, Vol. I, pp. 149–316.
I [1761]: ———, "Addition aux premières recherches...," *Miscell. Taurin.*, Vol. II, 1760–1761, pp. 323–344 = *Oeuvres*, Vol. I, pp. 317–332.
I [1762]: ———, "Solution de différents problèmes de calcul integral," *Miscell. Taurin.* Vol. III, 1762–1765 = *Oeuvres*, Vol. I, pp. 469–668.
II [1767]: ———, "Sur la résolution des équations numériques," *Mém. Acad. Sci. Berlin*, Vol. XXIII, 1767, pp. 311–352 = *Oeuvres*, Vol. II, 1868, pp. 539–578.
II [1770]: ———, "Additions au mémoire sur la résolution des équations...," *Mém. Acad. Sci. Berlin*, Vol. XXIV, 1768, pp. 111–180 = *Oeuvres*, Vol. II, pp. 581–652.
III [1771]: ———, "Sur le problème de Kepler," *Nouv. Mém. Acad. Sci. Berlin*, Vol. XXV, 1769, pp. 204–233 = *Oeuvres*, Vol. III, 1869, pp. 113–138.
III [1772]: ———, "Sur une nouvelle espèce de calcul relatif à la différentiation et à la intégration des quantités variables," *Nouv. Mém. Acad. Sci. Berlin*, Vol. XXVIII, 1772, pp. 186–221 = *Oeuvres*, Vol. III, pp. 441–476.
VI [1772]: ———, "Recherches sur la manière de former des tables des planètes d'après les seules observations," *Mém. Acad. Sci. Paris*, Vol. XXVI, 1772 = *Oeuvres*, Vol. VI, 1873, pp. 509–627.
IV [1775]: ———, "Recherches sur les suites récurrentes dont les termes varient de plusiers manières différentes, ou sur l'intégration des équations linéaires aux différences finies et partielles; et sur l'usage de ces équations dans la théorie des hasards," *Nouv. Mém. Acad. Sci. Berlin*, Vol. XXX, 1775 = *Oeuvres*, Vol. IV, 1869, pp. 151–251.
IV [1778]: ———, "Sur le problème de la determination des orbites des comètes d'après trois observations," *Nouv. Mém. Acad. Sci. Berlin*, Vol. XXXIII, 1778,

pp. 111–161 and Vol. XXXVII, 1783, pp. 296–332 = *Oeuvres*, Vol. II, pp. 439–532.

VII [1778] ———, "Sur les Interpolations" = "Über das Einschalten, nebst Tafeln und Beispielen," German transl. by Schulze, *Astron. Jahrbuch oder Ephemeriden f. das Jahr 1783*, Berlin, 1780, pp. 35–61 = Lagrange, *Mathematische Elementarvorlesungen*, Leipzig, 1880, German transl. by H. Niedermüller, pp. 535–553. The original French was reproduced from Lagrange's papers in the Bibliothèque de l'Institut de France.

V [1783]: ———, "Sur une méthode particulière d'approximation et d'interpolation," *Nouv. Mém. Acad. Sci. Berlin*, Vol. XXXVIII, 1783, pp. 279–289 = *Oeuvres*, Vol. V, 1870, pp. 517–532.

V [1792]: ———, "Mémoire sur la méthode d'interpolation," *Nouv. Mém. Acad. Sci. Berlin*, Vol. XLV, 1792–1793, pp. 276–288 = *Oeuvres*, Vol. V, pp. 663–684.

V [1793]: ———, "Mémoire sur l'expression du terme générale des séries récurrentes, lorsque l'équation génératrice a des racines égales," *Nouv. Mém. Acad. Sci. Berlin*, Vol. XLV ,pp. 247–257 = *Oeuvres*, Vol. V, pp. 627–641.

VII [1795]: ———, "Sur l'usage des courbes dans la solution des problèmes," in "Leçons élémentaires sur les mathématiques, données à l'École Normale en 1795, leçon cinquième," *Jour. Éc. Poly.*, 2, cah. 7–8, 1812, pp. 173–278 = *Oeuvres*, Vol. VII, 1877, pp. 271–287 (German transl. by H. Niedermüller, "Verwendung der Kurven bei der lösung der Probleme," in Lagrange, *Mathematische Elementarvorlesungen*, Leipzig, 1880, pp. 100–196.)

Lambert [1770]:

Johann H. Lambert, *Zusätze zu den logarithmen und trigonometrischen Tabellen*, Berlin, 1770.

Laplace

OC: Oeuvres complètes de Laplace, 14 Vols., Paris, 1787–1912.

O: Oeuvres de Laplace, 7 Vols., Paris, 1843–1847 = *Théorie analytique des probabilités*, 1st ed. Paris, 1812, plus *Traité de mécanique céleste*, 4 Vols., Paris, 1798–1825. (English transl. by N. Bowditch, 4 Vols., Boston, 1829–1839.) This reprinting was authorized by Louis Phillipe on June 15, 1842.

VIII [1772]: Pierre Simon Laplace, "Recherches sur le calcul intégral et sur le système du monde," *Mém. (Hist.) Acad. Sci. Paris*, 1772, IIe Partie 1776, pp. 267–376 and 533–554 = *OC*, Vol. VIII, 1891, pp. 369–477. (There are "Additions" on pp. 478–501.)

VIII [1774]: ———, "Mémoire sur les suites récurro-récurrentes et sur leur usage dans la théorie des hasards," *Mém. prés. divers sav. Acad. Sc. Paris*, Vol. VI, 1774, pp. 353–371 = *OC*, Vol. VIII, pp. 5–24.

VIII [1776]: ———, "Recherches sur l'intégration des équations différentielles aux différences finies et sur leur usage dans la théorie des hasards," *Mém. prés. divers sav. Acad. Sc.*, Vol. VII (1773) 1776, pp. 37–163 = *OC*, Vol. VIII, pp. 69–197.

IX [1777]: ———, "Mémoire sur l'usage du calcul aux différences partielles dans la théorie des suites," *Mém. (Hist.). Acad. Sci. Sc. Paris*, (1777) 1780, pp. 99–122 = *OC*, Vol. IX, 1893, pp. 313–335.

X [1779]: ———, "Mémoire sur les suites," *Mém. (Hist.) Acad. Sc. Paris*, (1779) 1782, pp. 207–309 = *OC*, Vol. X, 1894, p. 1–89.

X [1780]: ———, "Mémoire sur la détermination des orbites des comètes," *Mém. (Hist.) Acad. Sc. Paris*, (1780) 1784, pp. 13–72 = *OC*, Vol. X, pp. 93–146.

I [1805]: ———, *Traité de mécanique céleste*, Vol. I, 4th ed., Paris, 1798–1825 = *OC*, Vol. I, 1878.

II [1805]: ———, *Mécanique céleste*, Vol. II = *OC*, Vol. II, 1878.

III [1805]: ———, *Mécanique céleste*, Vol. III = *OC*, Vol. III, 1878.

IV [1805]: ———, *Mécanique céleste*, Vol. IV = *OC*, Vol. IV, 1880.

XII [1809]: ———, "Mémoire sur les approximations des formules qui sont fonctions de très grands nombres et sur leur application aux probabilités," *Jour. Éc. Poly.* 8 cah., Vol. XV, (1809) 1810, pp. 301–412 = *OC*, Vol. X, pp. 209–338.

XIV [1809]: ———, "Mémoire sur divers points d'analyse," *Jour. Éc. Poly.* 8 cah., Vol. XV, 1810, pp. 229–265 = *OC*, Vol. XIV, 1912, pp. 178–214.

VII [1820]: ———, *Théorie analytique des probabiliés, Revue et augmentée par l'Auteur*, Paris, 3rd ed., 1820 = *OC*, Vol. VII, 1820.

Legendre

[1785]: Adrien Marie Legendre, "Recherches sur l'attraction des sphéroïdes homogènes," *Mém. prés. divers sav., Paris*, Vol. X, 1785, pp. 411–434.

[1805]: ———, *Nouvelles méthodes pour la determination des orbites des comètes*, Paris, 1805. See also, "Appendice sur la méthode des moindres quarrés" (it is dated 15 ventose, an 13 = 6 mars 1805). After the Appendix there are two lengthy Supplements, the former of which appeared in 1806 and the latter in 1820.

[1809]: ———, "Recherches sur diverses sortes d'intégrales définies," *Mém. Inst. France*, 10, (1809) 1810, pp. 416–509.

[1811]: ———, *Exercices de calcul intégral sur divers ordres de transcendantes et sur les quadratures*, 1–3, Paris, 1811–1819.

[1817]: ———, "Sur une méthode d'interpolation employée par Briggs dans la construction de ses grandes Tables trigonométriques," *Connaissance des Temps*, 1817, Paris, 1815, pp. 219–222; *Additions*, pp. 302–331.

[1819]: ———, "Méthodes diverses pour faciliter l'interpolation des grandes Tables trigonométriques," *Connaissance des Temps*, 1819, Paris, 1816.

[1825]: ———, *Traité des fonctions elliptiques et des intégrales eulériennes avec des Tables pour les faciliter le calcul numérique*, 1–3, Paris, 1825–1828.

Leibniz

LMS: Leibnizens mathematischen Schriften, 7 Vols., ed. by C. J. Gerhardt, 1849–1863, Berlin–Halle. (Reprinted Hildesheim, 1962.)

XXVII: Gottfried Wilhelm Leibniz, "Symbolismus memorabilis calculi algebraici et infinitesimalis in comparatione potentiarum et differentiarum, et de Lege Homogeneorum transcendentali," *Misc. Berolinens* = *LMS*, Series 2, Vol. 1, 1855, pp. 377–382.

XVII: ———, Letter XVII in *Commercium epistolicum Philosophicum et mathematicum*, Lausanne and Geneva, 1745, pp. 90–100.

Le Verrier

Urbain J. J. Le Verrier, *Connaissance des Temps*, 1843, Paris, pp. 31ff.

Lidstone [1922]:

S. J. Lidstone, "Notes on Everett's Interpolation Formula," *Edinburgh Math. Soc.*, Vol. XL, 1922, pp. 21–26.

Lindelöf

[1894]: Ernst Lindelöf, "Sur l'application des méthodes d'approximations successive à l'étude des intégrales réelles des équations différentielles ordinaires," *Jour. de Math. pur. et appl.*, Sér. IV, Vol. X, 1894, pp. 117–128.

[1902]: ———, "Quelques applications d'une formule sommatoire générale," *Acta Soc. Sc. Fennicae*, Vol. 31, 1902, Nr. 3.

[1903]: ———, "Sur une formule sommatoire générale," *Acta math.*, Vol. 27, 1903, pp. 305–311.

[1905]: ———, *Le calcul des résidus et ses applications à la théorie des fonctions*, Paris, 1905.

Lipshitz

Rudolph O. S. Lipschitz, *Lehrbuch der Analysis*, 2 Vols., Bonn, 1877–1880.

Lobachevsky

Nikolai Ivanovich Lobachevsky, *Algebra*, Kazan, 1834, pp. 349–359 (in Russian).

Lohne [1965]:

J. A. Lohne, "Thomas Harriot als Mathematiker," *Centaurus*, Vol. 11, 1965–1966, pp. 19–45.

Lubbock [1829]:

J. W. Lubbock, "On the comparison of various tables of annuities," *Camb. Phil. Trans.*, Vol. 3, 1829, pp. 321–341.

Maclaurin

Fluxions: Colin Maclaurin, *A Treatise of Fluxions*, 2 Vols., Edinburgh, 1742; 2nd ed., London, 1801; French transl. Paris, 1749. References in the text are to the 1742 edition.

Account: ———, *An Account of Sir Isaac Newton's Philosophical Discoveries in Four Books*, published by P. Murdock from the author's Manuscript papers, 1st ed., London, 1748; 2nd ed., 1750; 3rd ed., 1775.

Treat: ———, *A Treatise of Algebra*, 1st ed., London, 1748; 2nd ed., 1756; 3rd ed., 1771; 4th ed., 1779; 5th ed., 1788; 6th ed., 1796. This is a posthumous work.

Malmsten [1884]:

C. J. Malmsten, "Sur la formula

$$hu'_x = \Delta u_x - \frac{h}{2}\Delta u'_x + \frac{B_1h^2}{1\cdot 2}\Delta u''_x - \frac{B_2h^4}{1\cdots 4}\Delta u^{\text{IV}}_x + \text{etc.,"}$$

Jour. f. reine u. angew. Math., Vol. XXXV, 1847, pp. 55–82 = *Acta Math.*, Vol. V, 1884, pp. 1–46.

Markoff [1896]:

Andrei Andraevitch Markoff, *Difference Equations*..., St. Petersburg, 1889–1891 (in Russian); *Differenzenrechnung*, German transl. by T. Friesendorff and E. Prümm, Leipzig, 1896.

Mascheroni

[1790]: Lorenzo Mascheroni, *Adnotationes ad Calculum Integralem Euleri. In quibus nonnulla Problemata ab Eulero proposita resolvuntur*, Ticino, 1790 = Euler, *Opera*, Ser. 1, Vol. XII, 1914, pp. 415–487.

[1792]: ———, *Adnotationum ad calculum integralem Euleri*..., *Pars Altera*, Ticino, 1792 = Euler, *Opera*, 1, Vol. XII, pp. 488–542.

Maurice [1844]:

F. Maurice, "Mémoire sur les interpolations, contenant surtout, avec une exposition fort simple de leur théorie dans ce qu'elle a de plus utile pour les applications, la démonstration générale et complète de la méthode de quinti-section de Briggs et de celle de Mouton, quand les indices sont équidifférents, et du procédé exposé par Newton, dans ses Principes, quand les indices sont quelconques," *C. R. Acad. Sc. Paris*, Vol. 19, 1844, pp. 81–85 = *Connaissance des Temps* 1847. Eng. transl. by T. B. Sprague and J. H. Williams, *Jour. Inst. Act. and Assur. Mag.*, Vol. XIV, 1867, pp. 1–36.

Mayer [1770]:

(Johann) Tobias Mayer, *Tabulae motuum solis et lunae, novae et correctae; auctore Tobia Mayer: quibus accredit methodus longitudinum promota, eodem autore*, London, 1770. This is in Latin and in English transl. by N. Maskelyne, astron. royal. (Eng. transl., *New and correct tables of the motions of the sun and moon*.... London, 1770). The *Tabulae Solares ex Theoria gravitatis, et observationibus praecipus Gottingensibus, deductae* appear on pp. I–CXXX, the equation of time on pp. II–III.

Mercator [1667]:

Nicolaus Mercator, *Logarithmo-technia; sive methodus construendi logarithmos nova, accurata, et facilis; scripto antehac communicata, anno sc.*, 1667..., London, 1668.

Milne
W. E. Milne, *Numerical Calculus. Approximations, Interpolation, Finite Differences Numerical Integration and Curve Fitting*, Princeton, 1949.

Milne–Thompson [1933]:
L. M. Milne–Thompson, *The Calculus of Finite Differences*, London, 1933.

Moigno [1840/61]:
François N. M. Moigno, *Leçons de calcul différentiel et de intégral, rédigées d'après les méthodes et les ouvrages publiés ou inédits de M. A. L. Cauchy*, 4 Vols., Paris, 1840–1861.

Moivre
[1722]: Abraham de Moivre, "De fractionibus algebraiciis radicalitate immunibus ad fractiones simpliciores reducendi...," *Phil. Trans.*, Vol. 32, (1722/23) 1724, pp. 162–178.
[1725]: ———, *A Treatise of Annuities on Lives;* London, 1st ed., 1725; 2nd ed., 1743; 3rd ed., 1750; 4th ed., 1752.
[1730]: ———, *Miscellanea Analytica de Seribus et Quadraturis...*, London, 1730.
[1756]: ———, *The Doctrine of Chances or, A Method of Calculating the Probabilities of Events in Play*, 3rd ed., London, 1756. The first edition appeared in 1718 but is not nearly as complete as the third. An Italian transl. appeared in Milan, 1776.

Moulton
NM: Forest Ray Moulton, *New Methods in Exterior Ballistics*, Chicago, 1925.
CM: ———, *An Introduction to Celestial Mechanics*, New York, 1931.
DE: ———, *Differential Equations*, New York, 1930.

Mouton [1670]:
Gabriel Mouton, *Observationes diametrorum Solis et Lunae apparentium meridianarumque aliquot altitudinum solis & paucarum fixarum. Cum tabula declinationum solis constructa...*, Lyon, 1670.

Muir [1906/23]:
Sir Thomas Muir, *The Theory of Determinants in the Historical Order of Development*, 4 Vols., London, 1906–1923.

Murray: *See* Goldstine

Napier
Descr: John Napier, *Mirifici Logarithmorum Canonis descriptio, Ejusque usu, in utraque Trigonometria; ut etiam in omni Logistica Mathematica, Amplissimi, Facillimi, & expeditissimi explicatio. Authore ac Inventore, Joanne Nepero, Barone Merchistonii, &c Scoto.*, Edinburgi, Ex officina Andreae Hart, Bibliopôlae, MDCXIV (1614). (4°.)
Const: ———, *Mirifici Logarithmorum Canonis Constructio; Et eorum ad naturales ipsorum numeros habitudines; unà cum Appendice, de aliâ eâque praestantiore Logarithmorum specie condenda. Quibus Accessere Propositiones ad triangula sphaerica faciliore calculo resolvenda: Unà cum Annotationibus aliquot doctissimi D. Henrici Briggii, in eas & memoratam appendicem. Authore & inventore Ioanne Nepero, Barone Merchistonii etc. Scoto.* Edinburgi, Excudebat Andreas Hart. Anno Domini 1619.
EC: ———, *The Construction of the wonderful Canon of Lagarithms; And their relations to their own natural numbers; with An Appendix as to the making of another and better kind of Logarithms. To which are added Propositions for the solution of Spherical Triangles by an easier method: with Notes on them and in the abovementioned Appendix by the learned Henry Briggs. By the Author and Inventor, John Napier, Baron of Merchiston, &c.*, in Scotland. Printed by Andrew Hart of Edinburgh, in the year of our Lord, 1619. Translated from Latin into English by William Rae MacDonald, 1888.

ED: ———, *A Description of the Admirable Table of Logarithmes, With A Declaration of the Most Plentiful, Easy, and speedy use thereof in both kinds of Trigonometrie, as also in all Mathematical calculations. Invented and published in Latin by That Honorable L. John Nepair, Baron of Marchiston, and translated into English by the late learned and famous Mathematician Edward Wright, with an Addition of an Instrumentall Table to finde the part proportionall, invented by the Translator and described in the end of the Booke by Henry Brigs, Geometry—reader at Greshamhouse in London. All perused and approved by the Author, and published since the death of the Translator.* London, printed by Nicholas Okes, 1616. (12°.)

NTY: ———, *Napier Tercentenary Memorial Volume*, ed. by C. G. Knott, London, 1915.

Napier, M. [1834]:
Mark Napier, *Memoirs of John Napier of Merchiston*, Edinburgh and London, 1834.

Naux [1966]:
Charles Naux, *Histoire des Logarithmes de Neper à Euler*, Vol. 1, Paris, 1966.

von Neumann: *See* Goldstine

Newton
Papers: The Mathematical Papers of Isaac Newton, 7 Vols., ed. by D. T. Whiteside, Cambridge, 1967–1976.

Works: The Mathematical Works of Isaac Newton, assemb. by D. T. Whiteside, 2 Vols., New York and London, 1967.

AU: Isaac Newton, *Arithmetica Universalis, sive de compositions et resolution arithmetica liber*, London, 1707. This appears in English translation as *Universal Arithmetic: or, a Treatise on Arithmetical composition and Resolution, written in Latin by Sir Isaac Newton and translated by the late Mr. Raphson and revised and corrected by M. Cunn*, London, 1728 = *Newton Papers*, Vol. V, 1972, pp. 538–621.

Principia: ———, *Mathematical Principles of Natural Philosophy and his system of the World*, transl. by F. Cajori, Berkeley, 1934.

Methodus: ———, *Methodus Differentialis* = Wm. Jones, *Analysis Per Quantitatum, Series, Fluxiones, ac Differentias: cum enumeratione Linearum Tertii Ordinis*, London, 1711 = *Newton Works*, Vol. 2, pp. 165–173.

See also Jones.

[1927]: ———, *Isaac Newton 1642–1727, A Memorial Volume*, ed. by W. J. Greenstreet, London, 1927, esp. D. C. Fraser, "Newton and Interpolation," pp. 45–69.

Nicole
[1717]: François Nicole, "Traité du calcul des différences finies," *Mém.* (*Hist.*) *Acad. Sc. Paris*, (1717) 1719, pp. 7–21.

[1723]: ———, "Seconde partie du calcul des différences finies," *Mém.* (*Hist.*) *Acad. Sci. Paris*, (1723) 1725, pp. 20–37.

[1724]: ———, "Seconde section de la seconde partie du calcul des différences finies où l'on traite des grandeurs exprimées par des fractions," *Mém.* (*Hist.*) *Acad. Sc. Paris*, pp. 181–198; "Addition aux deux mémoires sur le calcul des différences finies, imprimés l'année dernière," *Mém.* (*Hist.*) *Acad. Sc. Paris*, (1724) 1726 pp. 138–158.

[1727]: ———, "Méthode pour sommer une infinité de suites nouvelles, dont on ne peut trouver les sommes par les méthodes connues," *Mém.* (*Hist.*) *Acad. Sc. Paris*, 1727, pp. 257–268.

Nörlund [1924]:
Niels Erik Nörlund, *Vorlesungen über Differenzenrechnung*, Berlin, 1924, (reprinted Ann Arbor, 1945.)

Padé [1892]:
H. Padé, "Sur la réprésentation approchée d'une fonction par des fractions rationelles," *Ann. Sci. Éc. Norm. Sup. Paris*, Vol. 9, 1892, pp. 1–93.

Perron [1909]:
Oskar Perron, "Über einen Satz des Herrn Poincaré," *Jour. f. reine u. angew. Math.*, Vol. CXXXVI, 1909, pp. 17–37.

Picard
[1890]: Charles Emile Picard, "Mémoire sur la théorie des équations aux dérivées partielles et la méthode des approximations successives," *Jour. de Math. pur. appl.*, Sér. IV, Vol. 16, 1890, pp. 145–210.
[1890′]: ———, "Rectification au sujet du mémoire sur les équations aux dérivées partielles," *Jour. de Math. pur. et appl.*, Sér. IV, Vol. 16, 1890, p. 231.
[1891]: ———, *Traité d'Analyse*, 3 Vols., Paris, 1st ed., 1891–1896; 2nd ed., 1901; 3rd ed., 1922–1928; 4th ed., 1942.

Plana [1820]:
Giovanni A. A. Plana, "Sur une nouvelle expression analytique des nombres bernoulliens, propre à exprimer en termes finis la formule générale pour la sommation des suites," *Mém. Ac. Sc. Torino*, (1), Vol. XXV, 1820, pp. 403–418.

Poincaré [1885]:
Henri Poincaré, "Sur les équations linéaires aux différentielles ordinaires et aux différences finies," *Amer. Jour. Math.*, Vol. VII, 1885, pp. 213–217, 237–258. (These results first appeared in a note in the *Comptes Rendus*, 1883.)

Poisson [1823]:
Siméon–Denis Poisson, "Sur le calcul numérique des intégrales définies," *Mém. de l'Acad. des Sc. de Paris*, Vol. VI, 1823, pp. 571–602.

Raabe
[1848]: Joseph Ludwig Raabe, *Die Jacob Bernoullische Function*, Zürich, 1848.
[1851]: ———, "Zurückführung einiger Summen und bestimmten Integrale auf die *Jacob–Bernoulli*sche Function," *Jour. f. reine angew. Math.*, Vol. XLII, 1851, pp. 348–376.

Raphson [1690]:
Joseph Raphson, *Analysis Aequationum Universalis seu ad aequationes algebraicas...*, London, 1st ed., 1690; 2nd ed., 1702.

Riccati [1758]:
Jacopo Riccati, *Opere del Conte Jacopo Riccati*, 4 Vols., Treviso, 1758.

Rigaud [1841]:
S. P. Rigaud, *Correspondence of Scientific Men of the Seventeenth Century*, 2 Vols., Oxford, 1841–1862.

van Roomen [1593]:
Adrianus Romanus, *Ideae mathematicae pars prima, seu Methodus polygonorum qua laterum, perimetrorum et arearum eujuscunque poligoni investigandorum ratio...*, Antwerp, 1593.

Rosenhain [1846]:
Georg Rosenhain, "Neue Darstellung der Resultante der Elimination von z aus zwei algebraischen Gleichungen $f(z) = 0$ und $\varphi(z) = 0$ vermittelst der Werthe welche die Functionen $f(z)$ und $\varphi(z)$ für gegebene Werthe von z annehmen," *Jour. f. reine u. angew. Math.*, Vol. XXX, 1846, pp. 157–165.

Ruffini
[1804]: Paolo Ruffini, *Sopra la determinazione delle radici nelle equazioni numeriche di qualunque grado...*, Modena, 1804.
[1813]: ———, "Di nuovo metodo generale di estrarre le radici numeriche," *Mem. di Mat. e di Fis. d. Soc. It. d. Sc.*, Vol. XVI, Part I, 1813, pp. 373–429 and Vol. XVII, 1815, pp. 1–15.

Runge

RK: C. Runge and H. König, *Vorlesungen über numerisches Rechnen*, Berlin, 1924.

[1895]: Carl Runge, "Ueber die numerische Auflösung von Differentialgleichungen," *Math. Ann.*, Vol. 46, 1895, pp. 167–178.

[1900]: ———, *Praxis der Gleichungen*, 1st ed., Leipzig, 1900; 2nd ed., Berlin, 1921.

[1901]: ———, "Über empirische Funktionen und die Interpolation zwischen äquidistanten Ordinaten," *Zeit. Math. Phy.*, Vol. 46, 1901, pp. 224–243.

[1904]: ———, *Theorie und Praxis der Reihen*, Leipzig, 1904.

[1905]: ———, "Ueber die numerische Auflösung totaler Differentialgleichungen," *Nach. Ges. Wiss. Göttingen, Math.–Phy. Kl.*, 1905, pp. 252–257.

[1915]: C. Runge and Fr. Willers, "Numerische und grafische Quadratur und Integration gewöhnlicher und partieller Differentialgleichungen," *Enzyklopädie der math. Wiss.*, IIC 2, 1915, pp. 47–176.

St. Vincent

Gregory St. Vincent, *Opus Geometricum quadraturae circuli et sectionum coni*, 2 Vols., Antwerp, 1647.

Seidel [1874]:

Phillip L. Seidel, "Über ein Verfahren, die Gleichungen, auf welche die Methode der kleinsten Quadrate führt, sowie lineäre Gleichungen überhaupt, durch successive Annäherung aufzulösen," *Münch. Abh.*, Vol. II, 1874, Abt. 3, pp. 81–108.

Simpson

[1740]: Thomas Simpson, *Essays on several subjects in speculative and mixed mathematics*, London, 1740.

[1743]: ———, *Math. Dissertations on a Variety of physical and analytical subjects...*, London, 1743.

Smart [1953]:

W. M. Smart, *Celestial mechanics*, London, 1953.

Von Staudt

[1840]: Karl G. C. von Staudt, "Beweis eines Lehrsatzes, die Bernoullischen Zahlen betreffend," *Jour. für reine u. angew. Math.*, Vol. XXI, 1840, pp. 372–374.

[1845]: ———, *De numeris Bernoullianis*, 1 and 2, Erlangen, 1845.

Steffensen [1950]:

J. F. Steffensen, *Interpolation*, New York, 1st ed., 1927; 2nd ed., 1950.

Stifel

[1553]: Michael Stifel, *Die Coss Christoffs Rudolffs*, Königsberg, 1553.

[1554]: ———, *Arithmetica integra*, Nürnberg, 1544.

Stirling

[1719]: James Stirling, "Methodus Differentialis Newtoniana Illustrata," *Phil. Trans.*, Vol. 30, 1719, pp. 1050–1070.

[1730]: ———, *Methodus Differentialis:sive Tractatus de Summatione et Interpolatione Serierum Infinitarum*, London, 1730.

[1749]: ———, *The Differential Method: Or, a Treatise concerning Summation and Interpolation of Infinite Series by James Stirling, Esq., FRS.* Translated into English with the Author's Approbation By Francis Holliday, Master of the Grammar Free-School at Haughton–Park near Retford, Nottinghamshire, London, printed for E. Cave at St. John's Gate, M,DCC,XLIX (1749). References are to this translation.

Taylor

Methodus: Brook Taylor, *Methodus Incrementorum Directa & Inversa*, London, 1715.

Review: ——— *Phil. Trans.*, Vol. XXX, 1715, pp. 339–350.

Thiele

[1906]: T. N. Thiele, "Différences réciproques," *Overs. Dansk. Vid. Sel. For.*, Copenhagen, 1906, pp. 153–171.

[1909]: ———, *Interpolationsrechnung*, Leipzig, 1909.

Tisserand [1889]:

F. Tisserand, *Traité de Mécanique céleste*, 4 Vols., Paris, 1889–1896.

Tropfke *GEM*:

J. Tropfke, *Geschichte der Elementar-Mathematik*, 1st ed., 1902; 2nd ed., 1920; 3rd ed., 1930; refs. are to 3rd ed., 6 Vols.

Tukey: *See* Cooley

Turnbull

[1932]: H. W. Turnbull, "A Study in the Early History of Interpolation," *Proc. Edinburgh Math. Soc.*, Series 2, Vol. 3, 1932–1933, pp. 150–172.

[1945]: ———, *Mathematical Discoveries of Newton*, London, Glasgow, 1945.

Vandermonde [1772]:

Alexandre–Théophile Vandermonde, "Mémoire sur des irrationnelles de différens ordres avec une application au cercle," *Mém. (Hist.) Acad. sc. Paris*, (1772) 1775, premiére partie, pp. 489–498 = *Abh. aus der reinen Math.*, Berlin, 1888, pp. 65–81.

Ver Eecke [1960]:

Paul Ver Eecke, *Les Oeuvres complètes d'Archimède*, Vol. I, Paris, 1960.

Vetter [1930]:

Q. Vetter, "Sur l'équation de quarante-cinquième degré d'Adriaan van Roomen," *Bull. des Sci. Math.*, Series 2, Vol. 54, 1930, pp. 277–283.

Vieta

OP: Opera Mathematica, ed. by F. van Schooten, Leyden, 1646.

[1600]: François Vieta, *De numerosa Potestatum ad Exegesim Resolutione: Ex Opere resitutae Mathematicae Analyseos, seu Algebra nova*, Paris, 1600 = *De numerosa Potestatum purarum, atque adfectarum ad Exegesim Resolutione Tractatus, OP*, pp. 163–228.

Ad P: ———, *Ad Problema, Quod Omnibus Mathematicis Totius Orbis Construendum Proposuit Adrianus Romanus, Responsum, OP*, pp. 305–324.

Ad A: ———, *Ad Angulares Sectiones Theoremata ΚΑΘΟΛΙΚΩΤΕΡΑ, Demonstrata per Alexandrum Andersonum, OP*, pp. 287–304.

Vlacq

AL: Adrian Vlacq, *Arithmetica Logarithmica, sive Logarithmorum Chiliades Centum, Pro Numeris naturali serie crescentibus ab Unitate ad 100,000. Una Cum Canone Triangulorum, Seu Tabula Artificialium Sinuum, Tangentium & Secantium, Ad Radium 10,00000,00000. & ad singula Scrupula Prima Quadrantis. Quibus Novum Traditur Compendium, Quo Nullum nec admirabilius, nec utilius solvendi pleraque Problemata Arithmetica & Geometrica. Hos Numeros Primus Invenit Clarissimus Vir Iohannes Neperus Baro Merchistronij: eos autem ex ejusdem sententiâ mutavit, Eorumque ortum & usum illustravit Henricus Briggius, in celeberrimâ Academiâ Oxoniensi Geometriae Professor Savilianus. Editio Secunda Aucta per Adrianum Vlacq Goudanum. Deus Nobis Usuram Vitae Dedit et Ingenii, Tanquam Pecuniae, Nulla Praestituta Die.* Goudae, Excudebat Petrus Rammasenius. MDCXXVIII (1628). Cum Privilegio Illust. Ord. Generalium. (Folio.) (This is a Dutch edition of Briggs's *Arithmetica Logarithmica*. In it Vlacq completed the logarithmic table of Briggs.)

Wallis

[1668]: John Wallis, "Logarithmotechnia Nicolae Mercatoris. Concerning which we shall here deliver the account of the Judicious Dr. I. Wallis given in a Letter to the Lord Viscount Brouncker, as follow," *Phil. Trans.*, Vol. III, 1668, pp. 753–764.

OP: ———, *Arithmetica Infinitorum*, Oxford, 1656 = *Opera Mathematica*, Vol. 1, Oxford, 1695, pp. 355–478.

Waring [1770]:

Edward Waring, *Meditationes algebraicae*, Cambridge, 1770.

Weierstrass

Karl T. W. Weierstrass, *Mathematische Werke*, 7 Vols., Berlin, 1894–1927.

Whiteside

Patterns: Derek Thomas Whiteside, "Patterns of Mathematical Thought in the later Seventeenth Century," *Archive for History of Exact Sciences*, Vol. 1, 1960–1962, pp. 179–388.

See also Newton.

Whittaker

WR: E. T. Whittaker and C. Robinson, *The Calculus of Observations*, London and Glasgow, 1940.

WW: Whittaker and G. N. Watson, *A Course of Modern Analysis*, Amer. ed. New York, 1948.

Williamson [1889]:

B. Williamson, *An elementary treatise on the differential calculus*, New York, 1889.

Willers [1918]:

Fr. A. Willers, *Methoden der praktischen Analysis*, Berlin–Leipzig, 1928. Eng. transl. by R. T. Beyer, New York, 1948.

See also Runge.

Woolhouse [1888]:

W. S. B. Woolhouse, "On integration by means of selected values of the function," *Jour. Inst. Act.*, Vol. 27, 1889, pp. 122–155.

Wright [1599]:

Edward Wright, *Certaine errors in navigation...and tables of declination of the Sunne and fixed Starres...*, London, 1599.

Index

Entries followed by n refer to footnotes on the cited pages.

I.J. Bienaymé: Statistical Theory Anticipated
By **C.C. Heyde** and **E. Seneta**

The purpose of this book is to focus on the scientific work of I.J. Bienaymé (1796-1878), both for its intrinsic interest and for the perspective it gives on developments in probability and statistics (including demography and social statistics) in the 19th century. The book is addressed not only to the historian of science, but also to the working mathematician.

1977. approx. 175p. 1 illus. cloth

Sources in the History of Mathematics and Physical Sciences

Edited by **M.J. Klein** and **G.J. Toomer**

Diocles: On Burning Mirrors
With Text in Arabic and Greek
English translation and commentary by G.J. Toomer

The first edition of an important text from the most productive period of Greek mathematics will significantly alter previously accepted ideas on the early history of conic sections. This presentation, the first major addition to knowledge of mathematics during the Hellenistic period since Heiberg's work on Archimedes' "Method" in 1907, contains the Greek text of the extracts in Eutocius and the complete medieval Arabic text with English translations using modern notation and extensive commentary.

1976. ix, 249p. (64p. in Arabic, 12p. in Greek) 32 illus. 24 plates. cloth

Studies in the History of Mathematics and Physical Sciences

Edited by **M.J. Klein** and **G.J. Toomer**

Volume 1

A History of Ancient Mathematical Astronomy

By **O. Neugebauer**

"A totally different appreciation of Ptolemy is afforded by O. Neugebauer's new three-volume work on early astronomy. The inclusion of the word 'mathematical' is deliberate, for Neugebauer eschews the vague, speculative cosmologies of pre-Socratic philosophers . . . . for Ptolemy, it is the source par excellence.

Divided into six 'books,' this compendium distills much of a lifetime of scientific research into three volumes, and it is surely one of the landmark publications of this century in the history of astronomy."

Science

1975. xxxiii, 1456p. 619 illus. 9 plates.
1 foldout. cloth
(Also available in three separate parts)

Volume 3

The Origins of Digital Computers

Selected Papers
Second Edition
Edited by **B. Randell**

The Origins of Digital Computers brings together some of the more important and interesting written source material on the history of digital computers. The basic starting point is the Analytical Engine that Charles Babbage began to design in 1834. It ends with two papers that were presented at the inauguration of EDSAC in June, 1949. Each significant milestone from Babbage to EDSAC is covered.

Introductory and linking text is provided in order to place the work of the various pioneers into perspective, and to cover such topics as early calculating machines and sequence-control mechanisms, and the development of electromagnetic and electronic digital calculating devices. An annotated bibliography of over 350 items is also included.

The book is intended for computer science students or those employed in the computer science field who are interested in the history of their subject, and particularly in the technical details of the precursors of the modern electronic computer.

1975. xvi, 464p. 120 illus. cloth